INTRODUCTION TO WIRELESS SYSTEMS

P. M. SHANKAR
Professor of Electrical and Computer Engineering
Drexel University

JOHN WILEY & SONS, INC.

Acquisitions Editor   Bill Zobrist
Production Management   Jeanine Furino
Production Editor   Sandra Russell
Senior Marketing Manager   Katherine Hepburn
Designer   Harry Nolan
Production Management Services   Publication Services

This book was set in 10/12 Times Roman by Publication Services and printed and bound by Hamilton Printing Company. The cover was printed at The Lehigh Press, Inc.

The paper in this book was manufactured by a mill whose forest management programs include sustained yield harvesting of its timberlands. Sustained yield harvesting principles ensure that the number of trees cut each year does not exceed the amount of new growth.

This book is printed on acid-free paper ∞

Library of Congress Cataloging-in-Publication Data
Shankar, P. M.
    Introduction to wireless systems / P. M. Shankar.
        p. cm.
    Includes bibliographical references and index.
    ISBN 0-471-32167-2 (cloth : alk. paper)
    1. Wireless communication systems. 2. Mobile communication systems. I. Title.

TK5103.2 .S517 2001
621.382—dc21   2001026251

Printed in the United States of America
10 9 8 7 6 5 4 3 2 1

To my parents,
Padmanabha Rao and Kanakabai Rao,
both former teachers,
who instilled in me the values of education
and a strong commitment to teaching.

PREFACE

The field of wireless systems is expanding and moving toward becoming standard fare of electrical engineering programs at the undergraduate level. This book addresses the needs of the undergraduate students interested in learning the science and engineering of wireless communications. Keeping this in view, the book's purpose is to educate undergraduate and first-year graduate students by providing the analytical tools, fundamental physics, communication theory, and other essentials.

The book is based on the course I developed at the undergraduate level at Drexel University. Students taking a class based on this book are expected to have already taken an introductory-level course in probability/random variables and also a course in modulation techniques. These courses are required for students at the junior level of the electrical engineering programs at most colleges and universities. For those students who are not very familiar with this subject matter, some important basic topics in these areas are presented in the appendices.

For the benefit of students and readers, each chapter ends with a summary of the material covered in the chapter. The summary is detailed enough for students and the instructor to go back and reread to reinforce the concepts. Chapters 2–6 have a number of exercises that have to be completed using MATLAB. This strengthens the fundamental concepts and allows the students to practice the theory and techniques used in mobile and wireless communication systems by varying the parameters and exploring the resulting changes in the signals, waveforms, and other characteristic features of wireless signals. This project-based focus will help the students absorb the lectures through an interactive approach. These projects also demonstrate how and why mobile systems operate, bringing out the nuances and subtleties of communication systems. Chapters 2 through 6 also contain example problems to demonstrate some of the core concepts discussed. These examples are mostly numerical; the projects and exercises are both analytical and numerical.

A very comprehensive list of references allows the students and other readers to investigate and explore the topics discussed in the book.

Chapter 1 gives a brief overview and survey of the characteristic features of the different wireless systems available.

Chapter 2 introduces the students to the issues and problems in the transmission of wireless signals. The major topics of attenuation and fading are covered in detail. Diagrams illustrating attenuation and computation of received power are given. The topic of fading is covered in depth from a scattering approach, followed by the application of the probability theory to derive results for Rayleigh, Rician, and lognormal fading. Various statistical models of fading are described along with topics of hypothesis testing to validate statistical models. The concept of dispersion is introduced using impulse response so that frequency-selective and flat fading can be explained. The curves and figures were all generated using MATLAB. The origins of dispersion and distortion are detailed in the appendices.

Chapter 3 covers the topic of modems for wireless communications. Starting with an overview of analog communication formats such as AM, FM, and FDM, the area of digital communication is covered in detail. Topics of Nyquist pulse shaping, analysis of receivers, orthogonal functions, and error probability calculations based on signal constellations are presented. The concept of signal-to-noise ratio and the relationship to energy-to-noise ratio are explained in detail.

The different digital modulation techniques are presented. The two major techniques, forms of QPSK and MSK (including GMSK), are covered in great detail, providing a complete theoretical basis for these major formats of mobile and wireless communications. All the necessary derivations and explanations have been provided along with block diagrams and curves. The waveforms for these different modulation formats have been generated and are given. MATLAB was used for the generation of the waveforms, and these programs can be made available to instructors.

Other M-ary modulation schemes are also discussed. Comparison of the various modulation formats is undertaken using Shannon's theorem. The trade-off between power and bandwidth efficiency is studied.

A number of background topics are presented in two of the appendices, one dealing with topics in signals and systems, and the other dealing with topics in telecommunications.

Chapter 4 covers cellular aspects of wireless communications. The effect of co-channel interference (CCI) and techniques for its mitigation using sectored antennae are described. Trunking efficiency is discussed, and a table of Erlang B probabilities is provided. A detailed derivation of Erlang B and C probabilities is given in an appendix.

The effect of fading in wireless communication systems is covered in Chapter 5. The effect of randomness of the received signal and measures to overcome the increase in bit error rate brought on by fading are described in detail. Improvements brought on by different forms of diversity and diversity-combining methods are discussed. Detailed derivations and results for the effects of fading on the different modulation formats are presented. Methods to reduce the effects of frequency-selective fading using equalizers are discussed.

The various multiple-access techniques used in wireless systems are presented in Chapter 6. The three schemes, FDMA, TDMA, and CDMA, are described in terms of their characteristic advantages and disadvantages. CDMA techniques are explored in greater detail, going through both DS and FH techniques, along with the theoretical development necessary to understand and appreciate the importance of the concepts. Use of the RAKE receiver to achieve fading mitigation as a form of time diversity is introduced. All the different wireless and mobile systems currently in use are reviewed and summarized in tabular form. A number of examples have been provided to demonstrate the characteristics of CDMA-based systems.

Third-generation (3G) wireless communications are discussed with a review of the characteristics of IMT 2000. Specifically, the unique features of cdma2000 are presented along with comparisons with existing CDMA systems in North America and W-CDMA, the parallel IMT 2000–based systems in Europe and Asia. A brief overview of the new wireless paradigm, Bluetooth, is also given.

Three appendices are included. Appendix A covers topics in signals and systems such as Fourier transforms and random variables. Appendix B covers topics in telecommunications, such as trunking theory, source coding, channel coding, orthogonal frequency division multiplexing (OFDM), and eye patterns. Appendix C covers three topics in wireless communications: dispersion, attenuation in PCS systems, and macrodiversity.

The solutions manual contains MATLAB programs for the project-based exercises and will be available at the Wiley Web site, located on the back cover.

ACKNOWLEDGMENTS

I would like to express my sincere thanks to a number of my colleagues and associates who have provided encouragement and help with the preparation of this book. Of special mention here are two of my colleagues at Drexel, Dr. Stan Kesler and Dr. Maja Bystrom, who, despite their busy schedules, were kind enough to review some of the chapters of the book. I received quite a bit of support at home as well. A number of diagrams were drawn by my daughter Raji, who, along with my wife, Rajakumari, spent time reviewing the text for omissions, compiling the references, occasionally doing the typing, and keeping track of my commitment to Wiley to submit the manuscript on time. Indeed, the book has been a family project (as it should be), and I am very grateful to my wife and my daughter for their patience, support, encouragement, and willingness to help complete this project. Our family friend Ms. Maura Curran was generous with her time, going through the text thoroughly to correct typographical and grammatical errors.

The book would not have been possible without the strong support of Bill Zobrist, editor at Wiley, and his staff. Indeed, it was a pleasure working with Bill, who was responsive and enthusiastic about this project. I am grateful to the local Wiley representative, Dennis Laynor, for our initial discussion on the book proposal and for his support throughout this process. I also would like to express my sincere appreciation of the reviewers of this book, whose suggestions were invaluable:

Brian D. Woerner, Virginia Tech

Gregory J. Pottie, University of California, Los Angeles

Prashant Krishnamurthy, University of Pittsburgh

Gerald Mitchell, University of Colorado

Alexander Haimovich, New Jersey Institute of Technology

Thomas Robertazzi, State University of New York at Stony Brook

Victor S. Frost, University of Kansas

John M. Shea, University of Florida

Bruce A. Black, Rose-Hulman Institute of Technology

P. M. Shankar
Philadelphia, PA
January 2001

CONTENTS

HISTORICAL OVERVIEW OF WIRELESS SYSTEMS

1.0 OVERVIEW

The history of mobile communications is a long one. We can trace some of the earliest uses of wireless systems to military, fire, and police communications. The equipment used during these early days was relatively bulky; communication was in the simplex mode. This was followed by what was known as Improved Mobile Telephone System (IMTS), which allowed full-duplex provision. Owing to the very small number of available channels, the IMTS system was inadequate for providing service to customers waiting to gain access. The situation did not improve until the concepts of cellular systems were developed. Though cellular technology started with analog systems, mobile telephone providers have ushered in new developments incorporating digital technology and making it possible to cover large geographical regions and to provide access to a large number of users. At the time these developments were taking place in North America, Europe and Japan were also in the forefront of new developments in mobile and wireless systems and standards. This almost simultaneous progress across the world has made it possible to provide mobile telephone access to an ever-increasing population. But competing standards within North America and around the world, coupled with claims of superior performance of one standard over others, make it difficult for the ordinary customer to choose wisely. Even though these multiple standards and systems coexist in the United States, Europe has promoted a single standard, which is being increasingly adopted by the rest of the world. It is necessary for us to understand the natural growth of these multiple standards and systems to fully appreciate where we are today, where we came from, and how far we have come.

We will briefly look at the progress and deployment of cellular systems around the world, starting with the North American cellular systems.

1.1 NORTH AMERICAN CELLULAR SYSTEMS

There are several cellular systems operating within North America. These include the Advanced Mobile Phone System (AMPS) and the American Digital Cellular System, based on Interim Standard 54 (IS 54) and Interim Standard 95 (IS 95).

The idea of "cellular communications," conceived and developed at AT&T Bell Laboratories, was the first major step in what we now regard as the modern era of wireless communications. With the idea of splitting a geographical region into cells and reusing some of the channels over and over again, it was demonstrated that the

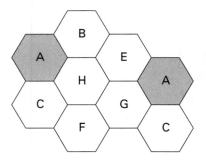

FIGURE 1.1 The various letters correspond to different radio channels. The shaded regions use the same radio channel. This pattern is a seven-cell reuse pattern.

capacity of wireless communications could be significantly increased. The concept of a cell structure is shown in Figure 1.1, where the shaded areas use and reuse the same frequencies. Thus without adding any new channels, it was possible to achieve an increase in capacity and allow subscribers to move from one location to another with ease, and become what are known as "mobile subscribers." The only requirement was that the distance between the regions be sufficient to have extremely low levels of signals coming from the faraway channels to the user in any given channel. The reuse was determined by the so-called co-channel interference, the contribution coming from areas using the same frequency.

Based on this cellular concept, AMPS was offered in the United States in 1983. Two providers shared the available spectrum in each market. The duplex provision, which is the ability to simultaneously transmit and receive calls, was accomplished by splitting the spectrum in two, using one band for transmission and the other for reception. Each radio channel spanned 30 kHz and was allocated for a single voice channel. Bidirectional communication was achieved through the use of two 30 kHz segments in each direction. This duplex operation using two simplex bands is shown in Figure 1.2. Note that this system operates in analog mode; it uses an analog modulation format, namely, frequency modulation (FM), for transmitting the voice signals. As we will see later, analog systems are inherently ill suited for providing many of the features of digital phones, such as voice messaging and multiparty calling. Analog systems also were nearing the limit to their capacity in terms of available channels, thus arresting the growth of the mobile systems.

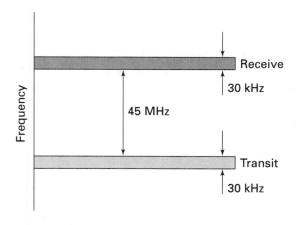

FIGURE 1.2 Concept of AMPS and similar analog systems employing frequency division multiple access (FDMA) for different users, with each user having a radio channel of bandwidth 30 kHz, one to transmit and another one to receive. The channel separation is 45 MHz.

To overcome the deficiencies of the existing analog cellular system, U.S. Digital Service was brought into being in 1991. The choice to usher in the digital revolution (as well as evolutions) was the time division multiple access (TDMA) scheme. In this mode, the radio channel is partitioned into bands, jointly "owned" or "shared" by a number of users. Each customer uses a particular time slot assigned to him or her so that the whole bandwidth of the radio channel is shared by all the users in a group, as shown in Figure 1.3. The digital technique also was compatible with the naturally evolving landline-based digital information transmission network.

The USDC (United States Digital Cellular) standard (IS 54) allowed cellular providers to gradually ease into the new service from the analog systems. This was possible because the radio channels of the analog systems and the bandwidth of the radio channels were exactly the same. Instead of a single voice channel for each radio channel, USDC allowed three voice channels to share the radio channel using time division multiple access. The USDC also kept the duplex provisions of the AMPS by separating the bidirectional transmissions over frequency bands separated by 45 MHz.

The modulation format used in USDC is $\pi/4$ differential quadrature phase shift keying ($\pi/4$-DQPSK), and the multiple-access method is TDMA. Frequency division duplex (FDD) is used in the same format as in AMPS. The use of the spectrally efficient digital modulation along with speech coding has made it possible to increase the capacity available in USDC over the analog AMPS system. With more advanced speech coding it may be possible to accommodate six-voice channels in the 30 kHz radio channel. Digital technology provides a number of advantages over the traditional analog schemes:

- Flexibility to transmit mixed signals, voice and data
- Increased capacity from efficient speech coding
- Reduced transmit power and hence longer battery life
- Possible digital encryption for privacy
- Ease of hand-off procedures such as mobile assisted hand-off
- A single transceiver capable of handling multiple users who share the particular radio channel
- Lessened demand on the duplexer (compared with FDMA-based analog systems) since transmit and receive slots are staggered

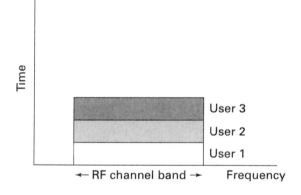

FIGURE 1.3 Concept of TDMA. Three users share the single radio channel, each using the bandwidth at a different time.

Another digital system offered recently is based on code division multiple access (CDMA). In this system, all the users share the same radio channel but use different codes. The digitized voice signal is spread using a long code (pseudo noise sequence) that increases the bandwidth to 1.25 MHz. Whereas AMPS and IS 54–based systems have what may be termed a "hard" capacity limit, CDMA-based systems (IS 95) allow more and more users to share the spectrum. As the number of users goes up, the interference increases, but users need not be turned away. Since every user is using the same channel, frequency planning, which is very important in AMPS and IS 54–based systems, is not necessary. This format also allows variable-rate transmissions. It is possible to reduce the effects of fading using a "RAKE" receiver.

While the AMPS operated at around 900 MHz, the digital systems, both TDMA-based and CDMA-based systems, operate in what is known as the PCS band, around 1900 MHz.

These are not the only systems available in North America. A number of companies are also offering service based on GSM in the 1900 MHz band. GSM is the pan-European system, used in Europe and most of Asia and Australia. GSM is discussed in the next section.

1.2 PAN-EUROPEAN CELLULAR SYSTEMS

While providers in North America were trying to improve their ability to supply better service and more capacity, Europeans were making great strides in the "digital revolution" of mobile communications. Different countries in Europe were using standards and systems that were not mutually compatible, and efforts were made to come up with a single, unified standard for Europe. The outcome of this effort was the birth of the pan-European system known as the Global System for Mobile (GSM), which is based on a completely digital format. This system was intended to provide a pan-European roaming capability that the previous analog systems lacked. It was also expected to provide lower costs and higher spectral efficiency than the analog systems.

In GSM, a radio channel of bandwidth 200 kHz is "time shared" by eight users. With eight users, pulse distortion is likely due to frequency-selective fading. This degradation can be reduced somewhat by the frequency hopping employed by GSM. The modulation format used is a modified digital form of frequency modulation known as GMSK (Gaussian-filtered minimum shift keying). This permits the use of low-cost nonlinear amplifiers in handsets, reducing the weight and cost of the units.

Though originally conceived for operation in the 900 MHz band, the format has now been adopted for 1800 MHz, known as the DCS 1800 (Digital Communication System 1800) system.

1.3 PACIFIC DIGITAL CELLULAR (PDC) SYSTEMS

PDC, also known as the Japanese Digital Cellular (JDC) system, is similar in principle to USDC in every respect expect for the lower radio channel bandwidth. Instead of a 30 kHz bandwidth, JDC uses a 25 kHz bandwidth for the three-voice channels. The frequency bands used are also slightly different from USDC.

1.4 UNIVERSAL MOBILE TELECOMMUNICATION SYSTEMS (UMTS)/INTERNATIONAL MOBILE TELECOMMUNICATIONS 2000 (IMT 2000)

Efforts had been under way to design telecommunication systems to provide universal access to wireless services with roaming capabilities to transmit and receive a wide range of services, including multimedia. These efforts have resulted in the development of parallel technologies in North America and elsewhere in the world, and in the preliminary testing and deployment of cdma2000 in North America and W-CDMA (wideband CDMA) in the rest of the world. The key features of these systems are the ability to transmit data at multiple rates, at multiple carrier frequencies, and at multiple bandwidths (narrower bandwidth in cdma2000 versus wider bandwidth in W-CDMA). These systems are operational anywhere, indoors as well as outdoors, in stationary and mobile environments. They operate in the 1.9–2 GHz range.

1.5 COMPONENTS OF A CELLULAR SYSTEM

The cellular systems, whether they are based on North American, European, or Pacific systems, have a number of common elements. Essential elements of any cellular systems are (1) mobile stations, (2) base stations, and (3) a mobile switching center. A mobile station is sometimes referred to as a mobile unit (MU), a portable unit (PU), or a handset. A mobile switching center (MSC) is often referred to as a mobile telephone switching office (MTSO).

A mobile unit is carried by the subscriber, and the term *mobile* is used to signify that the subscriber need not be stationary. A generic mobile unit used in today's digital cellular systems is shown in Figure 1.4. The antenna receives signals coming from the base station and transmits the signal from the subscriber. The unit typically has a duplexer (DUP) to separate the transmit and receive frequencies. A generic duplexer is shown in Figure 1.5. The duplexer separates the two frequency bands, the transmit and receive frequencies.

A base station (BS) is a fixed unit in a cellular system that communicates with the mobile units within a cell. It is located at the center or the edge of the cell and has

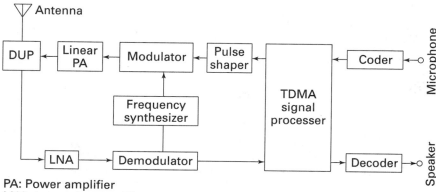

PA: Power amplifier
LNA: Low-noise amplifier

FIGURE 1.4 A generic mobile unit.

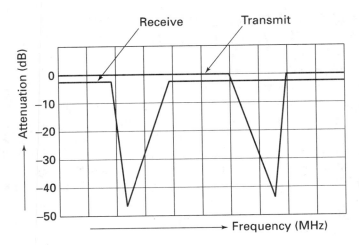

FIGURE 1.5 A duplexer. The two filters facilitate isolation of the two channels.

transmitter and receiver antennas mounted on a tower. A typical base station is shown in Figure 1.6. Note that a second receiving antenna is usually added to provide some form of diversity (the ability to receive multiple versions of the same signal, which can reduce the effects of fading).

The base station serves as the link between mobile units and the MSC, and facilitates connections to the public switched telephone network (PSTN). Note that each MSC will handle a number of base stations. This concept is illustrated in Figure 1.7.

A number of other terms are commonly used in mobile communication systems.

Cell. The cell is the smallest area covered by a base station. A geographical region may be composed of many cells. The same channel used in a cell will be reused in another cell or in cells that are separated from the cell.

Control channel. Control channels are radio signals used for transmission, setup, call request, call initiation, and other control purposes.

Duplex systems (full). Full-duplex systems allow simultaneous two-way communication.

Duplex systems (half). Half-duplex systems allow two-way communication. At any given time, the user can either transmit at a certain frequency or receive at the same frequency, but not both simultaneously.

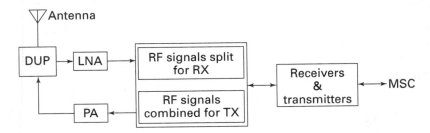

DUP: Duplexer   LNA: Low-noise amplifier   PA: Power amplifier
RF: Radio frequency   TX: Transmitter   RX: Receiver
MSC: Mobile switching center (or MTSO: Mobile telephone switching center)

FIGURE 1.6 A generic base station.

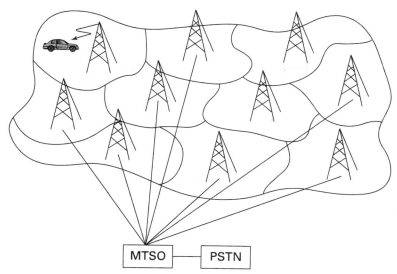

FIGURE 1.7 An overview of the cellular system. Each base station has an antenna, and all the base stations are connected to the mobile telephone switching office, which provides the link to the landline.

Forward channel. The forward channel is a radio channel used for transmission from the base station to the mobile unit.

Forward link (also known as *downlink*). The forward link is the connection from the base station to the mobile unit.

Hand-off. The hand-off is the process in which a mobile unit is transferred from one channel to another or from one base station to another.

Mobile station (also known as *mobile unit*). A mobile unit (MU) is carried by the subscriber. It may be handheld or vehicle mounted.

Mobile switching center (also known as *mobile telephone switching office*). The base stations in a large service area are connected to the MSC to facilitate the switchover to land-based telephone lines or to transfer calls from a land-based system to a mobile unit. In some systems, the MSC also makes decisions on hand-off, power levels of the handsets, and other issues.

Reverse channel. The reverse channel is a radio channel used for transmission from the mobile unit to the base station.

Reverse link (also known as *uplink*). The reverse link is the connection from the mobile unit to the base station.

Simplex systems. Simplex systems, unlike duplex systems, offer only a one-way communication link.

1.6 SUMMARY

This chapter has taken a brief look at the evolution of wireless communications. The transformation of the technology from the exclusive world of a select few users to the masses has been accomplished in a short period of time.

A number of books, monographs, and review papers trace this history and the technology of the wireless and mobile communication systems. A comprehensive bibliography is included below so that the reader can fully understand and appreciate the tremendous growth that is taking place.

BIBLIOGRAPHY

AKAIWA, Y. *Introduction to Digital Mobile Communication*. John Wiley & Sons, New York, 1998.

BARNES, F. S., EHLING, T. P., EPHREMIDES, A., EVANS, P. R., LOTOCHINSKI, E. B., & SHETH, J. N. (eds.). *Worldwide Wireless Communications*. International Engineering Consortium, Chicago, 1995.

CLARK, A. P. "Digital modems for land mobile radio," *IEE Proc.*, vol. 132, pt. F., no. 5, pp. 348–362, Aug. 1985.

CLARKE, R. H. "A statistical description of mobile radio reception," *Bell Syst. Tech. J.*, vol. 47, pp. 957–1000, July–Aug. 1968.

COOPER, G. R., NETTLETON, R. W., & GRYBOS, D. P. "Cellular land mobile radio: Why spread spectrum?," *IEEE Commun. Mag.*, vol. 17, no. 3, pp. 17–23, Mar. 1979.

COX, D. C. "Wireless personal communications: What is it?," *IEEE Pers. Commun.*, pp. 20–35, Apr. 1995.

FEHER, K. *Wireless Digital Communications: Modulation and Spread Spectrum Applications*. Prentice-Hall, Englewood Cliffs, NJ, 1995.

GIBSON, J. D. (ed.). *The Mobile Communications Handbook*. IEEE Press/CRC Press, Boca Raton, FL, 1996.

GIBSON, J. D. (ed.). *The Telecommunications Handbook*. IEEE Press/CRC Press, Boca Raton, FL, 1997.

GILBERT, E. N. "Energy reception for mobile radio," *Bell Syst. Tech. J.*, vol. 44, pp. 1779–1803, Oct. 1965.

GOODMAN, D. J. "Second generation wireless information networks," *IEEE Trans. Veh. Technol.*, vol. 40, no. 2, pp. 366–374, May 1991.

GOODMAN, D. J. *Wireless Personal Communications Systems*. Addison-Wesley, Reading, MA, 1997.

GUPTA, S. C., VISWANATHAN, R., & MUAMMAR, R. "Land mobile systems: A tutorial exposition," *IEEE Commun. Mag.*, vol. 23, no. 6, pp. 34–45, June 1985.

HAMMUDA, H. *Cellular Mobile Radio Systems*. John Wiley & Sons, Chichester, UK, 1998.

HANZO, L., & STEELE, R. "The pan European mobile radio system, Parts 1 and 2," *Eur. Trans. Telecommun.*, vol. 5, no. 2, pp. 245–276, Mar.–Apr. 1994.

HAUG, THOMAS. "Overview of GSM: Philosophy and results," *Int. J. Wireless Info. Networks*, vol. 1, no. 1, pp. 7–16, 1994.

HODGES, M. R. L. "The GSM radio interface," *Br. Telecommun. Res. Lab. J.*, vol. 8, no. 1, pp. 31–43, Jan. 1990.

HODGES, M. R. L., & JENSEN, S. A. "Laboratory testing of digital cellular systems," *Br. Telecommun. Res. Lab. J.*, vol. 8, no. 1, pp. 57–66, Jan. 1990.

IEEE VEHICULAR TECHNOLOGY SOCIETY COMMITTEE ON RADIO PROPAGATION. "Coverage prediction for mobile radio systems operating in the 800/900 MHz frequency range," IEEE *Trans. Veh. Technol.*, vol. 37, no. 1, pp. 3–44, Feb. 1988.

JAKES, W. C. (ed.). *Microwave Mobile Communications*. IEEE Press, Los Alamitos, CA, 1974.

KERR, R. "CDMA digital cellular," *Appl. Microwave Wireless*, pp. 30–41, Fall 1993.

KNISELY, D. N., KUMAR, S., LAHA, S., & NANDA, S. "Evolution of wireless data services: IS-95 to cdma 2000," *IEEE Commun. Mag.*, pp. 140–149, Oct. 1998.

KNISELY, D. N., LI, Q., & RAMESH N. S. "cdma 2000: A third generation radio transmission technology," *Bell Labs Tech. J.*, pp. 63–78, July/Sept. 1998.

LEE, W. C. Y. "Elements of cellular mobile radio systems," *IEEE Trans. Veh. Technol.*, vol. 35, no. 2, pp. 48–56, May 1986.

LEE, W. C. Y. "Smaller cells for greater performance," *IEEE Commun. Mag.*, vol. 29, pp. 19–23, Nov. 1991.

LEE, W. C. Y. *Mobile Communications Design Fundamentals*. John Wiley & Sons, New York, 1993.

LEE, W. C. Y. *Mobile Communications Engineering*. McGraw-Hill, New York,1997.

MACDONALD, V. H. "The cellular concept," *Bell Syst. Tech. J.*, vol. 58, no. 1, pp. 15–41, Jan. 1979.

MIKULSKI, J. J. "DynaT*A*C cellular portable radiotelephone system experience in the U.S. and the U. K.," *IEEE Commun. Mag.*, vol. 24, no. 2, pp. 40–46, Feb. 1986.

MONSEN, P. "Fading channel communications," *IEEE Commun. Mag.*, vol. 18, no. 1, pp. 16–25, Jan. 1980.

MOULY, M., & PAUTET, M.-B. "Current evolution of the GSM systems," *IEEE Pers. Commun.*, vol. 2, no. 5, pp. 9–19, Oct. 1995.

NOGUCHI, T., DAIDO, Y., & NOSSEK, J. A. "Modulation techniques for microwave digital radio," *IEEE Commun. Mag.*, vol. 24, no. 10, pp. 21–30, Oct. 1986.

OETTING, J. D. "A comparison of modulation techniques for digital radio," *IEEE Trans. Commun.*, vol. COM-27, no. 12, pp. 1752–1762, Dec. 1979.

OETTING, J. D. "Cellular mobile radio: An emerging technology," *IEEE Commun. Mag.*, vol. 21, pp. 10–15, Nov. 1983.

OKUMURA, Y., OHMURI, E., KAWANO, T., & FUKUDA, K. "Field strength and its variability in VHF and UHF land mobile radio service," *Rev. Electr. Commun. Lab.*, vol. 16, pp. 825–873, Sept.–Oct. 1968.

PADGETT, J. E., GUNTHER, C. G., & HATTORI, T. "Overview of wireless personal communications," *IEEE Commun. Mag.*, vol. 33, no. 1, pp. 28–41, Jan. 1995.

PADOVANI, R. "Reverse link performance of IS 95 based cellular systems," *IEEE Pers. Commun.*, vol. 1, no. 3, pp. 28–34, 1994.

PAHLAVAN, K., & LEVESQUE, A. H. *Wireless Information Systems.* John Wiley & Sons, New York, 1995.

PARSONS, D. *The Mobile Radio Propagation Channel.* John Wiley & Sons, New York, 1992.

PRASAD, N. R. "GSM evolution towards third generation UMTS/IMT 2000." In *1999 IEEE International Conference on Personal Wireless Communications*, Jaipur, India, pp. 50–54, Feb. 1999.

RAHMAN, T. A., BUROK, H., & GEOK, T. K. "The cellular phone industry in Malaysia: Toward IMT-2000," *IEEE Commun. Mag.*, pp. 154–156, Sept. 1998.

RAO, Y. S. & KRIPALANI, A. "cdma2000 mobile radio access for IMT." In *1999 IEEE International Conference on Personal Wireless Communications*, Jaipur, India, pp. 6–15, Feb. 1999.

RAPPAPORT, T. S. (ed.). *Cellular Radio & Personal Communications*, Volume 1. IEEE Press, Los Alamitos, CA, 1995.

RAPPAPORT, T. S. (ed.). *Cellular Radio & Personal Communications*, Volume 2. IEEE Press, Los Alamitos, CA, 1996.

RAPPAPORT, T. S. *Wireless Communications: Principles & Practice.* Prentice-Hall/IEEE Press, Englewood Cliffs, NJ, 1996.

RICCI, F. J. *Personal Communications Systems and Applications.* Prentice-Hall, Englewood Cliffs, NJ, 1997.

SAMPEI, S. *Applications of Digital Wireless Technologies to Global Wireless Communications.* Prentice-Hall, Englewood Cliffs, NJ, 1997.

SCHILLING, D. L., MILSTEIN, L. B., PICKHOLTZ, R. L., BRUNO, F., KANTERAKIS, E., KULLBACK, M., ERCEG, V., BIEDERMAN, W., FISHMAN, D., & SALERNO, D. "Broadband CDMA for personal communications systems," *IEEE Commun. Mag.*, pp. 86–93, Nov. 1991.

SCHOLTZ, R. A. "The spread spectrum concept," *IEEE Trans. Commun.*, vol. 25, pp. 748–755, Aug. 1977.

SCHOLTZ, R. A. "The origins of spread spectrum," *IEEE Trans. Commun.*, vol. 30, pp. 822–854, May 1982.

SCHWARTZ, M., BENNETT, W. R., & STEIN, S. *Communication Systems and Techniques.* IEEE Press, Los Alamitos, CA, 1996.

SHAFI, M., SASAKI, A., & JEONG, D.-G. "IMT-2000 developments in the Asia Pacific region," *IEEE Commun. Mag.*, p. 144, Sept. 1998.

SHUMIN, C. "Current development of IMT-2000," *IEEE Commun. Mag.*, pp. 157–159, Sept. 1998.

STEELE, R. "The cellular environment of lightweight handheld portables," *IEEE Commun. Mag.*, vol. 27, no. 6, pp. 20–29, June 1989.

STEELE, R. "The evolution of personal communications," *IEEE Pers. Commun.*, vol. 1, no. 2, pp. 6–11, 1994.

STEELE, R. *Mobile Radio Communications.* IEEE Press, Los Alamitos, CA, 1995.

TANTARATANA, S., & AHMED, K. M. *Wireless Applications of Spread Spectrum Systems.* IEEE Press, Los Alamitos, CA, 1998.

VITERBI, A. J. "Spread spectrum communication: Myths and realities," *IEEE Commun. Mag.*, vol. 17, pp. 11–18, May 1979.

WEAVER, C. S. "A comparison of several types of modulation," *IRE Trans. Commun. Syst.*, vol. CS-10, pp. 96–101, Mar. 1962.

WEE, K.-J., & SHIN, Y.-S. "Current IMT-2000 R&D status and views in Korea," *IEEE Commun. Mag.*, pp. 160–164, Sept. 1998.

WESEL, E. K. *Wireless Multimedia Communications: Networking Video, Voice, and Data.* Addison-Wesley, Reading, MA, 1998.

WHIPPLE, D. P. "The CDMA standard," *Appl. Microwave Wireless*, pp. 24–39, 1994.

WONG, D., & LIM, T. J. "Soft handoffs in CDMA mobile systems," *IEEE Pers. Commun.*, pp. 6–17, Dec. 1997.

YOUNG, W. R. "Advanced mobile phone service: Introduction, background, and objectives," *Bell Syst. Tech. J.*, vol. 58, no. 1, pp. 1–14, Jan. 1979.

YUE, ON-CHING. "Spread spectrum mobile radio, 1977–1982," *IEEE Trans. Veh. Technol.*, vol. 32, no. 1, pp. 98–105, Feb. 1983.

ZENG, M., ANNAMALAI, A., & BHARGAVA, V. K. "Recent advances in cellular wireless communications," *IEEE Commun. Mag.*, vol. 37, no. 9, pp. 128–138, Sept. 1999.

CHAPTER 2

PROPAGATION CHARACTERISTICS OF WIRELESS CHANNELS

2.0 INTRODUCTION

One of the major limitations on the performance of a mobile communication system is the attenuation undergone by the signal as it travels from the transmitter to the receiver (Gilb 1965, IEEE 1988, Deli 1985, Cheu 1998). The path the signal takes from the transmitter to the receiver may be line-of-sight (LOS), as shown in Figure 2.1, in which the case the signal loss may not be severe.

However, in a typical urban surrounding, the path between the transmitter and receiver is indirect and the signal reaches the receiver through the processes of reflection, diffraction, refraction, and scattering from buildings, structures, and other obstructions in the path (Jake 1974). These means of signal transmission are examples of non–line-of-sight (N-LOS) propagation mechanisms (Clar 1968). Reflection occurs when a propagating electromagnetic signal meets an object that is much larger than the signal's wavelength. This happens when the signal enters a building. Reflection may occur at the walls of the building, as shown in Figure 2.2. Note that depending on the angle of incidence and the impedance of the wall, reflection may or may not be accompanied by refraction.

Diffraction occurs when the surface encountered by the electromagnetic wave has irregularities such as sharp edges. This leads to bending of the wave, making it possible to receive the signal even when no direct path exists between the transmitter and the receiver, as shown in Figure 2.3.

Scattering occurs when the medium through which the electromagnetic wave propagates contains a large number of objects smaller than the wavelength. The wave is scattered in all directions, as shown in Figure 2.4. This happens when an electromagnetic wave passes through a medium containing vegetation, clouds, or street signs, for example.

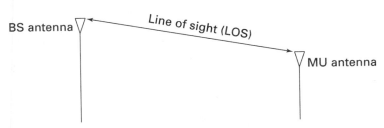

FIGURE 2.1 A direct path (line of sight) between two antennae.

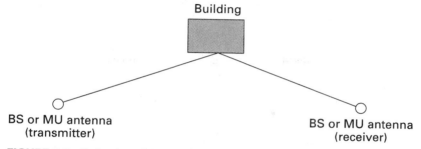

FIGURE 2.2 Reflection of the electromagnetic wave at a boundary.

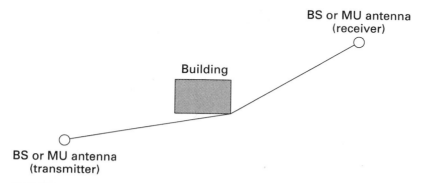

FIGURE 2.3 Diffraction of the electromagnetic wave at the edge of a building

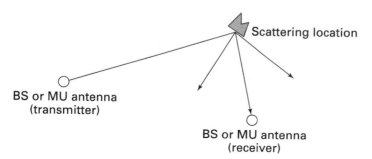

FIGURE 2.4 Scattering of the electromagnetic wave.

These non–line-of-sight (N-LOS) conditions (reflection, diffraction, and scattering) characterize most mobile communication transmissions. The free-space propagation models thus are not suited to calculate the attenuation undergone by the signal being received. The power detected by a receiver (MU or BS) is shown in Figure 2.5.

Observing the power at a separation of several kilometers, we see a steady decrease in power. This is the simple attenuation of power. It does not tell the whole story, however. If we zoom in to a distance of a couple of kilometers, we will see that the power fluctuates around a mean value and these fluctuations have a somewhat long period. This phenomenon is referred to as *long-term fading* or *large-scale fading*, and, as we shall see later, this can be described in terms of a *lognormal* distribution (Brau 1991, Bull 1977). If we zoom in further and examine the power over a few

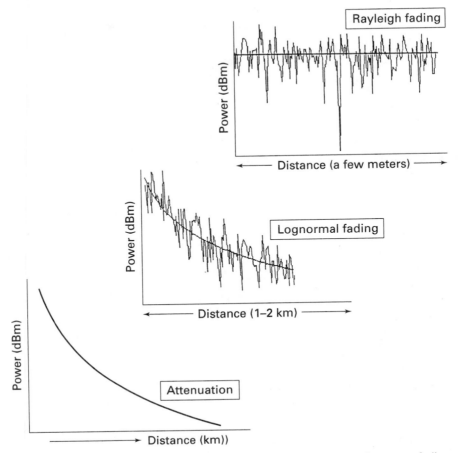

FIGURE 2.5 Power loss showing the three major effects: attenuation, long-term fading, and short-term fading.

hundred meters, we will see that the power is fluctuating more rapidly. The phenomenon giving rise to these fluctuations is referred to as *short-term fading* or *small-scale fading*, which, as we shall see later, can be described in terms of the *Rayleigh* distribution. We can thus see that the nature of the received signal is more complex than a simple description based on attenuation alone (Auli 1979, Akki 1994, Samp 1997). We will now look at these three phenomena associated with the propagation of wireless signals in greater detail.

2.1 ATTENUATION

Consider a very simple case where there is a direct path between the transmitter and receiver, as shown in Figure 2.6. In the absence of substantial obstacles in the path of the signal, the received signal power, P_r, follows the inverse square law (Gilb 1965, Jake 1974):

$$P_r \propto d^{-2},$$

(2.1)

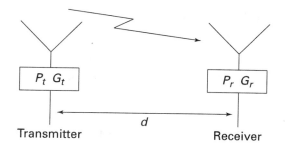

FIGURE 2.6 Free-space propagation geometry.

with d being the distance between the transmitter and the receiver. The received power, $P_r(d)$, is normally expressed as

$$P_r(d) = \frac{P_t G_r G_t \lambda^2}{(4\pi)^2 d^2 L}, \quad d > 0. \tag{2.2}$$

The parameters G_t and G_r are the transmitter and receiver gains, respectively (Jake 1974). The operating wavelength is λ, and L (≥ 1) represents any additional losses in the system not related to propagation losses, such as filter losses and antenna loss, etc. (i.e., hardware losses).

The transmitted power is P_t, in milliwatts. The product $P_t G_t$ is referred to as the *equivalent isotropic radiated power* (EIRP). Another parameter of interest is the *free-space* loss, L_{free}, given by

$$L_{\text{free}} = -20 \log_{10}\left(\frac{\lambda}{4\pi d}\right) \text{dB}. \tag{2.3}$$

The expression for EIRP can be used to estimate the received power at any distance from the transmitter using eq. (2.2),

$$P_r(d) = P_r(d_{\text{ref}})\left(\frac{d_{\text{ref}}}{d}\right)^2, \tag{2.4}$$

where d_{ref} is a reference distance. The reference distance must be smaller than the typical distances encountered in wireless communication systems and must fall in the far-field region of the antenna, so that losses beyond that point are purely distance-dependent effects (Rapp 1996b, Pahl 1995). This value is typically in the range of 100–1000 m. The power at a distance of d_{ref}, $P_r(d_{\text{ref}})$, is again in milliwatts. Making use of this approach, eq. (2.4) can be written as

$$P_r(d) \text{ dBm} = 10\log_{10}[P_r(d_{\text{ref}})] + 20 \log_{10}\left[\frac{d_{\text{ref}}}{d}\right]. \tag{2.5}$$

The actual loss suffered by signal at a frequency of f_0 (MHz) at a distance d (km) under the conditions of a flat, obstruction-free terrain, L_{free}, can also be obtained by rewriting the free-space propagation loss given in eq. (2.3):

$$L_{\text{free}} = -20 \log_{10}\left(\frac{c/f}{4\pi d}\right) \text{dB}, \tag{2.6}$$

where c is the free-space velocity of the electromagnetic wave, equal to 3×10^8 m/s, and f is the frequency. Expressing the frequency in megahertz, eq. (2.6) can now be expressed as

$$L_{\text{free}} = 32.44 + 20 \log_{10}(f) + 20 \log_{10}(d), \tag{2.7}$$

where d must be larger than 1 km, f is in megahertz, and d is in kilometers.

EXAMPLE 2.1

Consider an antenna transmitting a power of 10 W at 900 MHz. Calculate the received power at a distance of 2 km if propagation is taking place in free space.

Answer   The wavelength at 900 MHz is $3e^8/9e^6 = (1/3)$ m. Using eq. (2.2) with $L = 1$, $G_t = G_r = 1$, and $d_{\text{ref}} = 100$ m, the received power at d_{ref} is

$$\frac{10}{(4\pi)^2}\left(\frac{1}{3 \times 100}\right)^2 = 0.7\,\mu\text{W}$$

or -31.5 dBm. Now, using eq. (2.4), the power at a distance of 2 km is -57.5 dBm. Using eq. (2.6), the loss is $32.44 + 20 \log_{10}(900) + 20 \log_{10}(2) = 97.5$ dB. Therefore, the received power is $10 \log_{10}(10^4) - 97.5 = -57.5$ dBm. ∎

The case of free-space propagation discussed above, of course, is an ideal case, and the power often attenuates at a rate much higher than predicted by the inverse square law. Hence, the loss experienced in most cases will be considerably higher (Rapp 1995, Deli 1985). It is possible to explore whether, instead of the power decreasing as the inverse of the square of the distance from the transmitter, the power loss follows an exponent of a higher order, ν, so that the received power P_r can be expressed as

$$P_r \propto d^{-\nu}, \tag{2.8}$$

where the loss parameter, ν, has a minimum value of 2 in free space and takes a value larger than 2 when free-space propagation conditions do not exist. The received power under non–line-of-sight (N-LOS) conditions can now be written by combining equations (2.4) and (2.8) as

$$P_r(d)\ \text{dBm} = 10 \log_{10}[P_r(d_{\text{ref}})] + 10\nu \log_{10}\left[\frac{d_{\text{ref}}}{d}\right], \tag{2.9}$$

where d_{ref} is the reference distance (100 m). Typical plots of the received power for a few different values of ν are shown in Figure 2.7, indicating that losses tend to grow when N-LOS conditions prevail. The higher values of ν correspond to city and urban areas, and lower values of ν correspond to suburban or rural areas.

EXAMPLE 2.2

In Example 2.1, calculate the received power in dBm if the loss parameter ν is 2.5, 3, and 4.

Answer   The power at a short distance ($d_{\text{ref}} = 100$ m) is obtained from eq. (2.2) as -31.5 dBm. Now, using eq. (2.9), the power at a distance of 2 km is

-64 dBm for $\nu = 2.5$

-70.5 dBm for $\nu = 3$

-83.5 dBm for $\nu = 4$ ∎

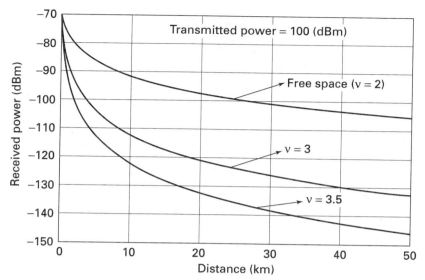

FIGURE 2.7 Received power for different values of loss parameter ν ($\nu = 2$ corresponds to free space). Increased loss is seen as ν goes up.

A number of models have been proposed to predict the loss of a signal as it travels to the receiver. These loss models combine empirical measurements in many cities and some physical models to account for the different ways in which radio-frequency signals travel. As mentioned earlier, the signals from the transmitter are reflected, scattered, refracted, diffracted, and absorbed by the terrain, which consists of buildings, vegetation, and other ground effects, before they reach the receiver. These various physical phenomena combine to produce significant attenuation in the signal. The loss of power is often accompanied by fluctuations, as shown in Figure 2.5, in the mean or median value of the power received, making predictions of received power (and design considerations based on received power) a little bit more difficult. A few different modes of transmission and reception of wireless signals (Meli 1993, Mehr 1999) are shown in Figures 2.8, 2.9, and 2.10.

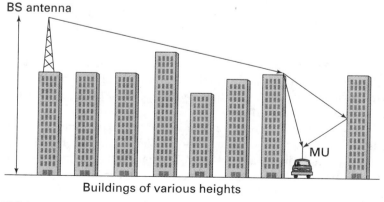

FIGURE 2.8 The signal reaches the receiver through reflection and diffraction.

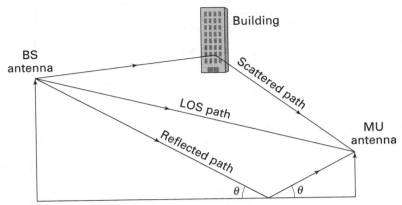

FIGURE 2.9   The signal reaches the receiver through reflection and scattering, as well as via a direct path.

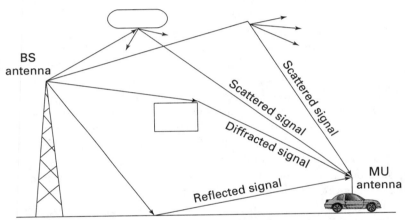

FIGURE 2.10   The most general case of signal reception, consisting of a direct path, a reflected path, a scattered path, and a diffracted path.

In Figure 2.8 the signal from the BS can reach the MU through the processes of diffraction and reflection. This is one of the common ways in which the transmission/reception of signals takes place in a city with tall buildings. In Figure 2.9 the signal from the BS takes three separate paths to reach the MU. In addition to the line-of-sight path, a path is provided by ground reflection and another path is provided by scattering from the building. The most general case is shown in Figure 2.10. In this case, the signal reaches the MU after undergoing reflection, scattering, diffraction, and possibly refraction after interaction with different structures in its path, or what may be described as "terrain-dependent" effects.

Combining all these different ways in which the signal can reach the receiver, Okumura and colleagues proposed a model to predict the median signal loss based on measurements conducted in and around Tokyo (Okum 1968). The model is based on the premise that it is possible to calculate the free-space loss between any two points for a base station antenna height of 200 m and a mobile antenna height of 3 m, with correction factors added to account for the terrain. Additional correction factors can be included to account for other factors, such as street orientation. These correction factors take into account the following features, among others:

Antenna heights and transmission frequency (or wavelength)

Suburban, quasi-open space; hilly terrain; etc.

Diffraction loss due to mountains

Lakes

Road shape

Even though this model is very reasonable in predicting the signal loss, it is not easy to use since correction factors must be incorporated for every conceivable scenario, or the results will have to be extrapolated. To overcome some of these problems, Hata proposed a reasonably simple model that fits Okamura's graphical results (Hata 1980, 1985). Another model that provides for transmission loss is Lee's model (Lee 1980, 1993, 1997).

Note that the Hata model and Lee's model merely provide a formula for path loss with respect to distance. But path loss alone is not sufficient to characterize the channel through which the signal is traveling. A typical path loss observed (Figure 2.5) shows the effects of fading, both long-term and short-term.

Before examining the relationship between attenuation and fading, we will look at the Hata model as well as Lee's model for calculating the propagation loss. We will also look at the differences between indoor and outdoor propagation, and examine ways of modeling and calculating indoor propagation losses.

2.1.1 Hata Model

The Hata model (Hata 1980) is a significant improvement over Okumura's model (Okum 1968) for the prediction of propagation losses. The propagation in different geographical regions is taken into consideration using correction factors that have been empirically derived. The starting point in the loss prediction is the propagation in an urban area. The loss models are generally given in terms of the median loss rather than the mean loss. The losses are given in terms of the effective height (h_b) of the BS antenna and the height (h_{mu}) of the MU antenna (measured above the ground). The approach for estimating the effective height of the BS antenna is shown in Figure 2.11. Typically, BS

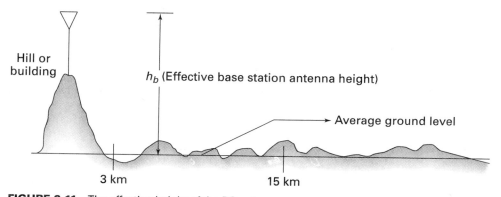

FIGURE 2.11 The effective height of the BS antenna.

antennae are mounted on top of existing buildings or other tall structures. Between 3 and 15 km from the base of the antenna to the MU antenna, the effective height is estimated to be above the average level of the terrain, as shown in the figure.

In the Hata model, median path loss, L_p (dB), in urban areas is given by

$$L_p \text{ (dB)} = 69.55 + 26.16 \log_{10}(f_0) + (44.9 - 6.55 \log_{10} h_b)\log_{10} d$$
$$- 13.82 \log_{10} h_b - a(h_{mu}), \tag{2.10}$$

where

f_0 = carrier frequency (MHz)

d = separation between base station and mobile unit (km)

h_b = height of the base station antenna (m)

h_{mu} = height of the mobile unit antenna (m)

$a(h_{mu})$ = correction factor for mobile unit antenna height.

For large cities, the correction factor $a(h_{mu})$ is given by

$$a(h_{mu}) = 3.2[\log_{10}(11.75 h_{mu})]^2 - 4.97 \quad (f_0 \geq 400 \text{ MHz}). \tag{2.11}$$

For small and medium cities, the correction factor is

$$a(h_{mu}) = [1.1 \log_{10}(f_0) - 0.7] h_{mu} - [1.56 \log_{10}(f_0) - 0.8]. \tag{2.12}$$

For suburban areas, the median loss, L_{sub}, is given by

$$L_{sub} \text{ (dB)} = L_p - 2\left[\log_{10}\left(\frac{f_0}{28}\right)\right]^2 - 5.4, \tag{2.13}$$

where L_p is the median loss in small-to-medium cities. For rural areas, the median loss, L_{rur}, is given by

$$L_{rur} \text{ (dB)} = L_p - 4.78[\log_{10}(f_0)]^2 + 18.33 \log_{10} f_0 - 40.94, \tag{2.14}$$

where L_p is the median loss in small-to-medium cities. The only limitation of the Hata model is the requirement that the distance, d, between the BS and MU exceed 1 km.

The typical loss as given by the Hata model is shown in Figure 2.12. Note that the difference in correction factors between a large and a medium-to-small city is only about 1 dB. This makes the loss curves for these two cases appear to be very close.

The Hata model can also be used to estimate the value of the loss parameter ν. The received power, $P_r(d)$ (dBm), for any separation, d (km), between MU and BS can be expressed as

$$P_r(d)(\text{dBm}) = P_t - P_{loss}(d), \tag{2.15}$$

where P_t (dBm) is the transmitted power and $P_{loss}(d)$ is the loss calculated from Hata's model. The received power can also be expressed as

$$P_r(d) \propto \left(\frac{1}{d}\right)^\nu, \tag{2.16}$$

where ν is the path loss exponent.

The loss at the two distances d_{ref} and d (>1 km) can be expressed as

$$P_{loss}(d_{ref}) \propto 10\nu \log_{10}(d_{ref}) \tag{2.17}$$

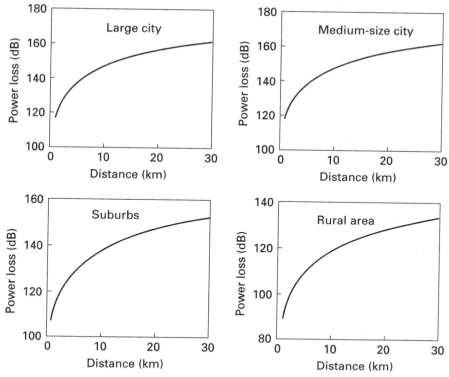

FIGURE 2.12 Loss calculations based on the Hata model for four different environments. Carrier frequency = 900 MHz, base station antenna height = 150 m, MU antenna height = 1.5 m.

and

$$P_{loss}(d) \propto 10\nu \log_{10}(d), \qquad (2.18)$$

and the expression for ν becomes

$$\nu = \frac{P_{loss}(d) - P_{loss}(d_{ref})}{10[\log_{10}(d) - \log_{10}(d_{ref})]}, \qquad (2.19)$$

where d_{ref} is 100 m and P_{loss} is given by

$$P_{loss}(d_{ref}) = 10 \log_{10}\left[\frac{(4\pi d_{ref})^2}{\lambda^2}\right], \qquad (2.20)$$

obtained from eq. (2.2).

EXAMPLE 2.3

Find approximate values of the loss parameter, ν, using the Hata model for the four geographical regions: large city, small-to-medium city, suburb, and rural area.

Answer Using Figure 2.12, the loss values at a distance of 5 km are 131.36 dB, 131.34 dB, 121.40 dB, and 102.8 dB, respectively, for large city, small-to-medium city, suburb, and rural area. Using eq. (2.19), the respective values of ν are 4.05, 4.04, 3.3, and 2.11. Note that these values are approximate, and d must be 2 km or more to get reasonably stable values of ν. ∎

Extension of Hata Model to PCS　The attenuation model proposed by Hata can be extended to PCS (personal communication system) environments (Meli 1993, Rapp 1996b, Garg 1997, Saun 1999). The median loss in urban areas, L_p (dB), can be expressed as

$$L_p(\text{dB}) = 46.3 + 33.93 \log_{10}(f_0) - 13.82 \log_{10}(h_b) - a(h_{mu})$$
$$+ [44.9 - 6.55 \log_{10}(h_b)] \log_{10} d + \text{Corr} \qquad (2.21)$$

where Corr is the additional correction factor given by

$$\text{Corr} = \begin{cases} 0 \text{ dB for medium city and suburban areas} \\ 3 \text{ dB for metropolitan areas} \end{cases} \qquad (2.22)$$

This model is valid for the following parameters only:

f_0: 1500–2000 MHz

h_b: 30–200 m

h_{mu}: 1–10 m

d: 1–20 km

The Hata model is applicable only for distances beyond 1 km, and thus cannot be used in microcells (see Chapter 4), where the distance between the transmitter and receiver may be only a few hundred meters. Newer models (the Har, Xia, and Bertoni model; and the Walfisch-Ikegami model) are available for loss prediction over short ranges, and are discussed in Section C.2 in Appendix C.

2.1.2　Lee's Model

Another model available for the prediction of path loss is the one proposed by Lee (Lee 1980, 1993). Based on measurements taken in three cities, including Philadelphia, the model provides the loss in signal strength on an area-to-area basis. These loss values can then be used as a set of initial values to get a point-to-point loss prediction. The median loss (area-to-area) at a distance d (km), $L(d)$, can be expressed as

$$L(d)\,(\text{dB}) = L_0 + 10\nu \log_{10}(d) + \alpha_c, \qquad (2.23)$$

where L_0 is the loss at 1 km, ν is the loss parameter, and α_c is a correction factor. The predictions were done for a carrier frequency of 900 MHz, a transmitting (BS) antenna of height 30.5 m, and a receiving antenna height of 3 m. The correction factor, α_c, is included to account for any change in the standard parameters used in the model and can be expressed as

$$\alpha_c = 10 \log_{10}(F). \qquad (2.24)$$

The parameter F is the product of various correction factors,

$$F = F_1 F_2 F_3 F_4 F_5, \qquad (2.25)$$

where

$$F_1 = \left[\frac{\text{actual base station antenna height (m)}}{30.5} \right]^2 \qquad (2.26a)$$

$$F_2 = \left[\frac{\text{actual transmitted power (W)}}{10} \right] \tag{2.26b}$$

$$F_3 = \left[\frac{\text{actual gain of base station antenna}}{4} \right] \tag{2.26c}$$

$$F_4 = \left[\frac{\text{actual mobile unit antenna height (m)}}{3} \right]^2, \quad \text{height} \geq 3 \text{ m} \tag{2.26d}$$

F_5 = differential antenna gain correction factor at the mobile unit. (2.26e)

The loss based on the point-to-point model, $L_p(d)$, can now be expressed as

$$L_p(d) \text{ (dB)} = L(d) + 20 \log_{10}\left(\frac{h_{\text{eff}}}{10}\right), \tag{2.27}$$

where h_{eff} is the effective height of the antenna, taking into account the location of the antenna in relation to the terrain, as shown in Figure 2.13.

Note that the Hata model is easier than most of the other models for predicting propagation losses and will be used for the computation of the cell radius in Chapter 4.

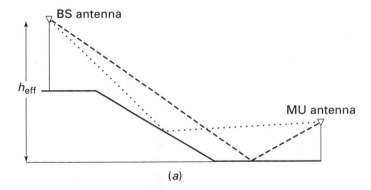

(a)

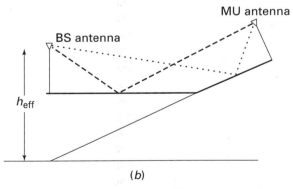

(b)

FIGURE 2.13 The concept of h_{eff}. (a) MU antenna on the ground. (b) MU antenna above the ground.

2.2 INDOOR PROPAGATION MODELS

The models discussed so far can be used to predict signal strength outdoors, but are not sufficient to predict signal strength indoors. Typical examples of indoor propagation are found in shopping malls, and in office buildings with different floors and room configurations. The variety of objects that reflect, scatter, and diffract the wireless signals include the different types of ceiling fixtures within a building, the different types of displays, and the different types of furniture. A model that can predict signal loss must take these different features into account (Bult 1987, 1989; Ders 1994; Molk 1991; Gane 1991). A building may have a very large room with no partitions and very few obstacles, or a very large room with a large number of obstacles. It is also possible for a building to have similar conditions in small rooms. The number of obstacles is not the only major factor in the determination of signal loss. The material used for partitioning the building also can adversely affect the strength of signal reaching the mobile user. This poses the challenge of coming up with a very general model that can predict the signal losses under all these conditions. The best approach for modeling propagation indoors is to classify these various environments into different "zone" configurations (Turk 1991b, Walk 1983, Samp 1997, Lots 1992, Rapp 1996b, Pars 1983). These configurations are based on the location of the base station and how the base station handles the traffic, and whether the base station is inside or outside the building.

2.2.1 Extra Large Zone

In an extra large zone there is a single base station outside the buildings that handles all the traffic in the buildings. This situation is ideal for a region that has a number of small offices or shops in adjoining buildings, as shown in Figure 2.14*a*.

Since the transmitting BS antenna is located outside the building, the signal will incur path-dependent losses as well as penetration-dependent losses because the signal must first traverse to the building boundary and then penetrate various floors and walls of the different buildings to reach the mobile unit. The median loss at a distance, d, from the transmitter in the extra large zone, $L_{ELZ}(d)$, can be expressed as

$$L_{ELZ}(d)\,(dB) = 10\log_{10}\left[L_d(d_0)\left(\frac{d}{d_0}\right)^{v_d}L_B(d_0)\left(\frac{d}{d_0}\right)^{v_B}A_B\right], \qquad (2.28)$$

where

$L_d(d_0)$ = attenuation due to propagation at $d = d_0$

$L_B(d_0)$ = attenuation due to building at $d = d_0$

v_d = attenuation factor of the propagation path loss with respect to distance

v_B = building attenuation factor

A_B = loss factor due to building penetration.

Note that quantities $L_d(d_0)$ and $L_B(d_0)$ are determined by the density of the obstacles present in the path as well as the frequency dependence of the losses. The distance-dependent loss parameter, v_d, is about 2 if there are very few obstacles (nearly free-space propagation), and if there are a lot of scattering centers v_d can be in the range of 3–6, as described in Section 2.1. The parameter v_B usually is in the range of 0.5–1.5. The quantity A_B depends on the difference between the heights of the transmitting and receiving antennae as well as on the materials in the building.

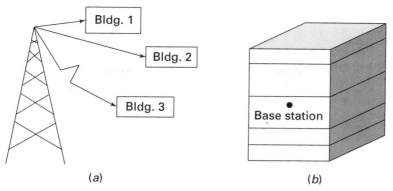

FIGURE 2.14 (*a*) Extra large zone. A single BS antenna is stationed outside the building. (*b*) Large zone. The base station is located in the building.

2.2.2 Large Zone

In a large zone, the building is very large but has a low population density. In this classification, a single base station is housed within the building itself. This situation is illustrated in Figure 2.14*b*.

The median loss in a large-zone system, $L_{LR}(d)$, can be expressed by simplifying eq. (2.28), for the loss in the extra large zone, as

$$L_{LR}(d)\,(\text{dB}) = 10\log_{10}\left[L_d(d_0)\left(\frac{d}{d_0}\right)^{\nu_0}\right], \tag{2.29}$$

where ν_0 is the loss parameter. It will be in the range 2–3 if the transmitter and receiver are on the same floor, and it will be greater than 3 if they are on different floors.

2.2.3 Middle Zone

In the middle zone, the building structure is large and also heavily populated. This is a situation that commonly exists in shopping malls.

In this classification, a number of base stations are located within the building to serve the mobile phone users. A typical scenario for a single base station located within the building structure is shown in Figure 2.15. The median loss in the middle-zone system, $L_{ML}(d)$, can be expressed as

$$L_{ML}(d)\,(\text{dB}) = 10\log_{10}\left[\left(\frac{4\pi f_0 d}{c}\right)^2 F(d)^{k_1} W(d)^{k_2} R(d)\right], \tag{2.30}$$

where

$f_0 = $ carrier frequency

$c = $ velocity of the electromagnetic wave

$F(d) = $ floor loss

$W(d) = $ wall loss

$R(d) = $ reflection loss

$k_1 = $ number of floors in the path

$k_2 = $ number of walls in the path.

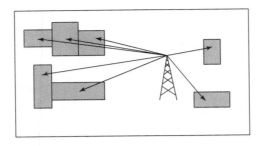

FIGURE 2.15 Middle zone. A base station is located within the structure to serve the MU in a portion of the structure. Similar base stations are present in other parts of the building.

2.2.4 Small Zone and Microzone

A building can have many partitions, with the penetration of the signal depending heavily on the material properties of the walls and partitions. This requires the provision of one base station for each room in the building. The loss models for the small zone can be obtained using the results for the large zone by incorporating the appropriate value of the loss parameter, ν, based on the number and types of obstacles between the transmitter and receiver. If LOS conditions exist, ν will be 2; for N-LOS conditions, ν will be close to 3.5.

The conditions of the building may be such that there is heavy traffic within each room, and this will require the use of several base stations within a single room. Path loss calculations for the microzone can be done similarly to those for the small zone, using slightly smaller values of the loss parameter ν.

2.3 FADING

As discussed earlier, the transmission characteristics are not determined by attenuation alone. The loss or attenuation observed may also fluctuate with distance and time, and this can be described in terms of fading.

When a signal leaves the transmitting antenna, it gets reflected, scattered, diffracted, or refracted by the various structures in its path (Arre 1973, Jake 1974, Gupt 1985, Fleu 1996, Hamm 1998, Stee 1999). We examined the signal loss arising from the presence of various obstacles in the channel. We also observed that the transmission loss fluctuates around a mean or median value. This aspect of the transmission loss curve, where the received signal loses its deterministic nature and becomes random in time and space, is described in terms of fading. In other words, fading is the process that describes the fluctuations in the received signal as the signal travels to the receiving antenna. Fading can be described either in terms of the primary cause (multipath or Doppler), the statistical distribution of the received envelope (Rayleigh, Rician, or lognormal), the duration of fading (long-term or short-term), or fast versus slow fading (Turi 1972, Stei 1987). We will look at these different characterizations of fading to understand their origin and specific consequences. We will also look at the various forms of fading to establish the relationships that exist among them.

2.3.1 Multipath Fading

Multipath fading, as the name suggests, arises from the existence of multiple paths between the transmitter and receiver. When a signal leaves the transmitting antenna, it can take a number of different paths to reach the receiver, as shown in Figure 2.16.

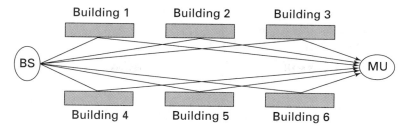

FIGURE 2.16 The multipath concept.

The mobile unit receives signal components that are scattered, reflected, diffracted, etc. by the buildings or by other artificial or natural structures, creating a number of different paths (Hodg 1990a, Ikeg 1980, Rumm 1986, Brau 1991, Rapp 1996b).

We assume that the signal components scattered by these different structures arrive at the receiving antenna independently of each other, as shown in Figure 2.17. Under this condition, the signal received at the antenna can be expressed as the vector sum of the components coming from these structures.

For the time being, we also assume that the receiver is stationary. The received signal $e_r(t)$ can then be expressed as a sum of delayed components:

$$e_r(t) = \sum_{i=1}^{N} a_i p(t - t_i), \qquad (2.31)$$

where a_i is the amplitude of the scattered component, $p(t)$ is the transmitted pulse shape, and t_i is the time taken by the pulse to reach the receiver. N is the number of different paths taken by the signal to reach the receiver. Note that we also make the assumption that no direct path exists between the transmitting and receiving antennae. Instead of writing the received signal amplitude as the sum of delayed components, we can also use phasor notation to represent the received signal:

$$e_r(t) = \sum_{i=1}^{N} a_i \cos(2\pi f_0 t + \phi_i), \qquad (2.32)$$

where f_0 is the carrier frequency. The ith signal component has an amplitude of a_i and a phase of ϕ_i. The difference between eq. (2.31) and eq. (2.32) is the assumption of a single signal of carrier frequency f_0 in eq. (2.32) versus the transmission of a pulse in eq. (2.31). We will go back to eq. (2.31) later.

Equation (2.32) can be rewritten in terms of *in-phase* and *quadrature* notation as

$$e_r(t) = \cos(2\pi f_0 t) \sum_{i=1}^{N} a_i \cos(\phi_i) - \sin(2\pi f_0 t) \sum_{i=1}^{N} a_i \sin(\phi_i), \qquad (2.33)$$

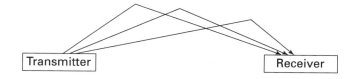

FIGURE 2.17 Conceptual "ray" diagram of the multipath between the transmitter and receiver.

where the first summation is identified as the *in-phase* term and the second summation is identified as the *quadrature* term.

If the locations of the structures are completely random, one can safely assume that the phase ϕ_i will be uniformly distributed (Papo 1991) in the range $(0, 2\pi)$. Under conditions of large N, the amplitude of the received signal can then be expressed as

$$e_r(t) = X\cos(2\pi f_0 t) - Y\sin(2\pi f_0 t), \tag{2.34a}$$

where

$$X = \sum_{i=1}^{N} a_i \cos(\phi_i), \quad Y = \sum_{i=1}^{N} a_i \sin(\phi_i). \tag{2.34b}$$

X and Y will be independent, identically distributed Gaussian random variables by virtue of the Central Limit Theorem (Papo 1991). Under these conditions, the envelope of the received signal, A, given by $(X^2 + Y^2)^{1/2}$, will be Rayleigh distributed. The envelope can be recovered through demodulation, which is discussed in Chapter 3. The probability density function $f_A(a)$ is given by

$$f_A(a) = \frac{a}{\sigma^2}\exp\left(-\frac{a^2}{2\sigma^2}\right)U(a), \tag{2.35}$$

where the parameter σ^2 is the variance of the random variable X (or Y) and $U(.)$ is the unit step function. Note that if the envelope of the signal is Rayleigh distributed, the power, P, will have an exponential distribution, given by

$$f_P(p) = \frac{1}{2\sigma^2}\exp\left(-\frac{p}{2\sigma^2}\right)U(p). \tag{2.36}$$

The Rayleigh and exponential probability density functions are shown in Figure 2.18. The Rayleigh-distributed envelope is characterized by a mean given by

$$E(A) = \sigma\sqrt{\frac{\pi}{2}} \tag{2.37}$$

and a variance, σ_A^2, given by

$$\sigma_A^2 = \sigma^2\left[2 - \frac{\pi}{2}\right]. \tag{2.38}$$

Note that the Rayleigh distribution is also unique in terms of its ratio of the mean to the standard deviation:

$$\frac{E(A)}{\sigma_A} = 1.91. \tag{2.39}$$

Plots of the radio-frequency (rf) signal and the corresponding envelope under Rayleigh fading are shown in Figure 2.19. We see that the received signal power is random even in the absence of noise introduced by the electronic system, namely, the additive white Gaussian noise. This is a consequence of the existence of multiple paths and the randomness of the phase (Auli 1979). Multipath fading thus leads to fluctuations in the received signal when the MU moves from place to place.

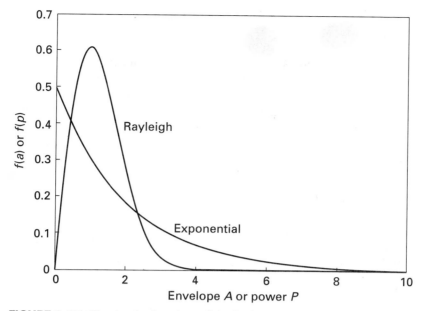

FIGURE 2.18 The density functions of the Rayleigh-distributed envelope and exponentially distributed power.

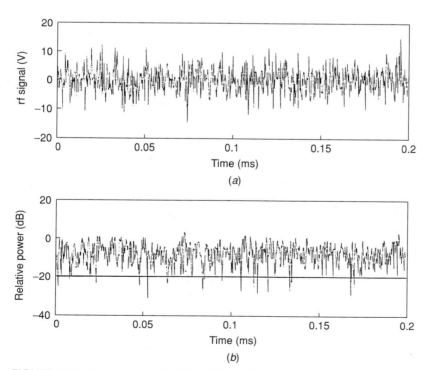

FIGURE 2.19 Rayleigh-faded rf signal (*a*) and its power (*b*). The plots were generated from 11 multiple paths. The envelope was obtained by demodulating the rf signal.

Note that we have not taken into account the fact that the MU may be moving at a certain speed.

To understand the implications of this fluctuation in the received power, consider the following scenario. If the receiver is designed to operate at an acceptable level only if a certain minimum power, P_{thr}, is being received, the receiver goes into outage whenever the power goes below this threshold value (Jake 1974, Ikeg 1980, Pale 1991). In Figure 2.19, we can clearly see that the system goes into outage if the threshold is set to -20 dB of relative power (indicated by the line drawn parallel to the frequency axis in Figure 2.19b). The *outage probability*, p_{out}, can now be calculated as

$$p_{\text{out}} = \int_0^{P_{\text{thr}}} f(p)\,dp = \int_0^{P_{\text{thr}}} \frac{1}{P_0}\exp\left(-\frac{p}{P_0}\right) dp = 1 - \exp\left(-\frac{p}{P_0}\right), \qquad (2.40)$$

where P_0 is the average power, given by $2\sigma^2$. One of the adverse consequences of fading is the existence of outage. When outage occurs, the performance of the wireless system becomes unacceptable.

EXAMPLE 2.4

Consider the case of a Rayleigh-fading channel. If the average power being received is 100 μW, what is the probability that the received power will be less than 50 μW?

Answer   Using eq. (2.40), the probability is $[1 - \exp(-50/100)] = 0.3935$.   ■

EXAMPLE 2.5

If the minimum required power for acceptable performance is 25 μW, what is the outage probability in a Rayleigh channel with an average received power of 100 μW?

Answer Using eq. (2.40), the probability of outage is $[1 - \exp(-25/100)] = 0.2212$, or 22.1%.   ■

2.3.2 Dispersive Characteristics of the Channel

The fluctuation of the received power is not the only effect of fading. Fading may also affect the shape of the pulse as it is being transmitted through the channel (Cox 1972, 1975; Hash 1979, 1989). Consider the diagram of multipath fading shown in Figure 2.20, corresponding to eq. (2.31). Only four different paths are shown. Because of the different paths taken, the replicas of the pulse will arrive at the receiver at four different times. If these pulses are not resolvable, the effect of the multipath is to produce a broadened pulse, as shown by the envelope of the overlapping pulses. In other words, the multipath effect can result in broadening of the transmitted pulse, leading to intersymbol interference (ISI) (see Section B.6, Appendix B).

The situation shown in Figure 2.20 can be simulated using MATLAB. The results are shown in Figure 2.21. A Gaussian pulse of width $\sigma_d = 14.14$ ms is being transmitted through a wireless channel. Ten different paths have been selected to replicate the scenario described in the previous paragraph. These pulses have different time delays, randomly chosen and having random power as they reach the receiver. A

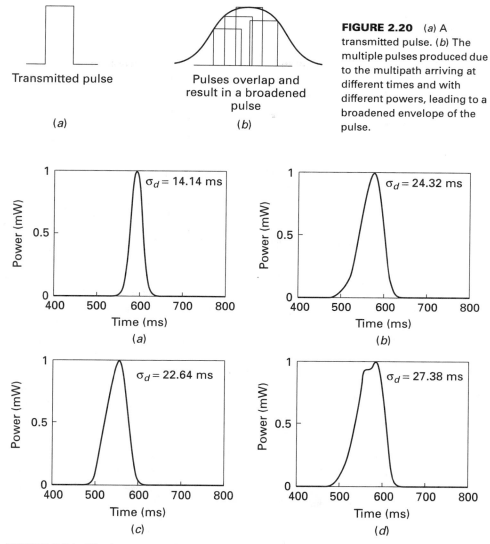

FIGURE 2.20 (*a*) A transmitted pulse. (*b*) The multiple pulses produced due to the multipath arriving at different times and with different powers, leading to a broadened envelope of the pulse.

Transmitted pulse

(*a*)

Pulses overlap and result in a broadened pulse

(*b*)

$\sigma_d = 14.14$ ms

(*a*)

$\sigma_d = 24.32$ ms

(*b*)

$\sigma_d = 22.64$ ms

(*c*)

$\sigma_d = 27.38$ ms

(*d*)

FIGURE 2.21 The frequency-selective fading channel simulated using MATLAB.

single pulse corresponding to one of the paths is shown in Figure 2.21*a*; this pulse is simply a delayed version of the transmitted pulse. In the absence of multiple paths, the pulse width remains the same as that of the transmitted pulse. Figures 2.21*b*, *c*, and *d* show the received pulse (the sum of the 10 multipath components) for three different simulations. The power has been normalized to unity. The standard deviation of each pulse is also indicated. It is obvious that the pulses have broadened. This is demonstrated by the increase in the pulse width of the received pulse compared with the transmitted pulse (Figure 2.21*a*). The situation depicted in Figure 2.20*b* is shown in Figures 2.21*b*, *c*, and *d*.

This dispersive behavior of the channel can be qualitatively described in the following manner. (The reasons for referring to this behavior of the channel as dispersive will be clear later.) Consider the transmission of a very narrow pulse

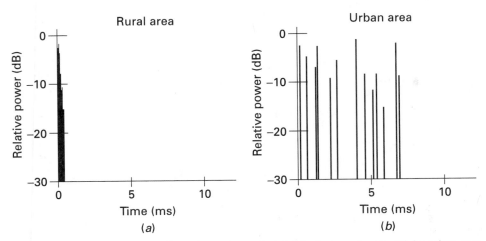

FIGURE 2.22 Impulse responses of two channels. (*a*) A typical rural area. (*b*) An urban area.

(delta function). The impulses corresponding to the multiple paths arrive at the receiver at different times and with different amounts of power depending on the nature of the scattering/reflection/refraction/diffraction that is responsible for the generation of the particular component. These multiple arrival times of signals with different powers can be used to define the impulse response of the channel as shown in Figure 2.22. For example, in a rural area, these impulses are likely to arrive at almost the same time and will most likely take a shorter time to reach the receiver. This is due to the fact that there are fewer tall structures, and therefore the paths are close to each other (Figure 2.22*a*). This means that the difference between arrival times of any information received will be too small to be observable or measurable. On the other hand, for an urban area (Figure 2.22*b*), the multiple paths will be more diverse and the received pulses will be spread out much more (Bult 1983, Hash 1993, Akai 1998, Hanz 1994). Under these conditions, information arriving in the form of finite-size pulses will overlap and result in a broadened pulse, as shown in Figure 2.21.

We can now write an expression for the average time taken by a pulse to reach the receiver. A typical impulse response is shown in Figure 2.23. The average delay, $\langle \tau \rangle$, experienced by the pulse as it traverses the channel is

$$\langle \tau \rangle = \frac{\displaystyle\sum_{i=1}^{N} p_i \tau_i}{\displaystyle\sum_{i=1}^{N} p_i}. \tag{2.41}$$

The quantity p_i represents the power coming along the ith path, and τ_i is the time taken by the ith component. The rms (root-mean-square) delay spread, σ_d, is given by

$$\sigma_d = \sqrt{\langle \tau^2 \rangle - \langle \tau \rangle^2}, \tag{2.42}$$

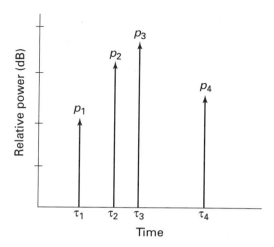

FIGURE 2.23 An impulse response of a wireless channel.

where $\langle \tau^2 \rangle$ is the mean square delay given by

$$\langle \tau^2 \rangle = \frac{\displaystyle\sum_{i=1}^{N} p_i \tau_i^2}{\displaystyle\sum_{i=1}^{N} p_i}. \tag{2.43}$$

In the ideal case, we would like to have

$$p_i = 0 \quad \text{if } i \neq 1,$$

indicating the existence of only a single path. In other words, if σ_d is zero, there will be no spreading of the pulse. Similarly, if σ_d is very high, we expect to see considerable pulse broadening. It is therefore possible to quantify the pulse broadening by defining the low-pass bandwidth of the channel to be inversely proportional to the r.m.s delay spread. The channel bandwidth, B_c, is approximately given by (Stei 1987, Rumm 1986, Rapp 1996b)

$$B_c = \frac{1}{5\sigma_d}. \tag{2.44}$$

The channel bandwidth can be identified as the coherence bandwidth of the channel, using the concept of envelope autocorrelation (see Appendix C, Section C.1).

It must be noted that this estimate of the channel bandwidth is only a means to characterize the channel by its frequency-dependent behavior, and not an exact representation of the bandwidth. If the channel bandwidth B_c is larger than the message bandwidth B_s, all the frequency components in the message will arrive at the receiver with little or no distortion, and ISI (intersymbol interference) will be negligible. This case is shown in Figure 2.24, and the channel is referred to as a *flat fading* channel. It is possible to characterize the rural areas as nearly "flat channels" since the likely values of σ_d are going to be very small, as shown in Figure 2.22a.

On the other hand, if the message bandwidth, B_s, is much larger than the channel bandwidth, B_c, different frequency components in the message will be

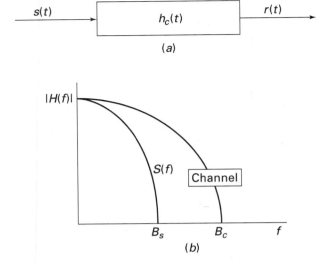

FIGURE 2.24 The baseband channel response (transfer function) of a "flat" channel along with the spectrum of the transmitted signal $S(f)$.

subjected to dispersive behavior, resulting in pulse broadening, and consequently the components will experience ISI (see Figure 2.25). The channel is thus classified as a *frequency-selective* channel (Bell 1963a, Jake 1974). The impulse response shown in Figure 2.22b can be viewed as an example of a frequency-selective channel. Note, however, that the distinction between flat and frequency-selective channels must be based on the relationship between the information bandwidth and σ_d, and not on the absolute value of σ_d. Thus, a flat channel becomes a frequency-selective one if the information is transmitted at a higher and higher data rate. We can now understand why we refer to a frequency-selective channel as a dispersive channel. The frequency-selective channel behaves as if different frequency components travel at different speeds (phenomenon of dispersion) and arrive at different times at the receiver, leading to pulse broadening.

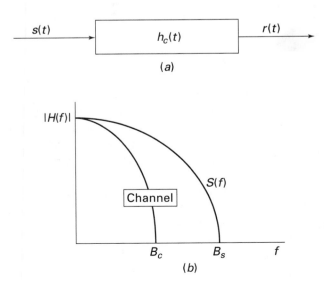

FIGURE 2.25 The baseband channel response (transfer function) of a "frequency-selective" channel (a) along with the spectrum of the transmitted signal (b), $S(f)$.

EXAMPLE 2.6

A typical impulse response of a wireless channel is given in Figure E2.6. If the data rate is 240 kbps, classify the channel as frequency-selective or flat.

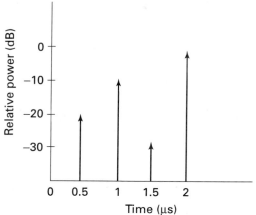

FIGURE E2.6

Answer $P(0.5) = 0.01, P(1) = 0.1, P(1.5) = 0.001,$ and $P(2) = 1,$ with $\sum_i P_i = 1.111.$

$$\langle \tau \rangle = \frac{0.5 \times 0.01 + 1 \times 0.1 + 1.5 \times 0.001 + 2 \times 1}{1.111} = 1.896\ \mu s$$

$$\langle \tau^2 \rangle = \frac{0.25 \times 0.01 + 1 \times 0.1 + 2.25 \times 0.001 + 4 \times 1}{1.111} = 3.695\ \mu s^2$$

$$\sigma_d = \sqrt{3.695 - 1.896^2} = 0.315\ \mu s$$

$B_c = 675$ kHz > 240 kHz. The channel is therefore "flat." ∎

We see that the two effects, the randomness of the received signal envelope and the frequency selectivity of the channel, are separate manifestations of the multipath propagation and can exist alone or in combination. However, in most practical cases, the received signal phases are random, and the Rayleigh distribution of the envelope will be exhibited irrespective of the frequency-selective nature of the channel.

One of the best ways to describe the frequency-dependent behavior of the channel is to use a two-ray model to represent the fading. In this model (Walk 1966, Clar 1968, Hash 1979, Ball 1982, Casa 1990), the impulse response, $h_c(t)$, of the channel is written as the sum of two Rayleigh fields having random phases and a delay of τ:

$$h_c(t) = a_1 \exp(j\psi_1)\delta(t) + a_2\exp(j\psi_2)\delta(t - \tau), \qquad (2.45)$$

where a_1 and a_2 are independent, identically distributed Rayleigh variables, and ψ_1 and ψ_2 are uniformly distributed in the range $(0, 2\pi)$. If a_2 is zero, we have a flat fading channel. By varying τ, it is possible to create channels with different bandwidths. Consider a simple case where a_1 and a_2 are scalars and deterministic with $b = a_2/a_1$. Assuming ψ_1 and ψ_2 to be deterministic and equal, eq. (2.45) for the impulse response can be rewritten as

$$h_c(t) = \delta(t) + b\delta(t - \tau), \qquad (2.46)$$

and the corresponding transfer function, $H_c(f)$, of the frequency-selective channel will be given by

$$H_c(f) = 1 + b \exp(-j2\pi f \tau). \tag{2.47}$$

The behavior of a typical channel may be observed by plotting the absolute value of the transfer function:

$$|H_c(f)| = \sqrt{1 + b^2 + 2b \cos(2\pi f \tau)}. \tag{2.48}$$

A plot of $|H_c(f)|$ is shown in Figure 2.26.

The transfer function has "notches" at intervals of $f\tau = 1$. For different values of τ, the bandwidth of the channel measured by the zero crossing will vary, causing the channel to go from being flat to being frequency selective. Note that in this simple description we have assumed the scaling factors to be deterministic. In practice, the scaling factors are random (Rayleigh distributed).

2.3.3 Time-Dispersive Behavior of the Channel

So far we have considered only the case of a stationary mobile unit. Consider now the case of a mobile unit traveling at speed v, as shown in Figure 2.27. The motion of the mobile unit will result in a Doppler shift in the frequency of the signal being received. The maximum Doppler shift, f_d, can be expressed as

$$f_d = f_0 \frac{v}{c}, \tag{2.49}$$

where c is the velocity of the electromagnetic wave in free space. Taking all possible directions into account, the instantaneous frequency, f_{in}, will be given by

$$f_{\text{in}} = f_0 + f_d \cos(\theta_i). \tag{2.50}$$

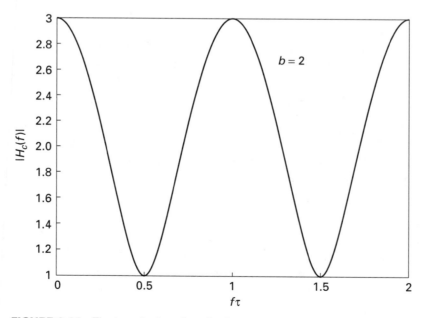

FIGURE 2.26 The transfer function of a "two-ray" model to describe the frequency-selective channel.

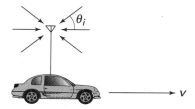

FIGURE 2.27 A mobile unit moving at speed *v.*

The received signal can once again be expressed as

$$e_r(t) = \cos(2\pi f_0 t) \sum_{i=1}^{N} a_i \cos[2\pi f_d \cos(\theta_i)t + \psi_i]$$

$$- \sin(2\pi f_0 t) \sum_{i=1}^{N} a_i \sin[2\pi f_d \cos(\theta_i)t + \psi_i], \tag{2.51}$$

where N is the number of multipaths available and ψ_i are the phases. If we make the assumption that N is sufficiently large, the envelope will be Rayleigh distributed as in the case of the stationary MU. A typical plot of the Doppler-faded signal is shown in Figure 2.28.

We can also calculate the power spectrum of the received signal (Gans 1972, Huan 1992, Pars 1992, Fleu 1996, Stee 1999). Equation (2.50) shows that the received signal will have a carrier frequency shifted by an amount $\pm f_d \cos(\theta_i)$. For a MU moving in a street, θ_i can be either zero or π, and the extreme values of the Doppler shift

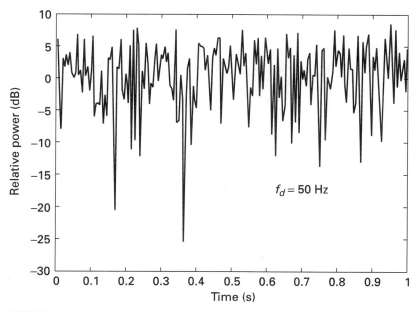

FIGURE 2.28 A Doppler-faded signal.

will be $\pm f_d$. If we make the assumption that the phases, ψ_i, and θ_i are uniform in the range $(0, 2\pi)$, the power spectrum, $S_d(f)$, of the received signal $e_r(t)$ can be expressed as (Gans 1972, Jake 1974)

$$S_d(f) = \begin{cases} \dfrac{\sigma}{\pi f_d \sqrt{1 - \left(\dfrac{f}{f_d}\right)^2}} & |f| \leq f_d \\ \\ 0 & \text{otherwise.} \end{cases} \tag{2.52}$$

A plot of the power spectrum is shown in Figure 2.29. It shows that most of the energy is concentrated around the maximum Doppler shift, f_d. Note that the power spectrum depends on the radiation pattern of the antenna and the polarization used.

Consider now the transmission of a short rf pulse as the vehicle is in motion. The motion of the MU will now introduce changes in the channel at a rate of f_d Hz. If the duration of the pulse is very short, the changes introduced by the motion will be very slow and will have very little or no impact on the transmission and, therefore, on the reception of the pulse. In other words, if the bandwidth of the signal measured in terms of the inverse of the pulse duration is much larger than the maximum Doppler shift, the channel will vary very slowly or will be a slow-fading channel. On the other hand, if the duration of the pulse is large, changes introduced in the channel from the motion of the mobile unit will be "fast" and thus will affect the transmission. In other words, for transmission at a very low data rate, a moving vehicle will introduce fast fading if the bandwidth of the signal is not much larger than the maximum Doppler shift.

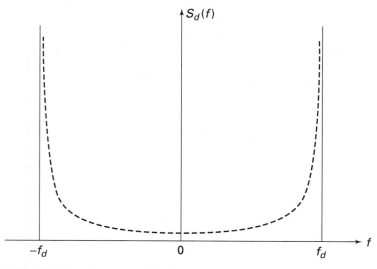

FIGURE 2.29 Spectrum of the Doppler-shifted signal.

The condition for slow versus fast fading can now be expressed in terms of the coherence time, T_c, of the channel (Stei 1987), measured in terms of the inverse of the maximum Doppler shift given by (see Section C.1, Appendix C)

$$T_c \approx \frac{9}{16 \pi f_d}. \tag{2.53}$$

If the pulse duration is smaller than T_c, the pulses are unlikely to undergo distortion (slow fading), and if the pulse duration is larger than the coherence time, the pulses undergo fast fading and will be distorted. Fast fading is thus a frequency-dispersive property of the channel brought on by the motion of the mobile unit. The difference between distortion and dispersion is explained in Appendix C.

EXAMPLE 2.7

Consider an antenna transmitting at 900 MHz. The receiver, a MU, is traveling at a speed of 30 km/h and is receiving/transmitting data at 200 kbps. Examine whether the channel fading is slow or fast.

Answer   The Doppler shift is given by

$$f_d = \frac{9 \times 10^8 \times 30 \times 1000}{3600 \times 3 \times 10^8} = 25 \text{ Hz}.$$

The coherence time is

$$T_c = \frac{9}{400 \pi} = 7162 \mu s \gg \frac{1}{200 \times 10^3}.$$

The channel is therefore a "slow-fading" one.   ∎

2.3.4 Level Crossing and Average Fade Duration

One of the important consequences of Doppler fading is that the signal will experience deep fades occasionally as the vehicle is in motion (Lee 1967, Kenn 1969, Bodt 1982, Adac 1988b, Pars 1992). The analysis of fading in terms of Rayleigh statistics does not allow a clear understanding of how often deep fades occur or how long they last; Rayleigh statistics merely provide information on the overall percentage of time that the signal goes below a certain level. Information is needed on the rate at which deep fades occur and their duration, so that system designers can choose specific approaches for appropriate data rates, word lengths, and coding schemes to mitigate the effects of deep fades.

Deep fades can be quantitatively expressed using the parameters level crossing rate, N_A, and average fade duration, τ_{av}. To understand the concept of level crossing, consider the envelope of a signal received from a moving vehicle as shown in Figure 2.30. The envelope of the signal is seen to fluctuate in time. The level crossing rate is defined as the expected rate at which the envelope crosses a specified signal level A in the positive direction, as shown in Figure 2.30, and is given by

$$N_A = \int_0^\infty \dot{a} p(A, \dot{a}) \, d\dot{a}, \tag{2.54}$$

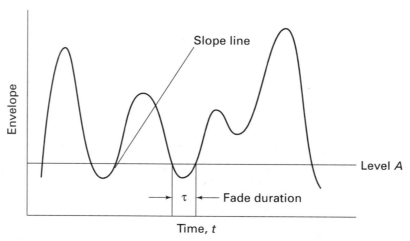

FIGURE 2.30 The concept of level crossing.

where $\dot{a}$ is the derivative of $a(t)$ and $p(A, \dot{a})$ is the joint probability density function of the level A and the rate of change of $a(t)$. Based on the results of Rice (Lee 1967, Kenn 1969), the expression for the level crossing rate, N_A, becomes

$$N_A = \sqrt{2\pi} f_d \alpha e^{-\alpha^2}, \qquad (2.55)$$

where α is the ratio A/A_{rms}. (A_{rms} is the rms value of the envelope.) The level crossing rate also depends on the maximum Doppler shift, f_d, and therefore on the speed of the mobile unit. By virtue of the factor $\alpha e^{-\alpha^2}$, there will be fewer crossings at low values of the signal level as well as at high values of the signal level. Certainly there will be more level crossings at higher speeds of the mobile unit.

Another parameter of interest is the average fade duration, τ_{av}, which is the average period of time the signal stays below a certain level A. For the case of Rayleigh fading, the average fade duration is given by

$$\tau_{\text{av}} = \frac{1}{N_A} \text{prob}\,(a \leq A), \qquad (2.56)$$

where $\text{prob}\,(a \leq A)$, the probability that the instantaneous signal is less than A, is given by

$$\text{prob}\,(a \leq A) = 1 - e^{-\alpha^2}. \qquad (2.57)$$

Using eqs. (2.56) and (2.57), the average fade duration τ_{av} is given as

$$\tau_{\text{av}} = \frac{e^{\alpha^2} - 1}{\sqrt{2\pi}\,\alpha f_d}. \qquad (2.58)$$

These two parameters are useful in enabling assessment of the instantaneous bit error rates, since the overall performance is determined not only by the error probabilities on a long-range basis but also by how the error rates vary on a short-term basis. The presence of deep fades as well as the number of such fades will change the instantaneous signal-to-noise ratio and hence, the bit error rates.

EXAMPLE 2.8

Continuing Example 2.7, calculate the average fade duration if $\alpha = 0.1$.

Answer

$$\text{Doppler shift} = 25 \text{ Hz}$$

$$\tau_{av} = \frac{e^{\alpha^2} - 1}{\sqrt{2}\,\pi\alpha f_d} = 1600 \ \mu s$$

2.3.5 Frequency Dispersion versus Time Dispersion

We have seen that fading can occur in the frequency domain or in the time domain. It is possible to treat fading in wireless communications systems as constituted by independent effects. The channel shows frequency-dispersive behavior when multipath phenomena are present. At the same time, the channel exhibits time dispersion if the mobile unit is moving. Even though these are independent effects, a relationship exists between these two dispersive attributes of the channel due to the uncertainty principle relating time and frequency dependence. This effect is demonstrated in Figure 2.31.

At very low data rates, the pulse duration is high and the channel is primarily slow and flat. If the data rate is very high and the MU is moving slowly, the channel will be slow but frequency selective. If, however, the data rate is high and the MU is moving at a very high speed, the channel will be fast and frequency selective. Channels falling in this category suffer from both time dispersion and frequency dispersion, and require additional corrective measures to overcome the distortion.

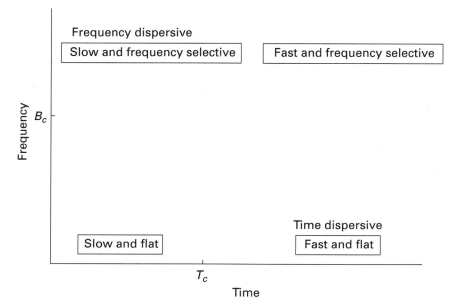

FIGURE 2.31 The "regions" of influence of different forms of fading.

2.4 OTHER FADING MODELS

2.4.1 Rician Fading

Rayleigh fading is not the only consequence of the multipath phenomenon (Stei 1964, Hash 1993). In addition to a number of random paths taken by the signal, it is possible to have a line-of-sight (LOS) propagation from the transmitter to the receiver. This LOS signal adds a deterministic component to the multipath signal. This is shown in Figure 2.32 and is compared with the case of Rayleigh fading.

The deterministic component makes the Gaussian random variable (eq. 2.32) one of nonzero mean, and consequently the envelope is Rician distributed. The pdf of the envelope can be expressed as (Papo 1991, Dave 1958)

$$f_A(a) = \frac{a}{\sigma^2} \exp\left(-\frac{a^2 + A_0^2}{2\sigma^2}\right) I_0\left(\frac{aA_0}{\sigma^2}\right), \tag{2.59}$$

where $I_0(.)$ is the modified Bessel function and A_0 is the component arising from the LOS signal. In the literature, it is customary to refer to the contribution of the randomly located scattering centers as the "diffuse" component and the contribution of the LOS component as the "steady" component. We therefore say that Rayleigh fading is the result of diffuse components and Rician fading is the result of the presence of a steady component along with the diffuse components.

The Rician probability density function is often characterized by the ratio of the power of the direct component to the power of the diffuse component, K (dB):

$$K(\text{dB}) = 10 \log_{10}\left(\frac{A_0^2}{2\sigma^2}\right). \tag{2.60}$$

For $K = -\infty$, we have no direct path and the Rician distribution becomes Rayleigh. For higher and higher values of K, the Rician distribution becomes almost Gaussian.

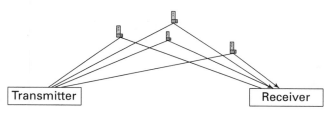

(a) Rayleigh fading (*no direct path*)

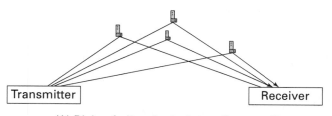

(b) Rician fading (includes a *direct path*)

FIGURE 2.32 Comparison of the conditions that exist for (*a*) Rayleigh fading and (*b*) Rician fading.

The Rician probability density function is shown in Figure 2.33 for different values of the parameter K.

The transition from Rayleigh to Gaussian can be seen clearly in Figure 2.33 as the value of K continues to increase. The LOS component of the Rician distribution provides a steady signal and serves to reduce the effects of fading.

EXAMPLE 2.9

In a wireless channel there is one direct path and six diffuse paths. The direct path has a power of 200% compared with the average power of any one of the diffuse paths. Find the Rician factor K.

Answer   Note that in eq. (2.60) the denominator is the sum of the contributions of the average powers of the diffuse paths. Therefore, $K = 2/6$, or $10 \log(2/6) = -4.7\,\text{dB}$. ∎

2.4.2 Lognormal Fading

The fading described so far falls under the category of "short-term" fading. However, as shown in Figure 2.5, the received signal also undergoes "long-term" fading. Consider the geometry of the scattering shown in Figure 2.34, where the propagation takes place in an environment with tall structures (trees, buildings).

Under these conditions, the signal reaching the receiver will not be the result of a single scattering effect; it will be the result of multiple scattering (Suzu 1977, Lee 1985, Hans 1977, Fren 1979). For example, the signal is likely to be multiply reflected or scattered before taking multiple paths to the receiver. The signal received, $r(t)$, by the MU or the base station receiver can now be expressed as

$$r(t) = \sum_{n=1}^{N} a_n \exp(-j\phi_n t),\tag{2.61}$$

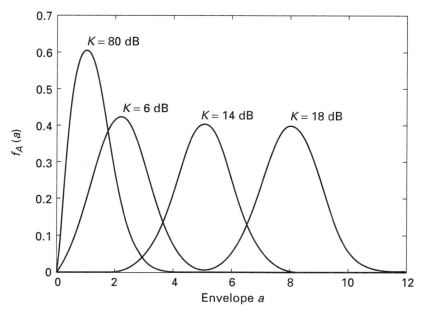

FIGURE 2.33   The Rician distribution for different values of K (dB).

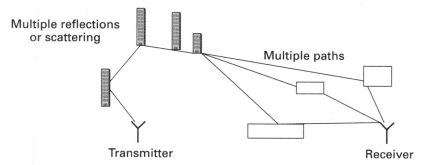

FIGURE 2.34 Geometry for lognormal fading, or shadowing. Note the existence of multiple reflections.

where the strengths of the multipath components are expressed as

$$a_n = \prod_{m=1}^{M} b_{mn}.$$ (2.62)

The quantities b_{mn} represent the scattering strengths of the multiply reflected components. Considering the fact that the region where the scattering takes place has a number of individual scatterers, the b_{mn} can be easily modeled as Rayleigh-distributed random variables. The impact of multiple scattering is to introduce further fluctuations in the received power, which will be manifested in the mean value of the received power, itself becoming random.

The mean value of the received power will be proportional to the variance of the Rayleigh-distributed envelope. However, the variance of the Rayleigh envelope will depend on $(b_{mn})^2$. The average signal power, P_{LT}, can then be expressed as

$$P_{LT} = \prod_{m=1}^{} p_m,$$ (2.63)

where

$$p_m = b_{mn}^2.$$ (2.64)

Applying the Central Limit Theorem for the product of random variables (Papo 1991), the density function $f(p_{dB})$ of P_{dB}, the logarithm of P_{LT}, given by

$$P_{dB} = 10 \log_{10}(P_{LT}) = \sum_{n=1}^{} 10 \log_{10}(p_n),$$ (2.65)

will be Gaussian and given by

$$f(p_{dB}) = \frac{1}{\sqrt{2\pi}\sigma_{dB}} \exp\left[-\frac{(p_{dB} - P_{av})^2}{2\sigma_{dB}^2}\right].$$ (2.66)

The parameters appearing in this equation are

P_{av} = average power (dBm)
σ_{dB} = standard deviation (dB)

The pdf of the signal power P_{LT} under lognormal fading can now be expressed as

$$f(p_{LT}) = \frac{1}{\sqrt{2\pi\sigma^2 p_{LT}^2}} \exp\left[-\frac{1}{2\sigma^2}\ln^2\left(\frac{p_{LT}}{p_0}\right)\right], \tag{2.67}$$

where p_0 is the average power in milliwatts and

$$\sigma = \frac{\sigma_{dB}\ln(10)}{10}. \tag{2.68}$$

Equation (2.67) is the lognormal pdf, and the multiple scattering leads to a lognormal distribution for the power received. The logarithm with respect to the base e is represented by ln. The lognormal pdf is shown in Figure 2.35. Lognormal fading is also referred to as *shadowing*, due to the fact that the shadowing seen in images can be modeled using an exponential transformation similar to the one given in eq. (2.63).

EXAMPLE 2.10

If the power received at the MU is lognormal with a standard deviation of 8 dB, calculate the outage probability. Assume that the average power being received is −95 dBm and the threshold power is −98 dBm. (Hint: Use the *erf* function. See Section B.7, Appendix B.)

Answer

$$P_{out} = \int_{-\infty}^{-98} \frac{1}{\sqrt{2\pi}\times 8} \exp-\frac{[x-(-95)]^2}{2\times 8^2}\, dx = 0.5 + 0.5\times \mathrm{erf}\left[\frac{-98+95}{8\sqrt{2}}\right] = 0.3538$$ ∎

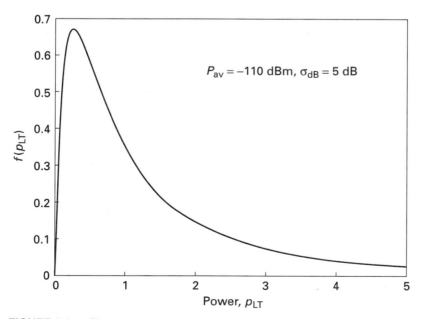

FIGURE 2.35 The lognormal probability density function.

Fluctuation of the received power from lognormal fading creates additional problems for designers of the wireless link. Including lognormal fading, the expression for the received power in any general area can be expressed by modifying eq. (2.9) as follows:

$$P_r(d) \text{ dBm} = 10 \log_{10}[P_r(d_{\text{ref}})] + 10\nu \log_{10}\left[\frac{d_{\text{ref}}}{d}\right] + Y_g, \qquad (2.69)$$

where Y_g is a zero-mean random variable with a standard deviation of σ_{dB}. In other words, the received power in dBm will be Gaussian distributed, with the median value of the received power (dBm) being the average and σ_{dB} being the standard deviation. A typical curve for the case $\nu = 3$ and standard deviation of lognormal fading of 6 dB is shown in Figure 2.36. It is possible to see that at any given location, the received power could be less than the median value of the power estimated from the Hata model or any other model. This necessitates the establishment of a power margin to account for fading at least as much as the standard deviation of fading, which reduces the maximum transmission distance as indicated in the figure.

We will reexamine the effects of lognormal fading on the coverage area in Chapter 4.

2.4.3 Nakagami Distribution

The Rayleigh and Rician models of fading assume that the amplitudes of the scattered components from the different paths are equal. The Nakagami model is very general and allows for the possibility of different strengths for the scattered components (Hash 1993, Brau 1991, Hoff 1960, Beck 1962). It can also work under conditions

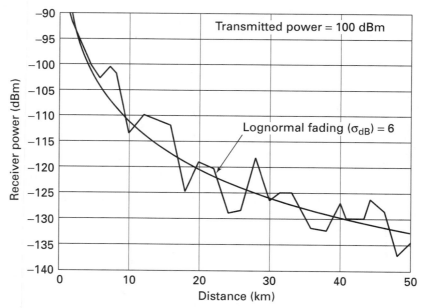

FIGURE 2.36 The lognormal long-term behavior (discontinuous line) versus pure attenuation. Compare this with Figure 2.5.

where the possibility of partial correlation exists between scattering elements. The density function of the envelope can be expressed as

$$f_A(a) = \frac{2m^m a^{2m-1}}{\Gamma(m)\Omega^m} \exp\left(-\frac{ma^2}{\Omega}\right) U(a), \qquad (2.70)$$

where $\Gamma(.)$ is the gamma function,

$$\Omega = \langle A^2 \rangle, \qquad (2.71)$$

and

$$m = \frac{\langle A^2 \rangle^2}{\langle (A^2 - \Omega)^2 \rangle}. \qquad (2.72)$$

There is an important restriction on the Nakagami parameter, that $m \geq 1/2$. For a value of $m = 1/2$, the Nakagami distribution becomes a single-sided Gaussian distribution. The Nakagami distribution becomes Rayleigh for $m = 1$, and for values of $m > 1$ the Nakagami distribution becomes Rician. The Nakagami probability density function is thus general enough to encompass both the Rayleigh and Rician distributions. The Nakagami probability density function is shown in Figure 2.37.

2.4.4 Suzuki Distribution

Rayleigh and lognormal fading have been considered to be two separate effects. However, the phenomena responsible for short-term fading (Rayleigh) and long-term fading (lognormal) occur concurrently (Suzu 1977, Fren 1979). The mean

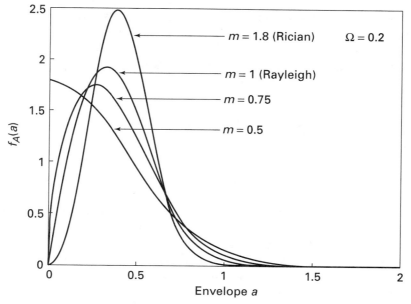

FIGURE 2.37 The Nakagami probability density function for a number of different values of m.

value of the power received under Rayleigh fading conditions typically has a lognormal distribution. In other words, the Rayleigh distribution essentially is not a marginal distribution, but a conditional one:

$$f(a|\sigma) = \frac{a}{\sigma^2}\exp\left(-\frac{a^2}{2\sigma^2}\right)U(a), \tag{2.73}$$

where σ has the lognormal probability density function. The density function of the envelope can then be obtained as

$$f_A(a) = \int_0^\infty f(a|\sigma)f(\sigma)\,d\sigma, \tag{2.74}$$

resulting in the Suzuki distribution for the envelope of the received signal, given by

$$f_A(a) = \int_0^\infty \frac{a}{\sigma^2}\exp\left(-\frac{a^2}{2\sigma^2}\right)\frac{1}{\sqrt{2\pi}\sigma\alpha}\exp\left[-\frac{(\ln\sigma - \mu)^2}{2\alpha^2}\right]d\sigma. \tag{2.75}$$

Once again, even though the Suzuki distribution is a more complete model, the fact that the pdf is not available in analytical form makes it a little difficult to work with.

2.4.5 Summary of Fading

The various fading mechanisms and the attenuation described can be summarized in a diagram as shown in Figure 2.38.

Note that Rician and Rayleigh fading arise out of multipath effects, and Nakagami fading can represent them both. This is not shown in the figure. For most cases, analyses based on Rayleigh or Rician fading are sufficient for understanding the nature of the mobile channel. A number of recent publications have suggested the use of Nakagami fading models to provide a generalized view of fading in wireless systems.

2.5 TESTING OF FADING MODELS

We have stated that the probability density functions of envelopes under various fading scenarios can be derived assuming certain fundamental conditions such as the existence of a multipath, the availability of a direct path, or the existence of multiple reflections. It is possible to conduct statistical tests to verify that the probability density function of the envelope of the faded signal follows a Rayleigh, Rician, or Nakagami distribution. One such test is the chi-square (χ^2) test (Papo 1991). The χ^2 test is a nonparametric (i.e., results are not dependent on the specific shape or parameters of the distribution) means of testing hypotheses (Papo 1991). Comparisons are made between theoretical populations based on assumed models and the actual data. The parameters of the expected theoretical probability density functions can be obtained from the data to estimate the theoretical probabilities.

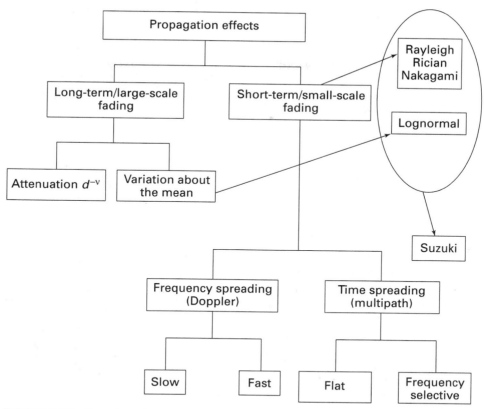

FIGURE 2.38 Overview of attenuation and fading. All forms of fading are shown along with their origins and relationships.

There are two major steps in χ^2 testing. First, a value is defined and calculated to estimate the difference between the expected theoretical frequency of occurrence and the experimentally observed frequency of occurrence. Second, this value is compared with a threshold to determine if it is too high. The threshold is determined by the significance level of the test selected by the investigator. The tests are conducted as follows:

1. Partition the observed sample space (N samples) into K disjoint intervals.

2. Calculate the number m_i of samples that fall in each of these intervals. This is a measure of the probability that the outcomes will fall in that interval.

3. Calculate the theoretical probability, p_i, that the outcomes would fall in the intervals. Thus, theoretically one expects Np_i samples to fall in the ith interval.

The χ^2 statistic is defined as the "weighted square error" and is given by

$$\chi^2 = \sum \frac{(\text{observed frequency} - \text{theoretical frequency})^2}{\text{theoretical frequency}}$$

$$= \sum_{i=1}^{K} \frac{(m_i - Np_i)^2}{Np_i}.$$

(2.76)

The test is good if the χ^2 statistic value is very small. The goodness of fit is quantified by setting a threshold X_T for the χ^2 statistic. We reject the hypothesis that the pdf indeed is the theoretical one if the value obtained in eq. (2.76) is larger than X_T.

The threshold is determined by the significance of the test. The test is based on the fact that when N is large, χ^2 has approximately a chi-square probability density function with $K - 1$ degrees of freedom. The chi-square probability density function, $f(x)$, of a random variable X is given by

$$f(x) = \frac{x^{(K-2)/2}}{2^{K/2}\Gamma(K/2)}\exp\left(-\frac{x}{2}\right)U(x). \qquad (2.77)$$

The pdf given in eq. (2.77) has $K - 1$ degrees of freedom. The threshold, X_T, is selected so that

$$\mathrm{prob}(X \geq X_T) = \alpha, \qquad (2.78)$$

where α is the significance level represented by the shaded area in Figure 2.39. Typical values of α are 1% and 5%.

If parameters such as the mean and standard deviation of the expected theoretical density functions are computed from the data, the degrees of freedom are reduced by the number of parameters computed. For example, if the mean is computed from the data, the number of degrees of freedom goes down from $K - 1$ to $K - 1 - 1$. If r parameters are computed from the data, the degrees of freedom of the chi-square distribution will be $K - r - 1$. In effect, each estimated parameter decreases the number of degrees of freedom by 1.

Table 2.1 gives values of the threshold for the chi-square test for significance levels of 5% and 1%.

We will now look at some examples of the use of the chi-square test.

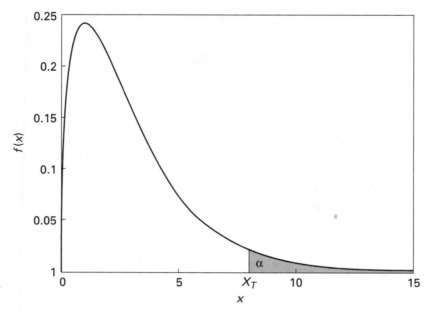

FIGURE 2.39 The χ^2 density function and the significance level.

TABLE 2.1 Chi-Square Values

K	X_T 5%	X_T 1%	K	X_T 5%	X_T 1%
1	3.84	6.63	15	25.00	30.58
2	5.99	9.21	16	26.30	32.00
3	7.81	11.34	17	27.59	33.41
4	9.49	13.28	18	28.87	34.81
5	11.07	75.09	19	30.14	36.19
6	12.59	16.81	20	31.41	37.57
7	14.07	18.48	22	33.90	40.30
8	15.51	20.09	24	36.40	43.00
9	16.92	21.67	25	37.65	44.31
10	18.31	23.21	26	38.90	45.60
11	19.68	24.73	28	41.30	48.30
12	21.03	26.22	30	43.80	50.90
13	22.36	27.69	40	55.80	63.70
14	23.68	29.14	50	67.50	76.20

EXAMPLE 2.11

Table 2.2 contains a set of numbers (100) given. Conduct a chi-square test to determine if these numbers are uniform in the range (0, 1).

TABLE 2.2 A Set of 100 Numbers

0.95	0.23	0.61	0.49	0.89	0.76	0.46	0.02	0.82	0.44
0.62	0.79	0.92	0.74	0.18	0.41	0.94	0.92	0.41	0.89
0.06	0.35	0.81	0.01	0.14	0.20	0.20	0.60	0.27	0.20
0.02	0.75	0.45	0.93	0.47	0.42	0.85	0.53	0.20	0.67
0.84	0.02	0.68	0.38	0.83	0.50	0.71	0.43	0.30	0.19
0.19	0.68	0.30	0.54	0.15	0.70	0.38	0.86	0.85	0.59
0.50	0.90	0.82	0.64	0.82	0.66	0.34	0.29	0.34	0.53
0.73	0.31	0.84	0.57	0.37	0.70	0.55	0.44	0.69	0.62
0.79	0.96	0.52	0.88	0.17	0.98	0.27	0.25	0.88	0.74
0.14	0.01	0.89	0.20	0.30	0.66	0.28	0.47	0.06	0.99

Let us pick a bin number of 10. If we now count the numbers in the bins (0 to 0.1, 0.1 to 0.2, etc.), bin 1 contains seven numbers, bin 2 contains seven numbers, and so on. If the numbers are uniformly distributed, $p_i = 0.1$. Hence $Np_i = 100 \times 0.1$. Note that for a uniform distribution, all the p_i are equal.

Therefore,

$$\chi^2 = \frac{(7-10)^2}{10} + \frac{(7-10)^2}{10} + \frac{(11-10)^2}{10} + \frac{(10-10)^2}{10} + \frac{(11-10)^2}{10}$$
$$+ \frac{(9-10)^2}{10} + \frac{(11-10)^2}{10} + \frac{(10-10)^2}{10} + \frac{(15-10)^2}{10} + \frac{(9-10)^2}{10} = 4.8.$$

(2.79)

From Table 2.1, $X_T(K-1) = X_T(9) = 16.92$ for $\alpha = 5\%$, which is obviously larger than the χ^2 value of 4.8 in this case. Therefore, the hypothesis that the numbers are uniform is accepted. ∎

EXAMPLE 2.12

Consider the set of numbers given in Table 2.3. We will test whether these samples follow a Rayleigh distribution.

TABLE 2.3 A Set of Numbers

2.53	5.52	1.99	1.19	2.38	2.43	2.38	2.01	2.00	0.83
2.40	2.56	3.17	4.37	2.45	0.24	3.11	2.70	0.56	2.53
0.64	2.98	2.74	3.38	1.48	2.97	3.64	3.69	2.88	1.72
1.80	1.46	2.57	3.03	2.64	3.25	3.29	2.77	2.58	0.35
3.28	1.76	2.34	4.11	1.95	2.00	0.66	2.27	4.67	2.41
2.04	1.24	1.40	3.59	1.22	1.55	0.78	3.67	1.21	2.70
0.94	1.92	2.19	3.95	1.37	2.11	1.52	4.27	0.28	3.46
2.33	3.20	0.77	1.49	3.59	2.51	0.62	1.06	2.89	1.02
1.56	1.76	2.38	4.67	2.75	2.10	2.89	1.68	3.51	0.68
3.20	1.43	4.12	1.51	3.83	2.02	2.76	1.21	2.89	2.29

Note that if these samples are Rayleigh distributed, the average of the squares of these samples should correspond to $2\sigma^2$, i.e.,

$$\langle A^2 \rangle = 2\sigma^2 \tag{2.80}$$

First we will compute this value. It is approximately equal to 6.6. This also means that the degrees of freedom will have to be reduced by 1. Once again, we will use 10 bins.

Since the largest value is 5.52, the bins will be in steps of 0.5 starting with 0.5. We can also compute the probabilities p_i ($i = 1, 2, ..., 10$) from the cumulative distribution function (CDF) of the Rayleigh distribution given by

$$F(a) = 1 - \exp\left(-\frac{a^2}{2\sigma^2}\right). \tag{2.81}$$

Table 2.4 gives the values of Np_i and m_i.

TABLE 2.4 Values of Np_i and m_i

Np_i	m_i
3.08	3
8.67	9
12.77	13
14.83	13
14.87	18
13.31	19
10.84	10
8.1	8
5.57	4
3.57	3

The chi-square test statistic can now be calculated using eq. (2.70), and we obtain a value of 3.93. From Table 2.1, the threshold value for acceptance for $\alpha = 5\%$ corresponds to $X_T(10 - 1 - 1) = X_T(8) = 15.51$, which is larger than 3.93, so we accept the hypothesis that the numbers are Rayleigh distributed. ∎

EXAMPLE 2.13

Let us see whether the numbers in Table 2.2 follow the Rayleigh distribution. We first estimate $2\sigma^2$. This is found to equal 0.358. Now we proceed as in Example 2.12, with the 10 bins going from 0.1 to 1, and calculate the theoretical probabilities and the frequency from the data. This information is provided in Table 2.5.

TABLE 2.5 Values of Np_i and m_i

Np_i	m_i
2.75	7
7.82	7
11.7	11
13.8	10
14.2	11
13.2	9
11.1	11
8.71	10
6.32	15
4.29	9

The test statistic can now be calculated using eq. (2.76); the value is equal to 27.5, which is larger than the threshold value of $X_T(8) = 15.51$ ($\alpha = 5\%$) from Table 2.1. We therefore reject the hypothesis that the numbers in Table 2.2 are Rayleigh distributed. ∎

2.6 POWER UNITS

A discussion of loss calculations would be incomplete without explaining the difference between dBm and dB. Power (P_0) in milliwatts (mW) can be expressed in terms of dBm as

$$P_0 \,(\text{dBm}) = 10 \, \log_{10}\left[\frac{P_0 \,(\text{mW})}{1 \,\text{mW}}\right]. \tag{2.82}$$

In other words, the power in dBm is an absolute measure of the power in mW. For example, 10 mW of power is 10 dBm, 1 W of power is 30 dBm, and 1 μW is −30 dBm. The unit dB, on the other hand, is the ratio of two powers in identical units. For example, if the average signal power is P_0 (mW) and the average noise power is P_n (mW), the signal-to-noise ratio (S/N) can be expressed as

$$(\text{S/N}) \, \text{dB} = 10 \, \log_{10}\left[\frac{P_0 \,(\text{mW})}{P_n \,(\text{mW})}\right]. \tag{2.83}$$

Thus, the signal-to-noise ratio expressed in dB carries information on how strong or how weak the signal is relative to the noise. For example, if the signal-to-noise ratio is 0 dB, the signal power and noise power are equal. If the signal-to-noise ratio is 20 dB, the signal power is 100 times stronger than the noise power. If the signal-to-noise ratio is −3 dB, the signal power is only 50% of the noise power. Note that dB expresses the ratio and therefore is not a measure of the absolute power. Because of this, we can write

$$\text{Transmit power (dBm)} - \text{receive power (dBm)} = \text{loss (dB)}. \qquad (2.84)$$

2.7 SUMMARY

This chapter has presented the problems associated with the propagation of a wireless signal. The effects of attenuation were described using the Hata model. The fading phenomena were examined to explain the fluctuations in the received signal power. Based on the physics of propagation, the fading can be described in different terms, namely, short-term and long-term fading; slow and fast fading; frequency-selective and flat fading; and Rayleigh, Rician, lognormal, Nakagami, and Suzuki fading. Statistical testing of the hypotheses to validate the appropriate probability density function was also explained.

- Attenuation is a result of reflection, scattering, diffraction, and refraction of the signal by natural and human-made structures.
- The received power, P_r, is inversely proportional to (distance)$^\nu$, or $P_r \propto 1/d^\nu$, where ν is the loss parameter.
- The loss parameter ν is equal to 2 for free space and is in the range of 2−4 for different environments, being higher for urban areas and lower for rural areas.
- The loss in outdoor areas can be modeled using the Hata model or Lee's model.
- Indoor propagation models are based on the characteristics of the interior of the building, building materials, and other factors and are described in terms of various zone models.
- The random fluctuations in the received power are due to fading.
- Multipaths and the Doppler effect contribute to short-term fading, and multiple reflections lead to long-term fading.
- Short-term fading can be described using Rayleigh statistics if no direct path exists between the transmitter and the receiver.
- Short-term fading can be described using Rician statistics if there is a direct path between the transmitter and the receiver.
- Rician and Rayleigh statistics can be encompassed by a single distribution, the Nakagami distribution.
- Short-term fading due to a multipath not only causes random fluctuations in the received power, but also distorts the pulses carrying the information.
- The multipath-fading channel can be modeled by treating the channel as a low-pass filter.
- If the bandwidth of the channel is higher than the bandwidth of the message, the signal is characterized by "flat fading" and no pulse distortion. If the bandwidth of the channel is less than the bandwidth of the message, the result is a "frequency-selective fading" channel. This flat or frequency-selective behavior can be quantitatively described in terms of the rms delay by taking the power of each multipath and the corresponding time delay.

- The bandwidth of the channel or the coherence bandwidth of the channel is inversely proportional to the rms delay.

- If there is relative motion between the transmitter and receiver (i.e., the MU is moving), the result is Doppler fading. If the maximum Doppler shift is less than the data rate, there is a "slow" fading channel. If the Doppler shift is larger than the data rate, there is a "fast" fading channel.

- In general, the worst performance occurs in fading channels that are fast and frequency-selective.

- In addition to fast and frequency-selective fading, a channel may also experience problems due to random frequency modulation. This arises from the $\cos(\theta_i)$ term in eq. (2.51), which makes the equation correspond to FM. By virtue of the fact that θ_i is random and fluctuates with time, random frequency modulation occurs, which will lead to increased bit error rates (see Chapter 5).

- Both short-term and long-term fading lead to outage. The system goes into outage when the signal-to-noise ratio or the received power falls below the threshold set for optimal performance.

- Long-term fading is modeled using lognormal distribution. If the received power is expressed in dBm, the long-term statistical fluctuations are Gaussian distributed.

- To prevent the system from going into outage, a margin is included in the power budget. This is the difference between the ideal threshold needed to maintain acceptable performance and a practical threshold that is set above the ideal threshold by a few dB. This power difference constitutes the *power margin*.

- The Suzuki model incorporates the short-term fading (Rayleigh) and long-term fading (lognormal) into a single distribution. This is based on the property that the average power in Rayleigh fading itself is random and can be described using the lognormal distribution.

- The statistics of fading may be verified using the chi-square test.

PROBLEMS

Most of these problems require MATLAB.
*** *Asterisks refer to problems better suited for graduate-level students.*

1. Using MATLAB, generate plots similar to the ones shown in Figure 2.6 to demonstrate the path loss as a function of the loss parameter for distances ranging from 2 km to 40 km. Calculate the excess loss (for values of $\nu > 2.0$) in dB.

2. The base station antenna is transmitting a power of 1 W. The transmitter antenna gain is unity while the receiver gain is 2. The system loss factor is unity (i.e., no loss). Find the received power in dBm at a distance of 5 km from the transmitter operating at 900 MHz in free space. (Hint: Use eq. (2.2))

3. Use MATLAB to plot the path loss predicted by the Hata model for the four separate environments. Assume a BS antenna height of 140 m, MU antenna height of 1.7 m, and a carrier frequency of 900 MHz.

4. Using the results of Problem 3 above, calculate approximately the loss exponent ν for the four separate cases.

5. Use MATLAB to generate the Rayleigh-faded signal shown in Figure 2.19. Use a carrier frequency of 900 MHz. (Hint: Use the concept of multiple paths. Take the number of multipaths to be equal to 10. Use random phase.)

6. Use MATLAB to generate a Rayleigh-faded signal with random delays. (Hint: Use the concept of multiple paths. Take the number of multipaths to be equal to 10. Calculate the mean delay for a certain distance. Generate uniformly distributed delays with the mean delay calculated.)

7. Use MATLAB to generate a set of Rayleigh random numbers and verify that the mean/std. dev ratio is 1.91. Calculate the mean/std. dev ratio for the set of random numbers obtained by squaring the Rayleigh number set.

8. For Problem 5, compute the outage probability. Set the threshold to 6 dB below the average power.

9. Use MATLAB to generate a Rician-faded signal. Use a carrier frequency of 900 MHz. Comment on the differences between these curves and those of Problem 5.

10. For the Rayleigh and Rician channels, compare the outage rates. (Hint: Use MATLAB to generate Rayleigh- and Rician-faded signals).

11. If the system goes into outage when the received signal-to-noise ratio falls 5 dB below the average, calculate the outage probability in Rayleigh fading with an average signal-to-noise ratio of 5 dB. Note that the signal-to-noise ratio is exponentially distributed when Rayleigh fading is present.

12. The instantaneous signal-to-noise ratio in a Rayleigh channel is given by A^2/N_0, where N_0 is the noise power. The average signal-to-noise ratio expected at the receiver is $\langle A^2 \rangle/N_0 = 3$ dB. What is the probability that the instantaneous SNR will be less than 0 dB?

13. If a wireless system goes into outage when the received signal-to-noise ratio falls 10 dB below the average, calculate the outage probability in Rayleigh fading. Assume an average signal-to-noise ratio of 25 dB.

14. The maximum outage rate accepted in a wireless system subject to fading is 0.02. If the threshold SNR is 5 dB, what must be the minimum value of the average SNR?

15. Plot histograms corresponding to the envelope of the Rayleigh and Rician fading channels (of Problems 5 and 9).

16. Calculate the mean delay and rms delay spread of a channel having the characteristics

shown in Figure P2.16. What is the coherence bandwidth $1/(5 \text{ rms delay})$?

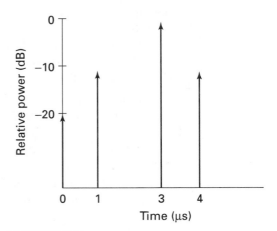

FIGURE P2.16

17. Compare the maximum data transmission capabilities of the two channels characterized by the impulse responses shown in Figure P2.17.

18. Generate a plot similar to the lognormal fading plot shown in Figure 2.30.

19. Consider two random variables X and Y that are independent and Gaussian with identical variances. One of them is of zero mean and the other is of mean A. Prove that the density function of $z = \sqrt{X^2 + Y^2}$ is Rician distributed.

20. For the Rayleigh-faded channel, verify that the Nakagami parameter is approximately equal to 1.

21. For the Rician channel, verify that the Nakagami parameter is greater than 1.

22. For a mobile unit moving at 80 km/h, calculate the Doppler spread. The carrier frequency is 900 MHz. What is the fade duration? What is the rate of level crossings? Assume that $\rho = 0.6$.

23. A MU is operating in a multipath environment. Measurements taken show that the coherence bandwidth is 10 kHz and the coherence time is 100 μs. The transmission of data is taking place at a rate of 5 kbps. Examine whether the channel is fast, slow, frequency selective, or flat.

24. Generate the frequency response of a frequency-selective faded signal using the equations given in the text. Instead of using Rayleigh-distributed random variables in the two-ray model, initially treat the

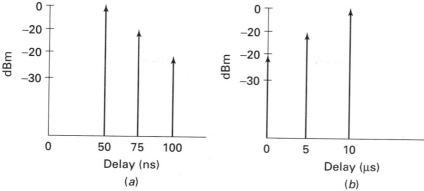

FIGURE P2.17

weights as deterministic and of ratio 2:1. Repeat the simulation two more times by using random variables in the same ratio, and in a ratio of 1:1. Discuss the results.

25. Generate two sets of Gaussian-distributed random variables and show that the density function of the sum of their squares is exponentially distributed. (Compare the histograms and verify that the mean/std. dev ratio of the exponentially generated set is 1.)

26. By considering a Gaussian pulse of rms width 5 ms, generate the output pulse of a frequency-selective channel. Compute the rms width of the output pulse.

27. Examine the Rician distribution. For what values of K can one approximate the Rician distribution to the Gaussian distribution?

28. Generate a set of Nakagami random variables and verify that when $m = 1$ the Nakagami distribution becomes Rayleigh, and when $m \geq 1$ the Nakagami distribution becomes Rician. (Hint: A Nakagami random variable is obtained by taking the square root of a gamma-distributed random variable. See Appendix A.)

29. Plot the outage probability (use MATLAB) as a function of the standard deviation of fading $(1-15$ dB$)$ for lognormal fading under the following conditions: average received power $= -90$ dBm; threshold power required $= -95$ dBm.

30. Use MATLAB to calculate the amount of lognormal fading tolerated by the system (i.e., standard deviation) under the following conditions: average power $= -85$ dBm; threshold power $= -90$ dBm; outage probability $= 3\%$.

31. Conduct a chi-square test to prove that the results of Problem 25 are correct. (Hint: Generate

two sets of 500 Gaussian random variables, and use them to generate the data.)

32. Generate five sets (500 each) of uniform random numbers (U_i) and create a new set X_k such that

$$X_k = \sum_{i=1}^{s} U_{ik}, \quad k = 1, 2, \dots, 500 \; .$$

Conduct a chi-square test to see whether the X_i are Gaussian (see the Central Limit Theorem in Appendix A).

33. Use the results of Problem 32 to justify Rayleigh fading.

34. Calculate the outage probabilities analytically for the Rayleigh fading (no motion). Compare the values with the MATLAB-generated ones.

35. Calculate the outage probabilities for the Rician fading case analytically. Compare the results with the Rayleigh fading case.

36. Using random number generation, examine whether the pdf of the product of a set of (10) of Rayleigh-distributed (or exponentially distributed) random variables will be lognormal. (Hint: Take the log of the product and see whether the histogram is Gaussian.)

37. Using the principle in Problem 36, generate a radio-frequency signal containing lognormal fading effects.

38. In Problem 37, examine whether the histogram appears to be lognormal.

39. Examine what happens if you increase the number of terms in the product from 5 to 10, 15, and 20. Comment on your results.

40. Conduct a chi-square test to see if the lognormal distribution is a fit based on the results of Problem 38 or Problem 39.

41. Compare the Hata model for loss prediction with the Walfish-Ikegami model (see Appendix C).

42. Explain the terms long-term fading and short-term fading. Compare Rayleigh and Rician fading channels, and explain why Rician channels perform better than Rayleigh channels.

43. Explain the terms *frequency-selective fading* and *flat fading*. Explain the meaning of the term *dispersive* (see Appendix C).

44. Use MATLAB to plot the frequency and time correlation functions (fading) shown in Figure C.1.1. (Appendix C). Consider now the case of an exponential delay model with a mean value of 0.5 ms. What is the rms delay? What is the maximum data rate that can be transmitted through this multipath channel without the use of an equalizer?

45. Repeat Problem 5 varying the number of multiple paths (3, 5, 10). Get histograms of the envelope and see whether the Rayleigh fading is taking place. (Hint: Superimpose a Rayleigh pdf with the same mean as the simulated one on the same diagram.) ***

46. Repeat Problem 45 by conducting a chi-square test to see when the Rayleigh model is an acceptable fit. ***

47. Generate a bit stream of $+1$s and -1s. Delay the bit stream by a fraction of the pulse duration. Create three such bit streams (delayed by different amounts), and add them and plot. Compare the plot with the plot of the original data stream. Examine what happens when you vary the delay. (Pulse distortion should be seen.)***

48. Take the pulse stream from Problem 47 and multiply it by a cosine wave. Choose an appropriate frequency so that you have a few cycles of RF in each of the pulses. Now create three such streams by adding a uniformly distributed random phase to the argument of the cosine. Plot the sum of these three streams. Comment on what you see.***

49. In Problem 47, for each time instant, scale the voltage by a Rayleigh-distributed random variable. Repeat the rest of the simulation. Comment on your results, and compare the results with those of Problem 47. (Hint: Effects of Rayleigh fading.)***

50. For each time instant in Problem 48, scale the voltage by a Rayleigh-distributed random variable. Repeat the rest of the simulation and compare the results with those of Problem 48. (Hint: Effects of Rayleigh fading.)***

CHAPTER 3

MODEMS FOR WIRELESS COMMUNICATIONS

3.0 INTRODUCTION

Modulation is the process in which the signal containing information is *modified* for transmission (Skla 1988, Toma 1998, Hayk 2001). This may involve the step of frequency translation, in which case the message signal appears as a passband around the carrier frequency. If the carrier frequency signal $c(t)$ is expressed as

$$c(t) = A_0\cos[2\pi f_0 t + \phi] , \tag{3.1}$$

frequency translation can be achieved in three ways: changing the amplitude of the carrier, A_0, changing the frequency of the carrier, f_0, or changing the phase of the carrier, ϕ, in accordance with the message signal or modulating signal. These modulation schemes are identified respectively as amplitude modulation (AM), frequency modulation (FM), and phase modulation (PM). Frequency and phase modulation techniques are also referred to as angle modulation schemes.

If $m(t)$ represents a message signal, the amplitude-modulated signal, $s_{AM}(t)$, can be expressed as

$$s_{AM}(t) = A_0[1 + k_a m(t)] \cos(2\pi f_0 t) , \tag{3.2}$$

where k_a is the modulation index. A zero phase ($\phi = 0$) has been assumed.

The frequency-modulated signal, $s_{FM}(t)$, can be expressed as

$$s_{FM}(t) = A_0\cos\left(2\pi f_0 t + 2\pi k_f \int_{-\infty}^{t} m(\tau)\,d\tau\right), \tag{3.3}$$

where k_f is the frequency deviation (Hz/V).

The phase-modulated signal, $s_{PM}(t)$, can be expressed as

$$s_{PM}(t) = A_0\cos(2\pi f_0 t + k_\phi m(t)), \tag{3.4}$$

where k_ϕ is the phase deviation (rad/V).

A casual examination of eqs. (3.2) and (3.3) indicates that AM is a form of linear modulation since $s_{AM}(t)$ is directly proportional to the message signal, $m(t)$. On the other hand, the envelopes of the signals in FM and PM do not vary with the message signal. This points to the fact that these modulation schemes are of the constant-envelope type and are nonlinear since the amplitude of the modulated signal does not vary with the message signal.

Analog modulation schemes are treated extensively in standard textbooks on communications theory (Shan 1979, Proa 1994, Hayk 2001, Couc 1997, Schw 1996). We will limit ourselves to an overview of these because analog modulation schemes are used less frequently in mobile or wireless communication systems. Most of the wireless communication systems use digital communication techniques, where the information is represented as a sequence of pulses (Oett 1979, 1983; Clar 1985; Taub 1986; Skla 1988; Pahl 1995; Fehe 1995; Stub 1996). Digital modulation techniques offer better noise immunity than the analog modulation methods. In addition, digital modulation techniques make it easy to multiplex different forms of information, such as voice, picture, and computer data. The digital technology is also compatible with the digital signal-processing methods, making it easy to implement.

3.1 ANALOG MODULATION

Even though analog modulation schemes are not commonly used in wireless communication systems, they provide a good starting point for a discussion of communication techniques. We will discuss two schemes, amplitude modulation and angle modulation.

3.1.1 Amplitude Modulation

Amplitude modulation is one of the simplest forms of modulation and is commonly used in some communication systems, such as radio. The information-bearing signal, $m(t)$, and the modulated signal, $s_{AM}(t)$, are related through

$$S_{AM}(t) = A_0[1 + k_a m(t)] \cos(2\pi f_0 t) . \tag{3.5}$$

If we choose the amplitude of the message signal such that

$$|m(t)| \leq 1, \tag{3.6}$$

the parameter k_a is positive and may be identified as the modulation index. This limitation on the amplitude of the message ensures that the envelope of the amplitude-modulated signal, $A(t)$, given by

$$A(t) = A_0[1 + k_a m(t)], \tag{3.7}$$

is always positive for modulation indices less than 1. Figure 3.1 shows the carrier signal, modulating signal, and modulated signal for the case of a sinusoidal modulating signal $m(t)$. Two modulation indices are shown.

If the modulation index exceeds 1, the envelope becomes negative and envelope distortion will occur, as shown in Figure 3.1d. The Fourier transform of the amplitude-modulated signal can be expressed in terms of the spectrum of the modulating signal as

$$s_{AM}(f) = \frac{A_0}{2}[\delta(f-f_0) + \delta(f+f_0)] + \frac{k_a A_0}{2}[M(f-f_0) + M(f+f_0)]. \tag{3.8}$$

In addition to the requirement on the modulation index, it is also possible to deduce the requirement on the carrier frequency, f_0, from the spectrum shown in Figure 3.2. For an undistorted envelope, the carrier frequency, f_0, must meet the condition

$$f_0 \gg W , \tag{3.9}$$

where W is the bandwidth of the message signal. We can see that the spectrum of the modulated signal contains an upper sideband (USB) and a lower sideband (LSB), and

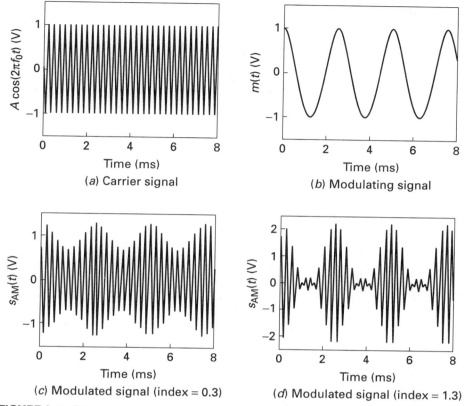

FIGURE 3.1 The carrier signal (a), single-tone modulating signal (b), and modulated signal (AM) for two values of the modulation indexes (c, d).

the bandwidth occupied by the AM signal is twice the baseband bandwidth. In other words, the bandwidth required for transmission, B_{AM}, is twice the baseband bandwidth and is normally expressed as

$$B_{AM} = 2W.$$ (3.10)

The presence of lower and upper sidebands accounts for the name *double-sideband amplitude modulation* (DSB-AM).

The power, P_{AM}, in the transmitted signal is

$$P_{AM} = \frac{1}{2}A_0^2\left[1 + k_a^2 S_m\right],$$ (3.11)

where S_m is the power in the message signal, $m(t)$, which is assumed to have no DC component, i.e.,

$$\overline{m(t)} = 0, \quad \overline{m(t)^2} = S_m,$$ (3.12)

where the upper bar denotes the average. We can now separate the power into two parts, power in the transmitted carrier, P_c, and power in the sidebands, P_{sb}:

$$P_{AM} = P_c + 2P_{sb}$$ (3.13)

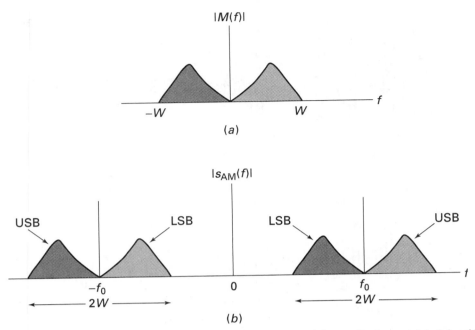

FIGURE 3.2 The spectra of a generic modulating signal (*a*) and the amplitude-modulated signal (*b*).

where

$$P_c = \frac{1}{2}A_0^2, \quad P_{sb} = \frac{1}{4}A_0^2 k_a^2 S_m = \frac{1}{2}P_c k_a^2 S_m. \tag{3.14}$$

This power calculation shows one of the major problems with amplitude modulation. Since the message signal and the modulation index values are limited by unity, more than half of the transmitted power (P_{AM}) resides in the carrier (P_c) as opposed to the message part of the transmitted signal (P_{sb}). On the other hand, the presence of the carrier in the transmitted signal and the lower power in the sidebands makes it possible for a simple envelope detector to recover the message signal.

Transmission and Reception of AM Signals   It is possible to transmit AM signals with suppressed carrier (DSB-SC) so that all the transmitted power resides in the message signal. The modulated signal, $s_{SC}(t)$, in this case is

$$s_{SC}(t) = A_0 m(t)\cos(2\pi f_0 t). \tag{3.15}$$

The spectrum of the DSB-SC signal will look similar to the spectrum of the DSB-AM signal except for the absence of the δ-functions that arose from the presence of the carrier wave. The drawback of the DSB-SC format is the need for a coherent or synchronous demodulator to recover the message signal, $m(t)$. The demodulator is shown in Figure 3.3. If there is a frequency mismatch between the local oscillator and the incoming wave, the output signal will be $m(t)\cos(2\pi\Delta ft)$, where Δf is the frequency mismatch. The presence of this cosine term reduces the signal power. Similarly, if there is a phase mismatch between the incoming signal wave and the local oscillator, the output signal will be $m(t)\cos(\Delta\phi)$, where $\Delta\phi$ is the phase mismatch. It is easily seen that for $\Delta\phi = \pi/2$, the output is zero.

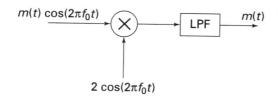

FIGURE 3.3 Block diagram of a coherent demodulator.

If we observe the spectrum of the DSB-AM or DSB-SC signal, we also note that there is no need to transmit both the upper and lower sidebands since they contain virtually the same information. We can therefore take the DSB-SC signal and pass it through a narrow bandpass filter so that the transmitted signal contains either the upper sideband or the lower sideband, resulting in *single-sideband modulation* (SSB-SC). Single-sideband modulation thus reduces the transmission bandwidth to the message bandwidth (and not twice the message bandwidth, as required for DSB-SC or DSB-AM). Even though SSB-SC is conceptually simple, as a practical matter, difficulty with filters having sharp cutoff has led to VSB-SC (*vestigial sideband*). VSB modulation allows the use of filters with a gradual roll-off, allowing retention of a small portion of the upper or lower sideband that was removed in SSB-SC. The spectra of SSB-SC and VSB-SC are shown in Figure 3.4, which indicates that the bandwidth of VSB-SC is about 25% higher than the bandwidth of SSB-SC.

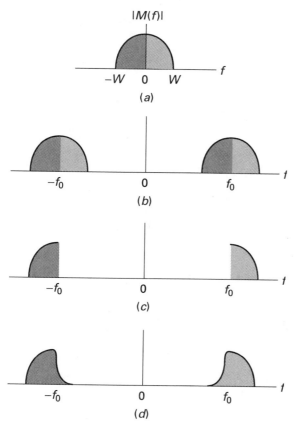

FIGURE 3.4 Spectra of the (a) message signal, (b) DSB-SC signal, (c) SSB-SC signal, and (d) VSB-SC signal.

Quadrature Amplitude Modulation To transmit more than one message signal, we need to use two separate carrier frequencies. Each of these signals at carrier frequencies f_{01} and f_{02} will be modulated by separate message signals, $m_1(t)$ and $m_2(t)$. The spectrum of the transmitted signal will be as shown in Figure 3.5. It is clear that to transmit the two signals, $m_1(t)$ and $m_2(t)$, the transmitted bandwidth will be twice the bandwidth required for the transmission of a single message.

It is possible to transmit two signals with a transmission bandwidth requirement of only a single message. This is accomplished using the quadrature amplitude modulation setup shown in Figure 3.6. The word *quadrature* refers to the simultaneous use of sine and cosine waveforms for the carrier. Block diagrams of the quadrature modulator and demodulator are shown in Figures 3.6a and b, respectively. Since we are only using a single carrier frequency f_0, the bandwidth required for transmission will be equal to the bandwidth required to transmit either $m_1(t)$ or $m_2(t)$. We will see modulators similar to this used in digital communications as well.

3.1.2 Angle Modulation

Frequency and phase modulation are generally identified as angle modulation schemes. Variation of either the frequency or the phase as a function of a modulating signal can be represented as a variation of the angle of the carrier wave. The carrier wave, $c(t)$, can be represented as

$$c(t) = A_0 \cos[2\pi f_0 t + \theta(t)], \tag{3.16}$$

where $\theta(t)$ is the instantaneous phase of the carrier given by

$$\theta(t) = \text{function of } m(t). \tag{3.17}$$

The frequency-modulated signal is one where the instantaneous frequency of the signal, $f_i(t)$, varies linearly with the baseband signal $m(t)$:

$$f_i(t) = f_0 + k_f m(t), \tag{3.18}$$

where k_f is the frequency deviation constant in units of Hz/V. If the modulating signal is a sinusoid of frequency f_m, i.e.,

$$m(t) = A_m \cos(2\pi f_m t), \tag{3.19}$$

the frequency-modulated signal becomes

$$s_{\text{FM}}(t) = A_0 \cos\left[2\pi f_0 t + \frac{A_m k_f}{f_m} \sin(2\pi f_m t)\right]. \tag{3.20}$$

The frequency modulation index, β_f, is given by

$$\beta_f = \frac{A_m k_f}{W} = \frac{\Delta_f}{W}, \tag{3.21}$$

where $W \, (= f_m)$ is the maximum frequency of the modulating signal and Δ_f is the peak frequency deviation. The transmission bandwidth, B_{FM}, is given by the Carson rule (Schw 1996, Hayk 2001):

$$B_{\text{FM}} = 2(\beta_f + 1)W. \tag{3.22}$$

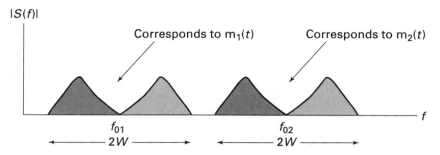

FIGURE 3.5 Spectrum of a frequency-division multiplexed signal.

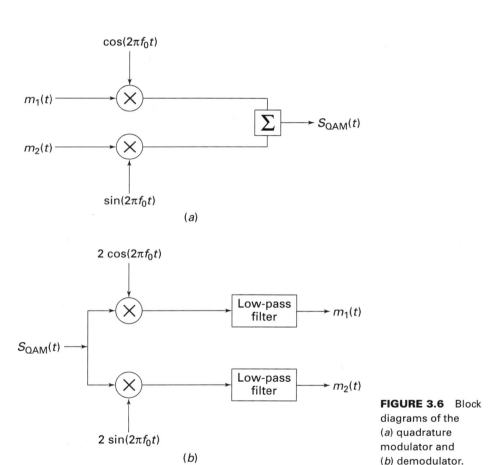

FIGURE 3.6 Block diagrams of the (a) quadrature modulator and (b) demodulator.

If we compare this with the transmission bandwidth of the AM signal, it is clear that $B_{AM} = B_{FM}$ if the FM index β_f is very small ($\beta_f \ll 1$). On the other hand, B_{FM} is much larger than B_{AM} if the FM index is high. The former case corresponds to narrowband FM (NBFM) and the latter to wideband FM (WBFM). It can also be seen from the expression for the FM index that as the amplitude of the modulating signal A_m goes up, the index and, consequently, the transmission bandwidth of FM also go up.

The FM signal with a modulation index of $\beta = 2$ is shown along with an AM signal of modulation index $k_a = 0.2$ in Figure 3.7. It is clear that the envelope of the FM signal does not vary with the modulation index, and hence the modulation technique is nonlinear. The linearity exists for the instantaneous frequency, which goes up with the amplitude of the modulating signal. The spectra of the FM and AM signals shown in Figure 3.7 demonstrate the impact of the modulation index on the transmission bandwidth of FM. Frequency modulation is the format employed in U.S. AMPS cellular systems. The voice signal is assumed to have the highest frequency of $f_m = 3$ kHz, and a modulation index $\beta_f = 3$ is employed.

The phase-modulated signal, $s_{PM}(t)$, can be expressed as

$$s_{PM}(t) = A_0 \cos[2\pi f_0 t + k_p m(t)] , \tag{3.23}$$

where k_p is the phase deviation constant in units of radians/volt. For the case of a pure sinusoidal modulating signal as given in eq. (3.19), the phase-modulated (PM) signal can be expressed as

$$s_{PM}(t) = A_0 \cos[2\pi f_0 t + k_p A_m \cos(2\pi f_m t)] . \tag{3.24}$$

The phase modulation index, β_p, is given by

$$\beta_p = k_p A_m. \tag{3.25}$$

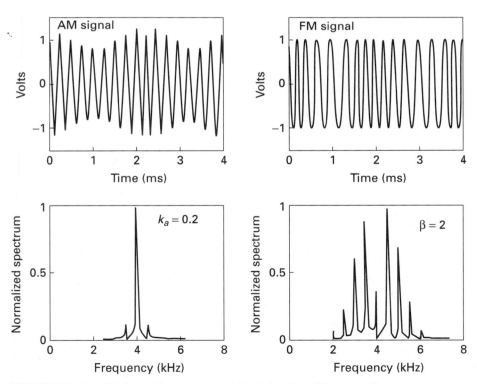

FIGURE 3.7   Amplitude- and frequency-modulated signals and the respective spectra are compared.

The relationship between FM and PM is clear from eqs. (3.20) and (3.23). By examining the arguments of the cosine function in these equations, it is easy to observe that for FM the instantaneous frequency is directly proportional to the message, while for PM the instantaneous phase is proportional to the message. In other words, with the help of integrating and differentiating networks, it is possible to generate FM from PM and vice versa, as shown in Figure 3.8.

Generation and Detection of Angle-Modulated Signals Frequency modulation can be accomplished using an indirect method. In this approach, we take advantage of the similarity that exists between amplitude modulation and narrowband FM. Let us go back and rewrite eq. (3.20) as

$$S_{FM}(t) = A_0 \cos(2\pi f_0 t)\cos[\beta_f \sin(2\pi f_m t)]$$

$$-A_0 \sin(2\pi f_0 t)\sin[\beta_f \sin(2\pi f_m t)]. \qquad (3.26)$$

For narrowband FM, with modulation index $\beta_f \ll 1$, the equation becomes

$$S_{FM}(t) = A_0 \cos(2\pi f_0 t) - A_0\beta_f \sin(2\pi f_0 t)\sin(2\pi f_m t). \qquad (3.27)$$

Equation (3.27) appears to be very similar to the amplitude-modulated signal given in eq. (3.5) for a sinusoidal modulating signal, except for two simple aspects: The positive sign is replaced by a negative sign, and the carrier waves in the two terms have a phase difference of $\pi/2$. Narrowband FM can therefore be generated using this approach and then converted to wideband FM using frequency scaling. A block diagram to accomplish this is given in Figure 3.9.

FM signals can be demodulated using a frequency discriminator. This setup is shown in Figure 3.10. The discriminator is preceded by a hard limiter. The hard limiter is necessary to remove the residual amplitude modulation that is typically present in the FM signal. In other words, A_0 in eq. (3.16) should in general be $A_0(t)$. The hard limiter removes the amplitude variations before the signal is fed to the discriminator. The discriminator involves two steps: first, differentiation of the signal, which can be undertaken using a slope detector, and second, envelope detection.

3.1.3 Comparison of Analog Modulation Schemes

The performance of the analog modulation formats can be compared in terms of the signal-to-noise ratio and the transmission bandwidth. While the transmission bandwidth of an amplitude modulation system is fixed (twice the message bandwidth), the transmission bandwidth of FM is always greater than or equal to the transmission bandwidth of AM systems. In essence, as the transmission bandwidth of the FM signal goes up, the signal-noise-ratio of the FM system also goes up. While such a trade-off between the transmission bandwidth and signal-to-noise ratio is absent in AM systems, it is characteristic of angle-modulated systems. In addition, the signal-to-noise ratio can be improved further by improving the level of modulation at the high end of the spectrum of the modulating signal (for example, the audio range). This technique is known as *preemphasis,* and the reverse process, *deemphasis,* is performed at the receiver. The preemphasis filter, where the high-frequency components of the modulating signal are enhanced, precedes the FM transmitter, and

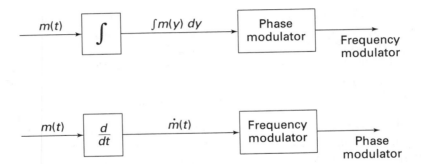

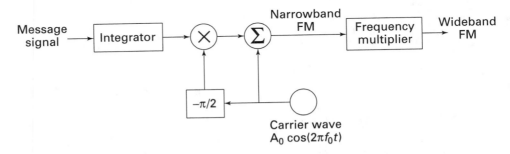

FIGURE 3.8   The interrelationship between FM and PM.

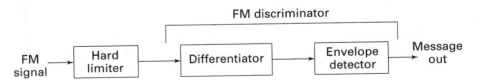

FIGURE 3.9   Generation of FM.

FIGURE 3.10   Frequency demodulator.

the deemphasis filter, which performs the proportionate reduction in the strength of the high frequencies, follows the FM receiver circuit. We will discuss these aspects when we look at the modulation/demodulation techniques used in analog cellular systems.

EXAMPLE 3.1

(a) A voice signal of bandwidth 4 kHz is to be transmitted by amplitude modulation. What is the minimum bandwidth required for transmission?

(b) If the voice signal is to be transmitted by frequency modulation, calculate the transmission bandwidth given that the modulation index is 2.

Answer

(a) $2 \times 4 = 8$ kHz

(b) Using the Carson rule, transmission BW $= 2(1 + 2) \times 4 = 24$ kHz. ∎

3.2 DIGITAL MODULATION

Even though analog communication schemes are simple and easy to implement, digital modulation schemes provide a much better alternative for information transmission. This is due not only to the enhanced noise immunity offered by the digital modulation schemes, but also to advancements in digital signal processing techniques and the availability of fast processors based on VLSI chips. Subject to cost and bandwidth factors, it is possible to encode the data appropriately to reduce errors at the receiver and prevent unauthorized use. The simple block diagram in Figure 3.11 shows a digital communication system. The transmitter portion contains an A/D converter, source encoder, channel encoder, and modulator. The A/D converter creates the digital information. The source encoder and A/D converter produce digital versions of the analog information for transmission. The source encoder encodes the information so that the overall bandwidth required for transmission can be kept to a manageable level. These coding algorithms depend on the various standards adopted by the wireless systems. The channel coding is necessary to mitigate the effects of noise and fading.

The key component in the digital communication system is the generation of data symbols. Symbols are, in effect, abstract quantities such as 0 (or −1) and 1, and they must be expressed in a tangible form for storage and transmission. This is done by representing these symbols using "pulses" in some fashion, and a number of such pulses form a "pulse train," which carries the information across the channel. If each of these symbols can take m possible values, the amount of information carried by the symbol is given by $n = \log_2 m$ bits/symbol. For example, if only two levels ($m = 2$) are used, there will be only a single bit per

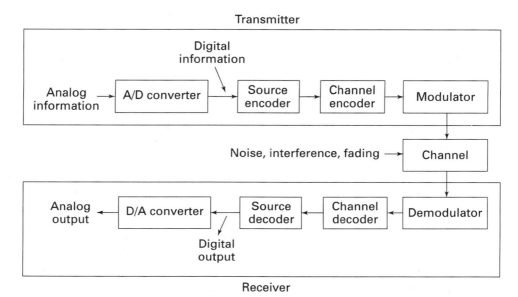

FIGURE 3.11 Block diagram of a generic digital communication system.

symbol. This is the binary case. Thus, in the binary case, one can use the terms *bit* and *symbol* interchangeably. If there are four possible values of *m*, we will have two bits/symbol (the quaternary case), and so on.

Once these pulses have been created, different modulation formats may be used to transmit them (Skla 1993). These formats differ in the ways in which data are encoded for transmission. It is also possible to have signals transmitted using one technique to be detected or demodulated using another method (Ande 1999). The different modulation and demodulation methods are chosen based on their efficiency and ease of implementation. The provision also exists for transmitting the pulses or symbols on a high-frequency carrier wave or in the baseband form itself. Between the transmitter and receiver, the information-bearing signal will be subject to fading (Stei 1987). Noise and interference from other sources add to the signal.

The discussion here is limited to the modulation techniques that employ a carrier wave. Before we study the modulation techniques, we will examine the different pulse shapes that can be used to represent the bits and symbols.

3.2.1 Pulse Trains and Pulse Shaping

One of the simplest pulses that can be used to represent bits or symbols is the rectangular pulse, given by

$$
\Pi\left(\frac{t}{T}\right) = \begin{cases} A & -\frac{T}{2} \le t \le \frac{T}{2} \\ 0 & \text{otherwise,} \end{cases}
\tag{3.28}
$$

where *A* is a constant. The duration of the pulse is *T*. Rectangular pulses can be used to form different types of binary codes, as shown in Figure 3.12. The commonly used NRZ (non–return-to-zero) format occupies the least bandwidth since the pulse occupies the whole bit duration *T* and does not change. This is shown in Figure 3.12*a*. The bipolar format (Figure 3.12*b*) offers greater separation between a "0" and a "1" since a "0" is represented by a −1. The NRZ code may offer low bandwidth capabilities, but it also offers poor synchronization capabilities since it is difficult to clearly identify the location of the pulses when one has a chain of either +1s or −1s. The RZ codes (Figure 3.12*c*), on the other hand, occupy half the bit duration and therefore require twice as much bandwidth as the NRZ pulses. The RZ formats, however, offer better synchronization capabilities. The Manchester format (Figure 3.12*c*) offers good synchronization capability since two short pulses are used to represent each bit. Each binary 1 is represented by a positive half period followed by a negative half bit period. The order of the half periods is reversed for the binary 0. The Manchester format occupies the largest bandwidth of the formats shown in Figure 3.12.

The pulse shapes in Figure 3.12 are all rectangular, and passage of these pulses through real physical channels will introduce distortion as a result of the band-limited nature of the channels. The distortion and the intersymbol interference (ISI) (see Section B.6, Appendix B) produced by the channel will increase the error rate in detecting a symbol. Since the bandwidth of a channel is limited, there is a need

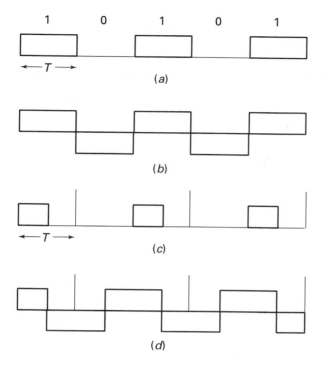

FIGURE 3.12 Various pulse shapes used in digital communications. (*a*) Unipolar NRZ. (*b*) Bipolar NRZ. (*c*) Unipolar RZ. (*d*) Manchester.

to explore other ways of eliminating or reducing ISI by appropriately shaping the pulses. Pulse shaping is essential in mobile and wireless communication systems, since these systems operate with the smallest possible bandwidth to accommodate a large number of users. The obvious technique here for shaping the pulses is to round off the corners so that sidelobes of the spectra of the pulses fall off faster. The other technique is to increase the width of the pulse so that overlapping pulses can be sent, thereby reducing the spectral occupancy of the pulse. It is important to note that overlapping pulses must also have amplitudes that drop off faster so that the tails of these pulses do not cause problems in the presence of jitter (sampling instant offset) and lead to lack of synchronization.

The techniques of pulse shaping to reduce ISI can be described and explained using the Nyquist criteria.

Nyquist Criteria Nyquist's first criterion addresses intersymbol interference (ISI). To understand ISI, consider the transmission of a rectangular pulse through a filter having a finite bandwidth. The output pulse shapes for a number of different bandwidths of the low-pass filter (LPF) are shown in Figure 3.13. If the bandwidth of the LPF is not sufficient, the pulse extends beyond its original duration of T, thus interfering with pulses in the adjoining slots. The interference of a signal with other pulses is referred to as ISI. It causes errors in the identification of the bits. Intersymbol interference can be studied using an "eye pattern" (see Section B.6, Appendix B).

It is essential that we choose pulse shapes that will result in no intersymbol interference or that will at least reduce ISI as they pass through communication channels (Ande 1986, 1999). Nyquist criteria help us to understand and develop a formulation to specify the nature of the pulses to achieve the goal of zero ISI.

A pulse satisfies the first Nyquist criterion if it passes through zero at $t = kT$, $k = \pm 1, \pm 2, \ldots$, except at $t = 0$. The pulse $s(t)$ that satisfies this criterion is referred to as a *Nyquist pulse*. Examples of pulses satisfying this criterion are shown in Figure 3.14. These pulse shapes are

A rectangular pulse

A sinc pulse

A raised cosine roll-off pulse

A spectrally raised cosine (RC) pulse

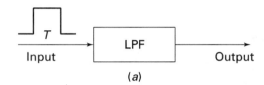

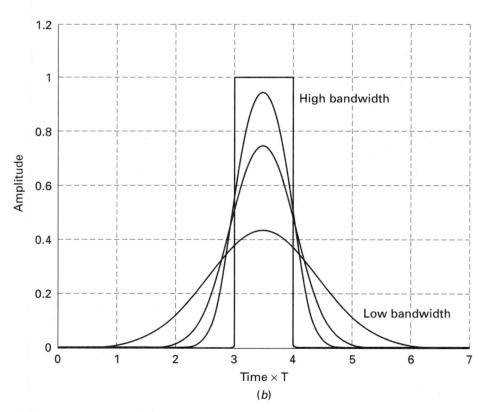

FIGURE 3.13 (*a*) A pulse of duration *T* as input to a low-pass filter. (*b*) Pulse shapes at the output of a low-pass filter having different bandwidths.

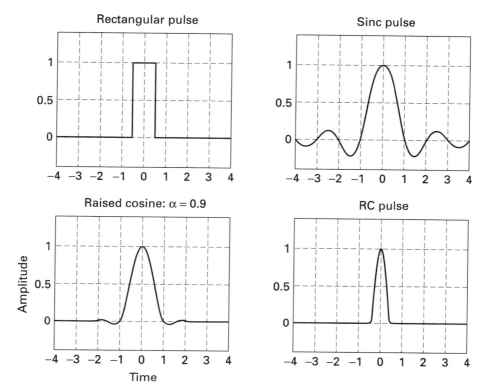

FIGURE 3.14 Different pulse shapes satisfying the first Nyquist criterion. The *x*-axis is normalized to the bit duration.

These pulses are defined by

$$s(t) = \begin{cases} \dfrac{1}{2T} & 0 \le t \le T \\[2em] \dfrac{\sin\left(\dfrac{\pi}{T}t\right)}{\left(\dfrac{\pi}{T}t\right)} & \\[2em] \dfrac{\sin\left(\dfrac{\pi}{T}t\right)}{\left(\dfrac{\pi}{T}t\right)} \dfrac{\cos\left(\dfrac{\pi}{T}\alpha t\right)}{1 - \left[\dfrac{4\alpha}{2T}t\right]^2} & 0 < \alpha < 1 \\[2em] \dfrac{1}{2}\left[1 + \cos\left(\dfrac{2\pi}{T}t\right)\right] & 0 \le t \le T \end{cases} \qquad (3.29)$$

Let us revisit the concept of intersymbol interference using a sinc pulse. A set of sinc pulses shifted by T is shown in Figure 3.15. These sinc pulses represent bits (or symbols) being transmitted. At the sampling instant, the previous bit and the next bit have zero values at $t = 0$ corresponding to the pulse (present bit). The sinc pulse thus satisfies the Nyquist criterion of zero contribution at the sampling instant from the pulses on either side of the present bit.

In essence, the effect of intersymbol interference can be eliminated if the impulse response of the overall system (transmitter, channel, and receiver) matches the pulse shapes shown in Figure 3.14. Since the rectangular pulse has no "intrusion" into the next symbol slot, it is an ideal shape for eliminating ISI. The sinc pulse also has no ISI, as seen in Figure 3.15. The parameter α is the roll-off factor, which determines the gradient of the spectral roll-off for the raised cosine pulse, a shape that also seems ideal in terms of the adjoining bits having zero amplitude at the sampling instant. If $\alpha = 0$, we have a sinc pulse. The corresponding spectra (magnitudes of the Fourier transforms) are shown in Figure 3.16.

Let us examine the pulses in the frequency domain and explore their suitability as practical pulse shapes with no ISI. Even though the rectangular pulse shape in Figure 3.14 seems to satisfy the Nyquist criterion, its spectrum has sidelobes that do not decay quickly, as seen in Figure 3.16. The presence of a significant amount of power in the sidelobes creates the following problem. If a LPF is designed to filter only the main lobe, then in case of inadequate filtering, these sidelobes will reappear, causing interchannel cross-talk since

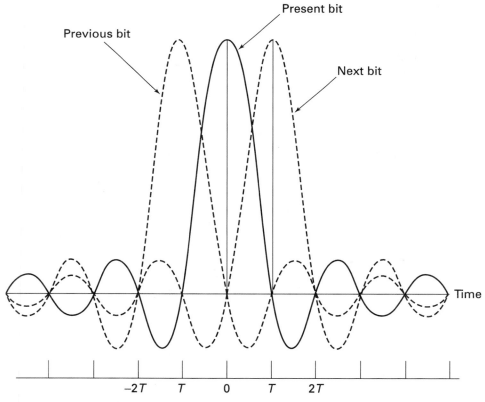

FIGURE 3.15 Three sinc pulses. At the sampling instant (the center of the "present bit"), the other two pulses have zero amplitude.

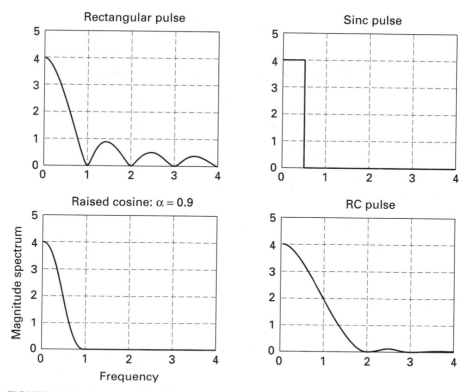

FIGURE 3.16 Spectra of the different pulse shapes shown in Figure 3.14. The frequency axis is normalized to the bit rate (1/T).

these sidelobes will fall in the passband of the next channel. The rectangular pulse, therefore, does not have a "tight" spectrum in which the energy is confined to a very narrow region of the spectrum. The sinc function of Figure 3.14 seems to meet the tightness requirement as far as the spectrum is concerned, and has the smallest spectral width or bandwidth. However, in the time domain, its sidelobes decay very slowly. Such a pulse will lead to synchronization problems and intersymbol interference in the presence of jitter (i.e., if the sampling instants are not exactly at $t = \pm kT$). The raised cosine pulse shape is ideal, since the decay of the sidelobes in the time domain can be controlled using the roll-off factor, α. Thus, the raised cosine pulse shape shown in Figure 3.14 seems to satisfy the requirements of faster roll-off in the time domain as well as tightness of the spectrum. Using the roll-off parameter, α, it is possible to strike a compromise between bandwidth and ISI.

Nyquist's second criterion refers to the bandwidth associated with the pulses. The bandwidth of a Nyquist pulse cannot be smaller than $1/2T$ Hz. This criterion is commonly referred to as Nyquist's sampling theorem. As we see from the spectra of the pulse shapes, the sinc pulse occupies the smallest bandwidth, and this corresponds to the Nyquist bandwidth. Since the data rate $R = 1/T$, the Nyquist bandwidth corresponds to $R/2$ Hz, the lowest possible bandwidth that is required for transmission.

The preceding criteria may not suffice to improve the performance of digital communication systems. When a number of symbols are sent, the detection schemes use correlation-type receivers. In this case, it is essential that pulses also be orthogonal under a shift of T s, i.e.,

$$\int_{-\infty}^{\infty} s(t)s(t - nT)\, dt = 0, \quad n = \pm 1, \pm 2, \dots. \tag{3.30}$$

From eq. (3.30), it is clear that if the pulses satisfy the Nyquist criteria, including the orthogonality criterion, there will be less ISI (zero values from other pulses at sampling instants), and minimal or no effect of the overlap integral (orthogonality).

Bandwidth   The bandwidth of a digital signal is a parameter that has several definitions, which is often confusing, leading to problems in understanding the spectral content of the signal (Amor 1980, Couc 1997). All these definitions are correct, but with different meanings. Since the design of communications systems is dependent on the bandwidth, it is absolutely essential that we understand the differences and relationships between different forms of bandwidth. To make the differences reasonably clear, we will use the example of binary phase-shift keying (BPSK), which can be represented essentially as

$$p(t) = p_T(t)\cos(2\pi f_0 t), \quad 0 \le t \le T. \tag{3.31}$$

The pulse train, $p_T(t)$ is expressed as

$$p_T(t) = \sum_{k=-\infty}^{\infty} m(t - kT), \tag{3.32}$$

with

$$m(t) = \pm 1, \quad kT \le t \le (k + 1)T. \tag{3.33}$$

The signal $m(t)$ is shown in Figure 3.17a, and the periodic signal, $p_T(t)$, is shown in Figure 3.17b. The power spectrum, $P(f)$, of $p(t)$ is shown in Figure 3.18a and is given by

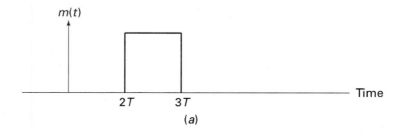

(a)

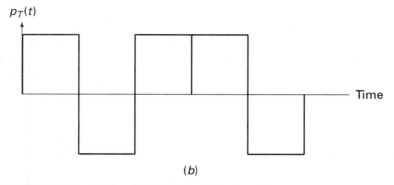

(b)

FIGURE 3.17   (a) Signal $m(t)$. (b) Periodic signal $P_T(t)$.

$$P(f) = \frac{T}{4}\left[\frac{\sin \pi T(f-f_0)}{\pi T(f-f_0)}\right]^2 + \frac{T}{4}\left[\frac{\sin \pi T(f+f_0)}{\pi T(f+f_0)}\right]^2. \tag{3.34}$$

The power spectrum, $M(f)$, of $m(t)$ is shown in Figure 3.18b. This corresponds to the spectrum of the baseband equivalent of the bandpass signal $p(t)$. The data rate is $R = 1/T$.

1. *Absolute bandwidth.* The absolute bandwidth of a signal is defined as the frequency range outside (positive frequencies) of which the power is zero. This is infinite both for the bandpass signal $p(t)$ and the baseband signal $m(t)$.

2. *3 dB bandwidth.* The 3 dB bandwidth, or half-power bandwidth, is the frequency range (positive frequencies f_1, f_2, with $f_2 > f_1$) where the power drops to 50% of the peak value. This bandwidth is given by $(f_2 - f_1)$.

3. *Equivalent bandwidth.* This is the extent of the positive frequencies occupied by a rectangular window such that the window has a height equal to the peak value of the positive spectrum and a power equal to the power contained within the positive frequencies. This is shown by the rectangular window in Figure 3.18a and has a value of R.

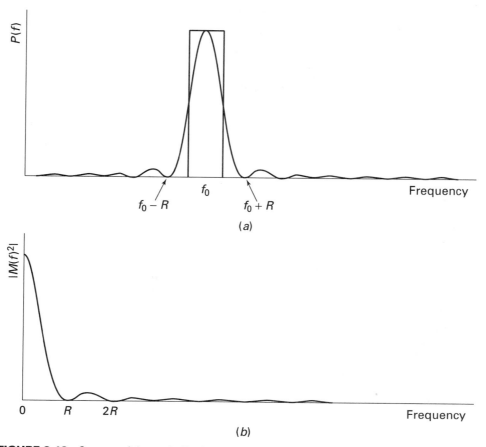

FIGURE 3.18 Spectra of the periodic signal (*a*) and a single pulse (*b*).

TABLE 3.1 Definitions of Bandwidth

Definition	Bandwidth
Absolute bandwidth	∞
3 dB bandwidth	$0.88R$
Equivalent bandwidth	$1.0R$
Null-to-null bandwidth	$2.0R$
Bounded-spectrum bandwidth	$201.5R$
Power (99%) bandwidth	$20.5R$

4. *Null-to-null bandwidth.* This corresponds to the frequency band between two "nulls" on either side of the peak. For the bandpass spectrum, this corresponds to $2R$.

5. *Bounded-spectrum bandwidth.* This is the value of $(f_2 - f_1)$ such that outside the band $f_1 < f < f_2$ the power spectrum is lowered by a significant amount at least (about 50 dB) below the peak value.

6. *Power (99%) bandwidth.* This is the range of frequencies $(f_2 - f_1)$ such that 99% of the power resides in that frequency band.

For the BPSK signal, these values are given in Table 3.1.

EXAMPLE 3.2

A voice signal is being sampled and transmitted. Assume that the voice signal is band-limited to 4 kHz. The modulation scheme is BPSK.

(a) Assuming a sinc pulse shape for the transmitted pulse, what is the minimum bandwidth required for transmission?

(b) If a raised cosine pulse shape of roll-off factor unity is used, what is the minimum bandwidth required for transmission?

Answer

(a) 4 kHz

(b) 8 kHz ∎

EXAMPLE 3.3

Consider transmission of data at the rate of 3 kbps using BPSK. Express the values of different bandwidths.

Answer

Definition	Bandwidth
Absolute bandwidth	∞
3 dB bandwidth	0.88×3 kHz
Equivalent bandwidth	3 kHz
Null-to-null bandwidth	6 kHz
Bounded-spectrum bandwidth	201.5×3 kHz
Power (99%) bandwidth	20.5×3 kHz

∎

Efficiency of Digital Modulation   Before we look at the digital modulation techniques employed in wireless and mobile communication systems, the factors that influence the choice of modulation scheme must be addressed. The two most important parameters are the *power efficiency* and the *bandwidth efficiency*. The power efficiency of a modulation system is a measure of the ability of the modulation scheme to preserve the quality or fidelity of the signal with minimal signal power (Amor 1980, Skla 1993, Glov 1998). A particular modulation technique that requires a lower signal-to-noise ratio to maintain a given bit error rate is considered to have a better power efficiency than another technique that requires a higher signal-to-noise ratio. After we describe some of the digital modulation techniques, we will compare the power efficiencies of these techniques.

The bandwidth or spectral efficiency of a modulation technique is a measure of the ability of the technique to transmit more data at a given bandwidth. In other words, if a modulation scheme is capable of transmitting data at a rate of R bps (bits per second) and B is the bandwidth occupied by the signal, the bandwidth efficiency (η_B) is defined as R/B bps/Hz. Again, we will compare the spectral efficiencies of the modulation schemes after we have described the modulation schemes in detail.

As we shall see, it is not possible to have high spectral efficiency and high power efficiency at the same time, and trade-offs have to be made in practical situations. Furthermore, the use of nonlinear amplifiers in communication systems may force us to examine the out-of-band power since inadequate filtering followed by nonlinear amplifiers results in spectral regrowth (Ariy 1989; Akai 1987, 1998). Once again, these aspects will be discussed after the specifics of digital modems are presented.

3.2.2 General Analysis of Receivers

The primary criterion used to characterize the different modulation and demodulation schemes is the bit error rate or the probability of error (Stei 1964, Taub 1986, Skla 1993). In a purely binary case with equal likelihood of transmitting a 1 or a 0, the probability of error, $p(e)$, can be written as

$$p(e) = \frac{1}{2}(p_{0/1} + p_{1/0}), \tag{3.35}$$

where $p_{0/1}$ (*miss*) is the probability of receiving a 0 when a 1 is transmitted, and $p_{1/0}$ (*false alarm*) is the probability of receiving a 1 when a 0 is transmitted. Before we examine ways to calculate these error probabilities, it is necessary to understand how the receiver processes the signals to make a decision on the choice of the two values.

Consider a case where we have transmitted two signals, $s_0(t)$ and $s_1(t)$, each of duration T, representing the two bits 1 and 0, respectively. These signals may be of constant voltages or two waveforms, such as $s_0(t) = A_0 \cos(2\pi f_0 t)$ and $s_1(t) = A_1 \cos[2\pi(f_0 + \Delta f)t]$, where f_0 is the carrier frequency and Δf is a frequency shift. These signals are corrupted by noise (see Section A.5, Appendix A), which is considered to be additive white Gaussian (AWG) noise, $n(t)$, with a noise spectral density of $G_n(f)$. This composite signal, consisting of a deterministic signal and noise, passes through a filter with transfer function $H(f)$ before being sampled to get a voltage of $v(T) = r_0(T) + n(T)$ or $r_1(T) + n(T)$, as shown in Figure 3.19. The quantity $n(T)$ is the noise component that is filtered and sampled, while $r_0(T)$ and $r_1(T)$ are the filtered and sampled values of the signals. Their probability density functions are

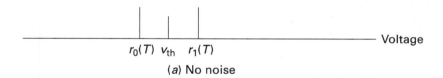

FIGURE 3.19 Input to the filter and the output of the filter.

shown in Figure 3.20. The goal is to design a filter so that the signal components will be emphasized and the effects of noise deemphasized. In the absence of noise, it is possible to set a threshold, v_{th}, equal to a value halfway between the two sampled values of the signal,

$$v_{th} = \frac{1}{2}[r_0(T) + r_1(T)] . \tag{3.36}$$

Now if we consider the fact that the signal is corrupted by the presence of noise, the decision will be in error if

$$n(T) + r_0(T) \geq \frac{1}{2}[r_0(T) + r_1(T)] \tag{3.37}$$

or

$$n(T) \geq \frac{1}{2}[r_1(T) - r_0(T)] . \tag{3.38}$$

The probability of error, $p(e)$, can therefore be expressed as

$$p(e) = \int_{\frac{1}{2}[r_1(T) - r_0(T)]}^{\infty} \frac{1}{\sigma_n \sqrt{2\pi}} \exp\left(-\frac{n(T)^2}{2\sigma_n^2}\right) dn(T) . \tag{3.39}$$

The quantity σ_n is the standard deviation of the noise $n(T)$. Note that the value of $p(e)$ given in eq. (3.39) is equal to either the probability of a "false alarm" (shaded area

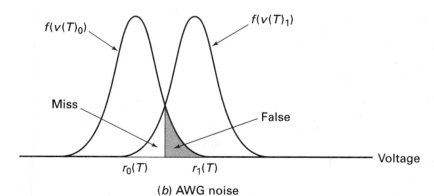

FIGURE 3.20 (a) The received signal constellation. (b) Probability density functions of the received signals when noise is present.

under the curve) or the probability of a "miss" (unshaded area under the curve) in Figure 3.20b. The functions plotted are the density functions of the received signal under the two hypotheses, H_1 (corresponding to the transmission of a 1) and H_0 (corresponding to the transmission of a 0). Using the properties of Gaussian random variables, the probability of error (Schw 1996, Peeb 1987, Proa 1995) can be expressed as

$$p(e) = \frac{1}{2}\text{erfc}\left[\frac{r_1(T) - r_0(T)}{2\sqrt{2}\sigma_n}\right], \qquad (3.40)$$

where erfc(.) is the complementary error function. Since erfc(x) decreases as x increases (Section B.7, Appendix B), the probability of error will be minimized if the quantity within the square brackets, Γ,

$$\Gamma = \frac{r_1(T) - r_0(T)}{2\sqrt{2}\sigma_n}, \qquad (3.41)$$

is maximum. The filter with a transfer function $H(f)$ that achieves this goal will be the *optimum filter*.

Optimum Filter   To obtain the form of the impulse response of the optimum filter, we start from the input-output relationship based on the quantity given in eq. (3.41). Noting that the quantity Γ will be maximized if the numerator is maximized along with minimization of the denominator, we can express the input to the filter, $m_{in}(t)$, as

$$m_{in}(t) = s_1(t) - s_0(t) \qquad (3.42)$$

and the output of the filter, $m_{out}(t)$, as

$$m_{out}(t) = r_1(t) - r_0(t). \qquad (3.43)$$

In the frequency domain, the relationship between the input and output can be expressed as

$$M_{out}(f) = H(f)M_{in}(f), \qquad (3.44)$$

where $M_{in}(f)$ and $M_{out}(f)$ are the Fourier transforms of $m_{in}(t)$ and $m_{out}(t)$, respectively. The sampled value of the output, $m_{out}(T)$, can now be expressed as

$$m_{out}(T) = \int_{-\infty}^{\infty} M_{out}(f) \exp(j2\pi fT)df$$
$$= \int_{-\infty}^{\infty} M_{in}(f)H(f) \exp(j2\pi fT)df. \qquad (3.45)$$

The output noise, $n(T)$, has a spectral density, $G_{out}(f)$, given by

$$G_{out}(f) = |H(f)|^2 G_n(f), \qquad (3.46)$$

and the noise power at the output, σ_n^2, becomes

$$\sigma_n^2 = \int_{-\infty}^{\infty} G_{out}(f)\,df = \int_{-\infty}^{\infty} |H(f)|^2 G_n(f)\,df. \qquad (3.47)$$

The condition for the optimum filter can now be expressed as

$$\Gamma^2 = \left[\frac{m_{out}(T)}{2\sqrt{2}\sigma_n}\right]^2 = \frac{\left|\int_{-\infty}^{\infty} M_{in}(f)H(f)\exp(j2\pi fT)\,df\right|^2}{\left(2\sqrt{2}\right)^2 \int_{-\infty}^{\infty} |H(f)|^2 G_n(f)\,df}. \tag{3.48}$$

The absolute value in the numerator does not affect the result since $m_{out}(T)$ is always positive (note that $r_1(t) > r_0(t)$). Applying the Schwarz inequality (Shan 1979, Skla 1988),

$$\Gamma^2 = \left[\frac{m_{out}(T)}{2\sqrt{2}\sigma_n}\right]^2 \leq \frac{\int_{-\infty}^{\infty} |M_{in}(f)|^2\,df}{8G_n(f)}, \tag{3.49}$$

the maximum value of Γ therefore is given by

$$\Gamma = \sqrt{\frac{\int_{-\infty}^{\infty} |M_{in}(f)|^2\,df}{8G_n(f)}}, \tag{3.50}$$

and from eq. (3.44), this value is obtained when

$$H(f) = KM_{out}^*(f)\exp(-j2\pi fT), \tag{3.51}$$

where K is a constant. In other words, the impulse response, $h(t)$, of the optimum filter is given by the Fourier transform of eq. (3.51), leading to

$$h(t) = K[s_1(T-t) - s_0(T-t)]. \tag{3.52}$$

Since $h(t)$ is a shifted version of the input signal itself, the filter is referred to as a *matched filter*.

The probability of error, $p(e)$, for the case of the matched filter can now be expressed as

$$p(e) = \frac{1}{2}\text{erfc}\sqrt{\left[\frac{m_{out}(T)}{2\sqrt{2}\sigma_n}\right]^2_{max}}, \tag{3.53}$$

and the right-hand side can be evaluated using Parseval's theorem (Section A.1, Appendix A):

$$\left[\frac{m_{out}(T)}{2\sqrt{2}\sigma_n}\right]^2_{max} = \frac{\int_{-\infty}^{\infty} |M_{in}(f)|^2\,df}{8G_n(f)} = \frac{1}{8}\left(\frac{2}{N_0}\right)\int_0^T m_{in}^2(t)\,dt$$

$$= \frac{1}{8}\left(\frac{2}{N_0}\right)\int_0^T [s_1(t) - s_0(t)]^2\,dt, \tag{3.54}$$

where $G_n(f) = N_0/2$ is the noise spectral density. The minimum probability of error corresponding to the maximum value of Γ becomes

$$p(e)_{min} = \frac{1}{2} \operatorname{erfc}\left[\sqrt{\left(\frac{m_{out}(t)}{2\sigma_n\sqrt{2}}\right)^2}\right]_{max} = \frac{1}{2} \operatorname{erfc}\sqrt{\frac{E_0 + E_1 - 2\rho\sqrt{E_1 E_0}}{4N_0}}, \quad (3.55)$$

where

$$E_0 = \int_0^T [s_0(t)]^2 \, dt \qquad (3.56)$$

$$E_1 = \int_0^T [s_1(t)]^2 \, dt \qquad (3.57)$$

$$\rho = \frac{1}{\sqrt{E_1 E_0}} \int_0^T s_0(t)s_1(t) \, dt. \qquad (3.58)$$

The matched filter–based receiver can be implemented as shown in Figure 3.21.

Let us now examine another receiver that will perform tasks similar to the matched filter, namely, the correlator.

Correlator versus Matched Filter A receiver configuration known as the correlator, or correlation receiver, also minimizes the probability of error. A block diagram of the correlator is shown in Figure 3.22. The input to the correlator is either of the signals $s_1(t)$ or $s_0(t)$ with noise, $n(t)$, added. This incoming signal is multiplied with a difference signal, $s_1(t) - s_0(t)$, and the outgoing signal is integrated and sampled at $t = T$.

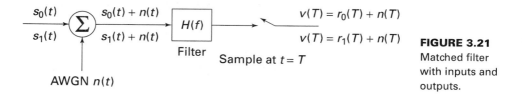

FIGURE 3.21 Matched filter with inputs and outputs.

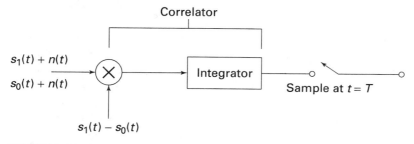

FIGURE 3.22 Schematic of the correlator.

This receiver is called a correlator because the output at the sampling instant is

$$
v_{out}(T) = \left. \begin{array}{l} \displaystyle\int_0^T s_0(t)[s_1(t) - s_0(t)]\ dt \\[12pt] \displaystyle\int_0^T s_1(t)[s_1(t) - s_0(t)]\ dt \end{array} \right\} \quad \text{signal} \tag{3.59}
$$

and

$$
n_{out}(T) = \int_0^T n(t)[s_1(t) - s_0(t)]\ dt \quad \text{noise} \tag{3.60}
$$

If we go back and examine the expression for the output of the optimum filter, we see that the output will be identical to the output of the matched filter (Skla 1988, Glov 1998). In other words, the performance of the two systems will be identical in terms of the bit error rate or probability of error. We can argue that the matched filter and correlator are two distinct ways of realizing the optimum filter, which minimizes the probability of error. Consider a simple sinusoidal pulse, $x(t)$, as the input to the matched filter or the correlator. The matched filter output, $x_{mf}(t)$, can be expressed as

$$
x_{mf}(t) = \int_0^t x(\tau)x(\tau + T - t)\ d\tau. \tag{3.61}
$$

The output of the correlator, $x_{co}(t)$, can be expressed as

$$
x_{co}(t) = \int_0^t x(\tau)^2 d\tau. \tag{3.62}
$$

The input sine pulse and outputs are plotted in Figures 3.23a and b. It can be seen that at $t = T$, the outputs of the matched filter and the correlator give the same value. This means that the bit error rate observed using either of these processors would be the same. However, if the timing of the decision is not the same, the outputs will be different and, consequently, the bit error rates will differ.

Note that with correlation, we are simply multiplying the two functions (or squaring the pulse if the functions are the same) and performing an integration. This process examines how close the functions essentially are. With the matched filter, we are sweeping the two functions past one another and integrating in steps. Thus, when these functions "match" as they slide past each other, the output will be the same as that of the correlator.

Geometric Representation of Signals and Orthonormal Functions

We have so far used $s_1(t)$ and $s_0(t)$ to represent the two signals for a particular modulation format. However, it is necessary to associate actual shapes and amplitudes to these signal functions so that the performances can be compared and decisions on the signal forms that optimize the performance of the modulation format can be realized. In addition, binary signaling is not the only possible form of information transmission. It is possible to transmit information using a higher-level modulation—M-ary modulation, where instead of just two signals, there are M signals ($M > 2$) to transmit the information. As we increase the number of levels, the amount of information we can transmit also goes up. For example, in a binary format we transmit one bit of information either as $s_1(t)$ or $s_0(t)$, while in

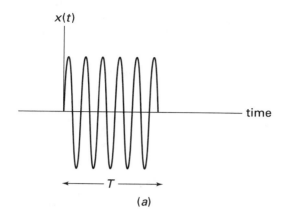

$x(t)$

time

T

(a)

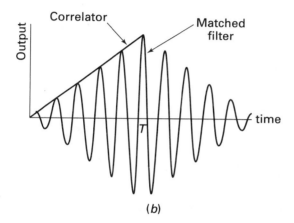

Output

Correlator

Matched filter

time

T

(b)

FIGURE 3.23 (a) Input sine pulse. (b) Comparison of the outputs of the correlator and matched filter.

M-ary format we have M possible signals, resulting in the transmission of $\log_2(M)$ bits of information. For $M = 4$, each signal or symbol contains two bits of information, for $M = 8$ each signal or symbol contains three bits of information, and so on.

If we present all M of these signals in a vector space, we can immediately observe the behavior of the modulation formats and compare the performances of the different formats as well as different levels of modulation. If S is the signal space, the signal space representing the M signals is

$$S = \{s_1(t), s_2(t), s_3(t), \ldots, s_M(t)\}. \tag{3.63}$$

We can represent any one of these signals, $s_1(t), \ldots, s_M(t)$, in terms of a set of orthonormal functions, i.e.,

$$s_i(t) = \sum_{k=1}^{N} a_{ik} \psi_k(t), \quad i = 1, 2, \ldots, M, \tag{3.64}$$

where the a_{ik} are the weighting factors and the $\psi_k(t)$ are the basis functions having the orthonormal property,

$$\int_{-\infty}^{\infty} \psi_k(t)\psi_l(t) \, dt = \begin{cases} 0 & k \neq l \\ 1 & k = l \end{cases}. \tag{3.65}$$

Thus each basis function or signal component can be thought of as having unit energy. Based on this formulation, we can represent the signal $s_1(t)$ or $s_0(t)$ as

$$s_1(t) = \begin{cases} \sqrt{\dfrac{2E}{T}} \cos(2\pi f_0 t) & 0 \leq t \leq T \quad \text{or} \\[2ex] \sqrt{\dfrac{2E}{T}} \sin(2\pi f_0 t) & 0 \leq t \leq T \end{cases} \tag{3.66}$$

$$s_0(t) = \begin{cases} -\sqrt{\dfrac{2E}{T}} \cos(2\pi f_0 t) & 0 \leq t \leq T \quad \text{or} \\[2ex] -\sqrt{\dfrac{2E}{T}} \sin(2\pi f_0 t) & 0 \leq t \leq T \end{cases} , \tag{3.67}$$

where $\psi(t)$ is given by

$$\psi(t) = \begin{cases} \sqrt{\dfrac{2}{T}} \cos(2\pi f_0 t) & 0 \leq t \leq T \quad \text{or} \\[2ex] \sqrt{\dfrac{2}{T}} \sin(2\pi f_0 t) & 0 \leq t \leq T \end{cases} , \tag{3.68}$$

and E is the energy contained in the bit of duration T and $a = \pm\sqrt{E}$, with the positive sign associated with the positive functions of eqs. (3.66) and (3.67) and the negative sign associated with the negative functions of eqs. (3.66) and (3.67).

Probability of Error in Terms of Signal Constellations   The error probability expression given in eq. (3.55) is useful for computing the performance of a number of commonly used digital modulation formats. These modulation schemes are typically described in terms of a plot of signal points referred to as a *signal constellation*. Consider the case of orthogonal signaling discussed in the previous section where the two waveforms are given by $s_i(t)$ and $s_j(t)$, with

$$\int_0^T s_i(t)s_j(t)\, dt = \begin{cases} E & i = j \\ 0 & i \neq j. \end{cases} \tag{3.69}$$

If we take the case of two orthogonal signals,

$$s_1(t) = \sqrt{\frac{2E}{T}} \cos(2\pi f_0 t) = \sqrt{E}\sqrt{\frac{2}{T}} \cos(2\pi f_0 t) = \sqrt{E}\,\psi_1(t) \tag{3.70}$$

$$s_0(t) = \sqrt{\frac{2E}{T}} \sin(2\pi f_0 t) = \sqrt{E}\sqrt{\frac{2}{T}} \cos(2\pi f_0 t) = \sqrt{E}\,\psi_0(t),$$

they can be shown in a constellation as in Figure 3.24a. The separation between these two signals is $\sqrt{2E}$. Let us now go back to eq. (3.55) for the probability of error. We now have $E_1 = E_0$, $\rho = 0$, and the expression for the probability of error becomes

$$p(e)_{min} = \frac{1}{2}\,\text{erfc}\,\sqrt{\frac{2E}{4N_0}} = \frac{1}{2}\,\text{erfc}\,\sqrt{\frac{d_{min}^2}{4N_0}} = \frac{1}{2}\,\text{erfc}\left[\frac{d_{min}}{2\sqrt{N_0}}\right], \qquad (3.71)$$

where d_{min} is the minimum separation between the two signals as shown in Figure 3.24a.

Now consider two signals that are antipodal, i.e.,

$$s_1(t) = -s_0(t), \qquad (3.72)$$

in which case the correlation coefficient $\rho = -1$. This situation is demonstrated in Figure 3.24b. The minimum separation between the two signals is given by $d_{min} = 2\sqrt{E}$. Going back to the equation for the error probability, the minimum error probability now becomes

$$p(e)_{min} = \frac{1}{2}\,\text{erfc}\,\sqrt{\frac{E_1 + E_0 - 2\rho\sqrt{E_1 E_0}}{4N_0}} = \frac{1}{2}\,\text{erfc}\,\sqrt{\frac{4E}{4N_0}} = \frac{1}{2}\,\text{erfc}\,\sqrt{\frac{d_{min}^2}{4N_0}}. \qquad (3.73)$$

It is possible to express the bit error rate in terms of the minimum separation between the two signals. Certainly the bit error rate or probability of error for the case of antipodal signaling is lower than the corresponding value for orthogonal signaling (Hayk 2001; Skla 1988; Ande 1986, 1999).

Another case we will be looking at refers to what is known as quaternary phase-shift keying (QPSK). In this case, each symbol consists of a pair of bits, as shown in Figure 3.24c. The error rate will be similar to the antipodal case.

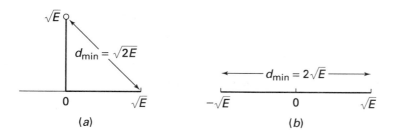

(a)

(b)

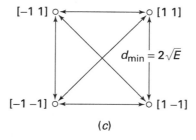

(c)

FIGURE 3.24
Signal constellations.
(a) Orthogonal signals.
(b) Bipolar signals.
(c) Multilevel signals (four-level PSK).

The concept of constellation and signal separation appears to be a simple way to calculate the probability of error. It is also intuitively obvious that the greater the separation between the two levels, the better the chances of detecting the bits with a lower error rate, since any decision criterion is based on a threshold level midway in the "separation" between the two signal points. In other words, a higher value of d_{min} will lead to a lower probability of error.

Signal-to-Noise Ratio in a Digital Communication System   The expression for the probability of error contains the parameter E/N_0. The probability of error for each of the different schemes can be expressed in terms of that single parameter. The unit of energy E (energy/bit) can be expressed in terms of the signal power, S, and the bit duration, T, as

$$E = ST, \tag{3.74}$$

and the parameter E/N_0 can be expressed as

$$\frac{E}{N_0} = \frac{ST}{N_0}. \tag{3.75}$$

Noting that the data rate R is equal to $1/T$, eq. (3.75) can be rewritten as

$$\frac{E}{N_0} = \frac{S}{N_0 R}. \tag{3.76}$$

If the signal bandwidth is B Hz, eq. (3.76) can be rewritten as

$$\frac{E}{N_0} = \frac{SB}{N_0 RB} = \left(\frac{S}{N}\right)\left(\frac{B}{R}\right), \tag{3.77}$$

where $N = N_0 B$ and R/B has units of bps/Hz. We see that E/N_0 and S/N are linearly related and, therefore, are sometimes used interchangeably. The quantity that relates these two ratios, energy/bit-to-noise and signal-to-noise, is the inverse of the bandwidth efficiency (Skla 1993, Ande 1999), defined in bps/Hz. If a certain modulation format can transmit more bps/Hz of available bandwidth for the same signal power at the same performance level, the format that provides a higher value of R/B (bps/Hz) will be more bandwidth efficient.

The performance of the modulation formats can be compared in terms of the value of E/N_0 required to maintain a fixed probability of error. A modulation scheme that requires a lower value of E/N_0 to maintain a certain bit error rate has better power efficiency than a scheme requiring a higher value of E/N_0 to maintain the same bit error rate.

We will revisit the issue of the relationship between the signal-to-noise ratio and bandwidth efficiency once we have looked at various modulation formats.

3.3   MODEMS

We will now examine the various modulation and demodulation schemes used in wireless communications (Sull 1972, Hira 1979a, Clar 1985, Nogu 1986, Aghv 1993, Fehe 1995, Samp 1997, Akai 1998). These techniques are distinctive because of the available bandwidth, cost of handsets, and other factors. We will look initially at simple modulation/demodulation (modem) techniques that may be used and then study the specific modems used exclusively in wireless and mobile communication systems. We will not derive expressions for the error probability. Instead, we will invoke the concept of signal constellations, discussed in the previous section, to obtain the error probability.

3.3.1 Amplitude Shift Keying (ASK)

The simplest form of digital modulation is amplitude shift keying (ASK). In this case, the transmitted carrier wave takes two amplitude values during the duration of the pulse:

$$S_{ASK}(t) = \sqrt{\frac{2E_i}{T}} \cos(2\pi f_0 t), \quad i = 1, 2. \tag{3.78}$$

Typically, E_i takes a value of E or 0; thus ASK is known as on-off keying (OOK). Demodulation of the on-off keying may be undertaken coherently (using a local oscillator) or incoherently (using an envelope modulator). The coherent demodulator takes the form described in the section on matched filters. The probability of error can be expressed (Taub 1986) as (see Problem 10)

$$p(e) = \frac{1}{2} \text{erfc} \sqrt{\frac{E}{2N_0}} \tag{3.79}$$

Amplitude shift keying is not used in wireless communications, since the bit error rate in ASK systems is higher than in most of the other digital modulation formats such as BPSK, QPSK, MSK, and GMSK, as we shall see in later sections.

3.3.2 Binary Phase-Shift Keying

In binary phase-shift keying (BPSK), the information is contained in the phase. During the transmission of a 1 the carrier phase is zero, and the carrier phase takes a value of π during the transmission of a 0. The BPSK signal can be expressed as

$$S_{PSK}(t) = \begin{cases} \sqrt{\dfrac{2E}{T}} \cos(2\pi f_0 t) & 1 \\[3mm] -\sqrt{\dfrac{2E}{T}} \cos(2\pi f_0 t) & 0 \end{cases} \quad 0 \le t \le T. \tag{3.80}$$

A typical BPSK signal is plotted in Figure 3.25. It shows that the waveform is discontinuous when the polarity of the bits changes (i.e., from 1 to -1 and vice versa). These abrupt changes in the waveform lead to sidelobes in the spectrum of the BPSK signal, which may create problems when inadequate filtering is present in the system. We will discuss these aspects in detail when we look at QPSK system.

Spectrum and Bandwidth of BPSK In the absence of additional pulse shaping, the power spectral density (PSD) of the BPSK signal is obtained from the Fourier transform of a rectangular pulse as shown in Section 3.2.1 (Pasu 1979, Amor 1980). The PSD can be expressed as

$$P_{\text{LP}}(f) = \text{const}\left(\frac{\sin \pi f T}{\pi f T}\right)^2, \tag{3.81}$$

where the subscript LP refers to the low-pass spectral density. The bandpass power spectral density, $P(f)$, of the BPSK signal at radio frequency is obtained through frequency translation to give

$$P(f) = \text{const}\left[\left(\frac{\sin \pi (f-f_0)T}{\pi (f-f_0)T}\right)^2 + \left(\frac{\sin \pi (f+f_0)T}{\pi (f+f_0)T}\right)^2\right]. \tag{3.82}$$

The power spectrum for BPSK (pulse shape) along with the spectra of a few other modulations formats are shown in Figure 3.26. The required bandwidth for transmission, defined on the basis of the first zero crossing of the PSD of the low-pass spectral density, is $1/T$. The bandwidth efficiency of the BPSK system modulation format is 1 bps/Hz. The sidelobes are clearly seen. These sidelobes will be detrimental in the presence of inadequate filtering since they are likely to be regenerated and amplified when nonlinearities are present in the amplifiers.

BPSK Receiver The BPSK signal can be demodulated using a matched filter or correlator. A typical block diagram of the demodulator with correlator is shown in Figure 3.27. This synchronous modulation approach requires that information on the phase and frequency of the carrier be available at the receiver. The carrier recovery circuit precedes the matched filter/correlator. The squaring device followed by the frequency divider provides the carrier frequency and phase to allow coherent or

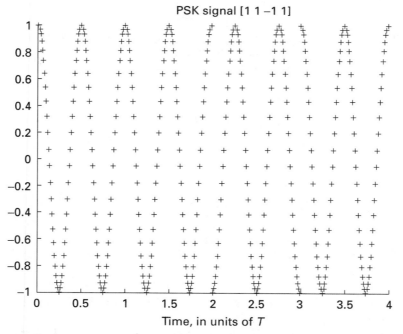

FIGURE 3.25 A BPSK waveform. Abrupt changes in the waveform are seen at $2T$ and $3T$.

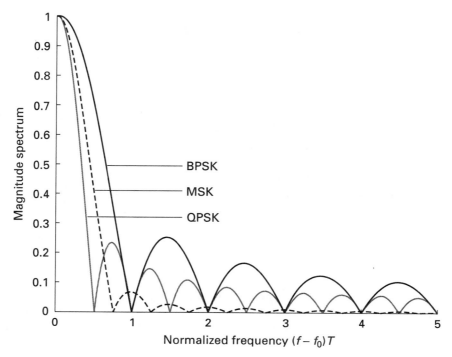

FIGURE 3.26 Spectra of BPSK, MSK, and QPSK.

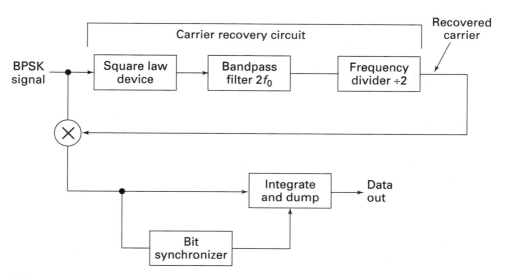

FIGURE 3.27 A coherent demodulator for BPSK.

synchronous detection using the correlator. The integrate-and-dump circuit is a correlation receiver. The integrator computes the area (integrates the product), and the output is sampled at $t = T$. After $t = T$, the integrator is immediately "dumped," or emptied, and restored to its initial condition. The bit synchronizer (Section B.4, Appendix B) makes correct sampling possible at the end of the bit period. The output is processed using a threshold decisionmaker to recreate the 0s and 1s.

The probability of error for BPSK can be obtained by noting that the BPSK signal has the constellation shown in Figure 3.24b. Hence, the error probability is

$$P_{\text{BPSK}}(e) = \frac{1}{2}\text{erfc}\left(\sqrt{\frac{E}{N_0}}\right). \tag{3.83}$$

The error probability plot for different values of E/N_0 (signal-to-noise ratio) is shown in Figure 3.28. Comparing eq. (3.83) with the bit error rate for ASK systems indicates that the performance of coherent BPSK is 3 dB better than that of coherent ASK. Note, however, that perfect carrier (phase) recovery is assumed.

Consider the case where the locally generated carrier is $\cos(2\pi f_0 t + \Delta\phi)$, with $\Delta\phi$ being the phase mismatch between the incoming signal and the regenerated carrier. The correlator removes $\sin(2\pi f_0 t)\sin(\Delta\phi)$, the amplitude of the correlator output is scaled by $\cos(\Delta\phi)$, and the energy at the output becomes $E\cos^2(\Delta\phi)$. The probability of error becomes

$$P_{\text{BPSK}}(e) = \frac{1}{2}\text{erfc}\left[\sqrt{\frac{E}{N_0}}\cos\Delta\phi\right]. \tag{3.84}$$

The phase error or mismatch may be caused by drift of the master oscillators at the transmitter and receiver if a separate signal source (master oscillator) is used at the receiver. It may also be caused by the imperfection of the carrier/phase recovery approach. The probability of error in the presence of phase mismatch is shown in Figure 3.29 for two values of phase mismatch.

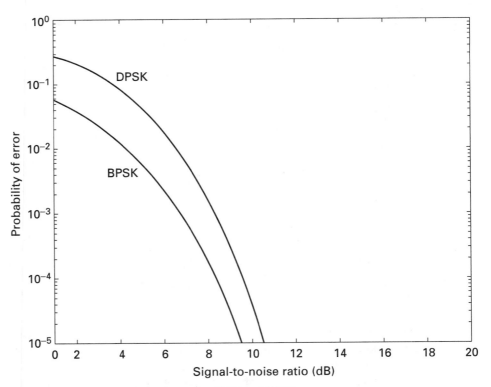

FIGURE 3.28 Probabilities of error for BPSK and DPSK.

The increase in the probability of error with phase mismatch can be easily seen in Figure 3.29, and it appears that for a mismatch of 20° or less, the increase in error probability at a given SNR is extremely small. From eq. (3.84), it can be easily shown that if

$$\cos^2(\Delta\phi) > 0.8, \tag{3.85}$$

we would require only a 1 dB additional increase in signal-to-noise ratio to maintain the same probability of error as in a perfectly coherent BPSK.

EXAMPLE 3.4

In a BPSK system, what is the average SNR required to maintain a bit error rate of 0.5×10^{-3}? (Hint: Use the MATLAB function *erfinv* making use of the identity $erf(x) + erfc(x) = 1$.)

Answer   Using eq. (3.83), we find the required SNR to equal 7.3 dB.   ■

EXAMPLE 3.5

Compare the average values of SNR required to maintain a bit error rate of 3×10^{-3} for a coherent OOK and a coherent BPSK system.

Answer   OOK, 8.77 dB; BPSK, 5.77 dB.   ■

EXAMPLE 3.6

Verify that eq. (3.85) holds true (left as as exercise).   ■

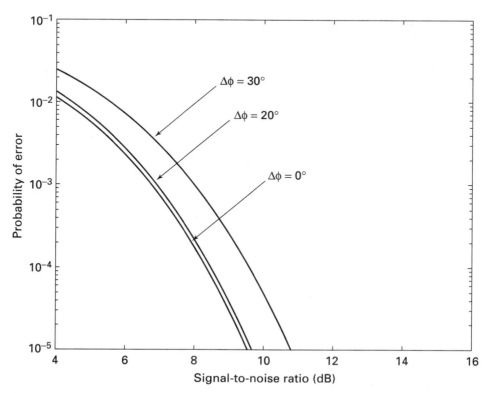

FIGURE 3.29   The effect of phase mismatch on the probability of error for two values of phase mismatch.

In eq. (3.84), it was assumed that the phase mismatch is deterministic. The error probability can also be evaluated under conditions of random phase errors. Quite often the phase of the local oscillator or the recovered carrier will be random because of noise introduced in the recovery process. Under these conditions, the phase mismatch $\Delta\phi$ will be a random variable. If we assume that the phase mismatch $\Delta\phi$ is Gaussian distributed with a standard deviation of σ_ϕ, the average error probability, $p_{av}(e)$, is

$$P_{av}(e) = \int_{-\infty}^{\infty} p_{BPSK}(e|\Delta\phi)f(\Delta\phi)\,d\phi$$

$$= \frac{1}{2}\int_{-\infty}^{\infty} \mathrm{erfc}\left(\sqrt{\frac{E}{N_0}}\cos\Delta\phi\right)\exp\left(-\frac{\Delta\phi^2}{2\sigma_\phi^2}\right)d\Delta\phi. \tag{3.86}$$

where $p_{BPSK}(e|\Delta\phi)$ is given in eq. (3.84) and $f(\Delta\phi)$ is the Gaussian probability density function of the phase mismatch.

The probability of error given in eq. (3.86) is plotted in Figure 3.30 for two values of σ_ϕ. It shows that as the standard deviation increases, the error not only goes up but also reaches an error floor (stays almost always at a high value). Such a trend is characteristic of digital systems subjected to random fluctuations in carrier frequency (or phase). The effects seen in Figure 3.30 are observed in fading channels when random FM is present, as we shall see in Chapter 5. The degradation in performance can be compensated only through the use of diversity techniques, a topic discussed in that chapter.

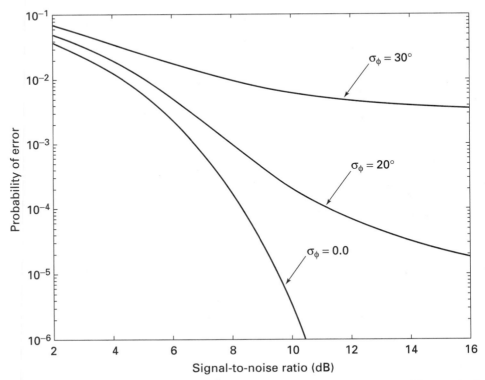

FIGURE 3.30   The average error probability given in eq. (3.86) for two values of the standard deviation of the phase mismatch, $\Delta\phi$.

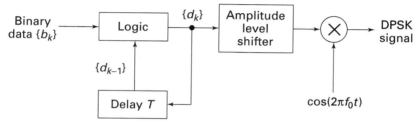

FIGURE 3.31 Block diagram of the DPSK transmitter.

DPSK Transmitter and Receiver The synchronous or coherent approach is not the only means of demodulating a BPSK signal. It is also possible to demodulate a BPSK signal using a differential approach that does not require the generation of the carrier information at the receiver. One of the simplest noncoherent demodulation schemes is the DPSK (differential phase-shift keying). The DPSK scheme uses a two-step process: In the first step, the data are differentially encoded, and in the second step, they are phase encoded. A block diagram of the DPSK transmitter is shown in Figure 3.31. The amplitude shifter produces the values of 1 and -1. The steps involved in the generation of DPSK are as follows.

The incoming bits are represented by $b_k(k = 1, 2, \ldots)$ and the encoded bits are represented by $d_k(k = 0,1,2, \ldots)$. The encoded bit d_0 corresponds to the reference bit chosen as 1. A modulo-2 addition $\oplus$ is performed between b_k and d_{k-1} ($0 \oplus 0 = 1 \oplus 1 = 0$; $0 \oplus 1 = 1 \oplus 0 = 1$). The encoded bit is the result of taking the complement of $b_k \oplus d_{k-1}$, i.e., $\overline{b_k \oplus d_{k-1}}$. These differentially encoded bits are used to phase-shift a carrier, with phase of angles 0 and π representing states 1 and 0, respectively. The differential encoding process is illustrated in Table 3.2.

The information can be recovered in a noncoherent fashion without the need for a local oscillator. A block diagram of the DPSK demodulator is given in Figure 3.32a. It looks like a coherent demodulator except for the absence of a carrier recovery circuit. In fact, the mixing takes place between the input bit stream and the bit stream delayed by T. This feature appears to make the demodulator shown in Figure 3.32a mirror the DPSK modulator shown in Figure 3.31.

The DPSK transmission/reception scheme shown in Figure 3.32a has the advantage of not requiring the generation of a phase-coherent source. The price of this simplicity is an increase in the probability of error (Biya 1995, Robe 1994).

The receiver in Figure 3.32a operates by correlating the received signal (transmitted signal plus any noise) with a delayed version of the same. The output of the correlator is examined in reference to zero. A decision is made in favor of 1 if the output is positive, and a decision of 0 is made if the output is negative. The correlation and decision procedures are shown in Table 3.3.

TABLE 3.2 DPSK Encoding: $d_k = \overline{m_k \oplus d_{k-1}}$, Where $\oplus$ Indicates Modulo-2 Addition and the Upper Bar Denotes the Complement

			1	0	0	1	0	0	1	1
Input data sequence $\{b_k\}$			1	0	0	1	0	0	1	1
Differentially encoded seqence $\{d_k\}$		1	1	0	1	1	0	1	1	1
Transmitted phase (radians)		0	0	π	0	0	π	0	0	0

Arbitrary bit

TABLE 3.3 DPSK Decoding

Transmitted phase (radians) (same as transmitted when no noise is present)	0	0	π	0	0	π	0	0	0
Phase comparison output		+	−	−	+	−	−	+	+
Output bits		1	0	0	1	0	0	1	1

This ideal recovery of data is achieved only in the absence of noise. Even though it is simple to implement, the receiver shown in Figure 3.32*a* is not the optimal receiver (Park 1978; Simo 1978, 1992; Proa 1995; Hayk 2001). An optimal receiver for DPSK uses a quadrature structure, shown in Figure 3.32*b*. The input to the receiver consists of the DPSK signal, $\pm A\cos(2\pi f_0 t)$, and additive white Gaussian noise $n(t)$ of spectral density $N_0/2$. Note that $A = \sqrt{2E/T}$. The input is split into two parts, one multiplied by $\cos(2\pi f_0 t)$ and the other multiplied by $\sin(2\pi f_0 t)$. The decisions are made on the basis of the test statistic h,

$$h = x_k x_{k-1} + y_k y_{k-1}. \tag{3.87}$$

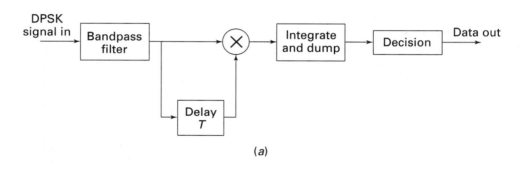

(a)

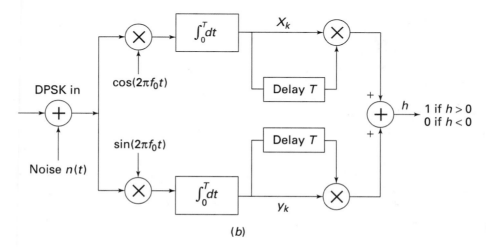

(b)

FIGURE 3.32 (*a*) Block diagram of a DPSK demodulator. (*b*) Block diagram of an optimal receiver for DPSK.

A decision in favor of 1 is made if $h > 0$. If $h < 0$, we conclude that a 0 has been received. The probability of error, $p_{DPSK}(e)$, is given by

$$p_{DPSK}(e) = \text{prob } \{x_k x_{k-1} + y_k y_{k-1} < 0 | 1 \text{ was sent}\}. \qquad (3.88)$$

We have assumed that 0s and 1s are equally likely and the error in the detection of 0 is equal to the error in the detection of 1. Let us now obtain expressions for x_k, x_{k-1}, y_k, and y_{k-1}.

Looking at the outputs of the correlators (k is taken to be unity), we can write

$$x_0 = \int_{-T}^{0} A \cos^2(2\pi f_0 t)dt + \int_{-T}^{0} n(t) \cos(2\pi f_0 t)dt \qquad (3.89a)$$

$$x_0 = A\frac{T}{2} + n_{x0} \qquad (3.89b)$$

$$x_1 = \int_{0}^{T} A \cos^2(2\pi f_0 t)dt + \int_{0}^{T} n(t) \cos(2\pi f_0 t)dt \qquad (3.90a)$$

$$x_1 = A\frac{T}{2} + n_{x1} \qquad (3.90b)$$

$$y_0 = \int_{-T}^{0} A \cos(2\pi f_0 t) \sin(2\pi f_0 t)dt + \int_{-T}^{0} n(t) \sin(2\pi f_0 t)dt \qquad (3.91a)$$

$$y_0 = 0 + n_{y0} = n_{y0} \qquad (3.91b)$$

$$y_1 = \int_{0}^{T} A \cos(2\pi f_0 t) \sin(2\pi f_0 t)dt + \int_{0}^{T} n(t) \sin(2\pi f_0 t)dt \qquad (3.92a)$$

$$y_1 = 0 + n_{y1} = n_{y1}. \qquad (3.92b)$$

The filtered noise components, n_{x0}, n_{x1}, n_{y0}, and n_{y1}, are all identically distributed independent Gaussian random variables, each with a zero mean and a variance of $N_0(T/4)$ (Skla 1988). This ensures that x_0 and x_1 are independent and identically distributed Gaussian random variables (Section A.6, Appendix A), each with a mean of $A(T/2)$ and a variance of $N_0(T/4)$. The equation for the probability of error, eq. (3.88), can now be rewritten as

$$p_{DPSK}(e) = \text{prob } \{x_0 x_1 + y_0 y_1 < 0\}. \qquad (3.93)$$

We can rewrite the product terms in the brackets (omitting a constant $1/4$), resulting in

$$p_{DPSK}(e) = \text{prob } \{(x_0 + x_1)^2 + (y_0 + y_1)^2 - (x_0 - x_1)^2 - (y_0 - y_1)^2 < 0\}. \qquad (3.94)$$

We can express the condition for the error to occur as

$$(x_0 + x_1)^2 + (y_0 + y_1)^2 > (x_0 - x_1)^2 + (y_0 - y_1)^2. \qquad (3.95)$$

Instead of comparing the squares, we can compare the positive square roots, leading to the condition under which an error occurs as

$$\sqrt{(x_0 + x_1)^2 + (y_0 + y_1)^2} > \sqrt{(x_0 - x_1)^2 + (y_0 - y_1)^2}. \qquad (3.96)$$

If we define

$$U_1 = x_0 + x_1$$

$$U_2 = y_0 + y_1$$

$$V_1 = x_0 - x_1 \qquad (3.97)$$

$$V_2 = y_0 - y_1,$$

the condition for the error to occur becomes

$$\sqrt{U_1^2 + U_2^2} > \sqrt{V_1^2 + V_2^2}, \qquad (3.98)$$

where

$$\langle U_1 \rangle = AT \qquad (3.99a)$$

$$\left.\begin{array}{c} \langle U_2 \rangle \\ \langle V_1 \rangle \\ \langle V_2 \rangle \end{array}\right\} = 0 \qquad (3.99b)$$

and

$$\left.\begin{array}{c} \text{var}(U_1) \\ \text{var}(U_2) \\ \text{var}(V_1) \\ \text{var}(V_2) \end{array}\right\} = \sigma_{ds}^2 = N_0 \frac{T}{2}. \qquad (3.100)$$

The probability density function of the random variable $R_1 = \sqrt{U_1^2 + U_2^2}$ is given by (Papo 1991)

$$f(r_1) = \frac{r_1}{\sigma_{ds}^2} \exp\left(-\frac{A^2 T^2 + r_1^2}{2\sigma_{ds}^2}\right) I\left(\frac{AT r_1}{\sigma_{ds}^2}\right) U(r_1), \qquad (3.101)$$

where I_0 is the modified Bessel function (Grad 1979).

The probability density function of the random variable $R_2 = \sqrt{V_1^2 + V_2^2}$ is given by

$$f(r_2) = \frac{r_2}{\sigma_{ds}^2} \exp\left(-\frac{r_2^2}{2\sigma_{ds}^2}\right) U(r_2). \qquad (3.102)$$

These are the Rician (3.101) and Rayleigh (3.102) density functions, which were seen in Chapter 2 in connection with fading. Equation (3.93) for the probability of error (Gagl 1988, Proa 1995, Simo 1992) now becomes

$$P_{\text{DPSK}}(e) = \text{prob}\{R_1 > R_2\} = \int_{r_1=0}^{\infty} f(r_1)\left[\int_{r_2=r_1}^{\infty} f(r_2)dr_2\right]dr_1 \qquad (3.103)$$

Using the table of integrals (Grad 1979), the error probability becomes

$$p_{\text{DPSK}}(e) = \frac{1}{2} \exp\left(-\frac{A^2 T^2}{4\sigma_{\text{ds}}^2}\right) = \frac{1}{2} \exp\left(-\frac{A^2 T^2}{4N_0\left(\dfrac{T}{2}\right)}\right) \tag{3.104}$$

Since the energy $E = A^2 T / 2$, eq. (3.104) becomes

$$p_{\text{DPSK}}(e) = \frac{1}{2} \exp\left(-\frac{E}{N_0}\right). \tag{3.105}$$

The probability of error for the DPSK scheme is shown in Figure 3.28, along with that for the coherent BPSK scheme. The performance appears to be worse by about 1 dB compared with that of a coherent BPSK.

It is clear that the BPSK scheme is simple to use in either coherent mode or differential mode. However, the bandwidth required for transmission may not be small enough for applications in wireless communications, where bandwidth is at a premium. It is therefore necessary to look at other modulation schemes that require bandwidths lower than that associated with the transmission of BPSK.

EXAMPLE 3.7

Calculate the average energy/bit-to-noise ratio required to maintain a bit error rate of 2×10^{-4} in coherent BPSK and DPSK.

Answer   BPSK, 6.915 dB; DPSK, 9.3 dB. ◾

3.3.3 *M*-ary Modulation Schemes

While the BPSK scheme is simple, its bandwidth efficiency is very poor, which makes it less suitable for wireless applications. On the basis of a null-to-null bandwidth criterion, BPSK requires a transmission bandwidth of $2B$ Hz for a data rate of B bits/s. Thus, we need to look for schemes that will reduce the bandwidth required for transmission. One of the ways in which this can be accomplished is through the use of M-ary modulation schemes, where each symbol has more than one bit. We will now look at a number of techniques based on M-ary modulation.

Quaternary Phase-Shift Keying (QPSK)   Whereas in BPSK the bits take either of two phase values, 0 or π, in quaternary phase-shift keying (Ande 1986, 1999; Dava 1989; Jenk 1972; Mase 1985a; Samp 1997; Bene 1999; Nogu 1986; Oett 1979) the phase can take any one of the four values 0, $\pi/2$, π, or $3\pi/2$. Each symbol consists of two bits, an *in-phase* component and a *quadrature* component. Thus a pair of bits will correspond to one of the four unique phase values.

The QPSK signal can be expressed as

$$S_{\text{QPSK}}(t) = \sqrt{\frac{2E}{T_s}} \cos(2\pi f_0 t + \phi_n) \quad 0 \le t \le T_s, \tag{3.106}$$

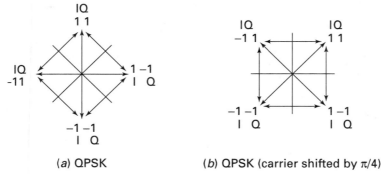

(a) QPSK   (b) QPSK (carrier shifted by π/4)

FIGURE 3.33 The two QPSK constellations. Note that they differ by $\pi/4$. When going from (1, 1) to (−1, −1), the phase is shifted by π. When going from (1, −1) to (1, 1), the phase shifts by $\pi/2$. Thus, depending on the incoming symbol, transitions from (1, 1) can occur to (1, 1), (1, −1), (−1, 1), or (−1, −1), or vice versa, leading to phase shifts of 0, $\pm\pi/2$, or $\pm\pi$ in QPSK. I and Q represent the in-phase and quadrature bits, respectively. Arrows show all possible transitions.

where the phase ϕ_n can take any one of the four phase values depending on the bit pairs. The symbol duration, T_s, is $2T$. The phase constellation associated with QPSK is shown in Figure 3.33a.It is easy to see that another phase constellation, shown in Figure 3.33b, will be created if the carrier phase is shifted by $\pi/4$ degrees, resulting in phases of $\pi/4$, $3\pi/4$, $5\pi/4$, and $7\pi/4$. The waveform resulting in this shifted version of QPSK can be expressed as

$$S_{\text{QPSK}}(t) = \sqrt{\frac{2E}{T_s}} \cos[(2\pi f_0 t + \pi/4) + \phi_n], \quad 0 \le t \le T_s. \quad (3.107)$$

Equation (3.107) can be rewritten to result in a simpler form for the QPSK signal as

$$S_{\text{QPSK}}(t) = \sqrt{\frac{2E}{T_s}} \cos[(2\pi f_0 t + \pi/4)] \cos(\phi_n)$$

$$(3.108)$$

$$-\sqrt{\frac{2E}{T_s}} \sin[(2\pi f_0 t + \pi/4)] \sin(\phi_n),$$

where the first term can be identified as the in-phase (I) term and the second term can be identified as the quadrature (Q) term. The phase values ϕ_n, are given in Table 3.4

TABLE 3.4 Phase Encoding in QPSK

Data		ϕ_n
1	1	$\pi/4$
−1	−1	$3\pi/4$
−1	−1	$-3\pi/4$
1	1	$\pi/4$

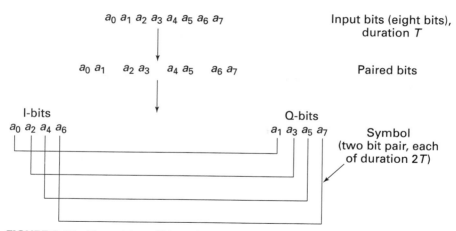

FIGURE 3.34 The pairing of bits to form symbols.

Consider the bit stream $a_0\, a_1\, a_2\, a_3\, a_4\, a_5\, a_6\, a_7$ (duration T s) where the a_i can take a value of $+1$ or -1. The various steps involved in the generation of QPSK signals are given in Figure 3.34 and explained below.

1. The bits are paired ($a_0\, a_1$, $a_2\, a_3$, $a_4\, a_5$, $a_6\, a_7$).

2. The even and odd bit streams are separated to create I and Q bits each of duration $T_s\ (=2T)$ seconds.

3. The even stream is multiplied by the *cosine* wave and odd stream is multiplied by the *sine* wave. These components are then added to get the QPSK signal. (Note that multiplication of the even and odd streams by cosine and sine waves creates two BPSK signals in quadrature. The only difference is that the pulses are of duration $2T$ in this arrangement, while in simple BPSK the pulses are of duration T.)

A QPSK signal for the bit pattern [1 1 -1 -1 -1 1 1 1] is shown in Figure 3.35. Let us examine the waveform in Figure 3.35*a*. One immediate observation is that the signal undergoes phase changes at intervals of $2T$. The phase difference between successive symbols may be 0, $\pm\pi/2$, or $\pm\pi$. These sharp changes in phase (particularly $\pm\pi$), resulting in discontinuities, make the modulation scheme less suited for some applications where nonlinear amplifiers may be used (Divs 1982, Ariy 1989, Rapp 1996b). The changes in phase are explained in Figure 3.36. Figure 3.35*b* shows another QPSK signal generated using a carrier wave that is shifted by $\pi/4$. The similarities between these two waveforms (Figures 3.35*a* and *b*) are obvious.

EXAMPLE 3.8

For the following bit stream to be transmitted as QPSK, what are the phase shifts between symbols? 0 0 1 0 1 0 1 1 0 1. (Note: 0 is the same as -1.)

Answer   The paired bits are 00   10   10   11   01. The respective phases are $-3\pi/4$, $-\pi/4$, $-\pi/4$, $\pi/4$, $3\pi/4$. The shifts are $-\pi/2$, 0, $-\pi/2$, $-\pi/2$. ∎

QPSK Transmitter and Receiver   The various steps in generating the QPSK signal are shown in Figure 3.37*a*. The data rate is R bits/s and the symbol rate is $(R/2)$ symbols/s. The input bits are paired and phase-encoded as shown in the figure, based on the methodology presented in the previous sections. The resulting phase terms,

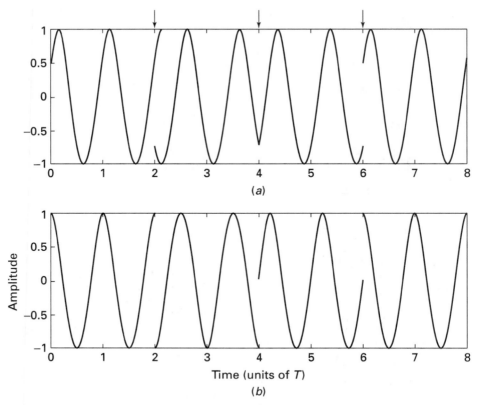

FIGURE 3.35 QPSK waveforms. (*a*) QPSK. (*b*) $\pi/4$-shifted QPSK. Arrows indicate phase changes.

What happens to the phase at intervals of $2T$ in QPSK?

Phases of the symbols: All possible combinations:

$1\ 1 \longrightarrow \pi/4$ $-1\ 1 \longrightarrow 3\pi/4$ $-1\ -1 \longrightarrow 5\pi/4$ $1\ -1 \longrightarrow 7\pi/4$

The bit stream is $1\ 1\ -1\ -1\ -1\ 1\ 1\ 1$.

π		$\pi/2$		$-\pi/2$			$\longrightarrow$ Phase shifts between symbols in QPSK

| $\pi/4$ | | $5\pi/4$ | | $3\pi/4$ | | $\pi/4$ | $\longrightarrow$ Phase of the symbols in QPSK |

| 1 | 1 | −1 | −1 | −1 | 1 | 1 | 1 |
| I | Q | I | Q | I | Q | I | Q |

FIGURE 3.36 Explanation of the phase shifts observed in QPSK, indicating the phases of the symbols and the phase difference between symbols.

$\cos(\phi_n)$ and $\sin(\phi_n)$, are the inputs to two equivalent BPSK modulators, one operating with a cosine carrier wave and the other with a sine carrier wave.

The coherent receiver for QPSK is very similar to the BPSK receiver if we imagine using two BPSK receivers in quadrature and appropriate filters. The receiver structure is shown in Figure 3.37*b*. The carrier recovery circuit provides precise information on the frequency (f_0) and phase of the incoming QPSK signal. A parallel-to-serial converter (P/S) provides the output data.

Offset QPSK (Staggered QPSK)   The problems with the phase reversal of π can be eliminated if the I and Q streams are offset by one bit period. This leads to offset QPSK(OQPSK) or staggered QPSK (Pasu 1979, Skla 1988). The block diagram of the OQPSK modulator is shown in Figure 3.38.

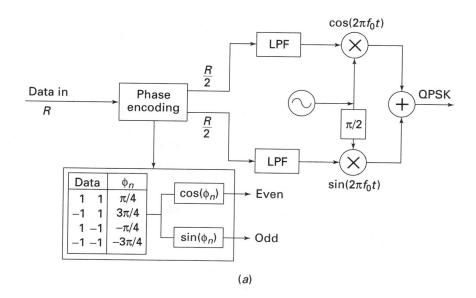

(a)

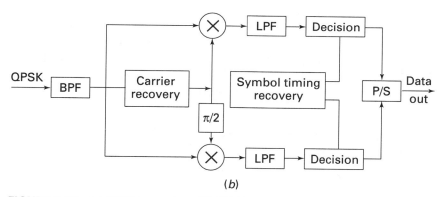

(b)

FIGURE 3.37   (a) QPSK transmitter. (b) QPSK receiver.

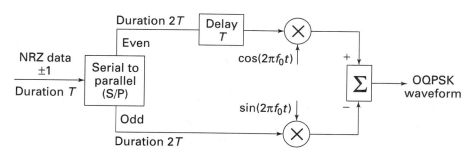

FIGURE 3.38   Block diagram of the OQPSK modulator.

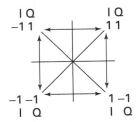

FIGURE 3.39 Phase constellation of OQPSK.

Once the bit streams are staggered, at intervals of T, only one of the bits in the two symbols can change at any given time, leading to a phase change of 0 or $\pi/2$. The phase constellation associated with OQPSK is shown in Figure 3.39 along with the phase transitions for the bit stream. Note that the phase transitions occur twice as many times (at intervals of T) as one would observe for QPSK. The elimination of the phase transition of π reduces spectral regrowth, making the modulation technique welcome in environments where nonlinear amplifiers are used. This is shown in Figure 3.40.

In OQPSK, the I pattern is delayed by T. Thus, at intervals of T, only one of the bits can change, not both. In other words, it is possible only to go from (1, 1) to $(-1, 1)$ or $(1, -1)$, and not to $(-1, -1)$. This limits the phase shifts to 0 or $\pm\pi/2$. *Note that the arrows in the constellation showing transitions go up, down, or sideways. There are no arrows along diagonals.*

An OQPSK waveform is shown in Figure 3.41. The phase differences at the transitions indicated by the arrows are either $\pm\pi/2$ or 0. A receiver similar to the one used for QPSK can be built to take the staggered nature of the bits into account.

$\pi/4$-QPSK and $\pi/4$-DQPSK   One way to alleviate the problems associated with QPSK is through the use of the two QPSK constellations described by eqs. (3.106) and (3.107). This modulation scheme, referred to as $\pi/4$-QPSK, uses both constellations (Bake 1962; Yang 1971; Fehe 1995; Gibs 1996, 1997). The constellations are switched for every other symbol, thus eliminating the phase transitions of $\pm\pi$ as well as $\pm\pi/2$. Instead, the phase transitions are now limited to $\pm\pi/4$ or

What happens to the phase at intervals of T in OQPSK

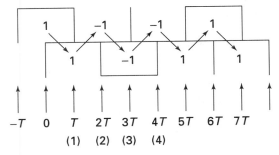

At (1), the symbols are (1, 1) and (−1, 1): The phase change is $\pm\pi/2$.
At (2), the symbols are (−1, 1) and (−1, −1): The phase change is $\pm\pi/2$.
At (3), the symbols are (−1, −1) and (−1, −1): The phase change is 0.
At (4), the symbols are (−1, −1) and (−1, 1): The phase change is $\pm\pi/2$, and so on.
Thus in OQPSK, the phase changes are limited to 0 or $\pm\pi/2$.

FIGURE 3.40 Explanation of the phase transitions in OQPSK.

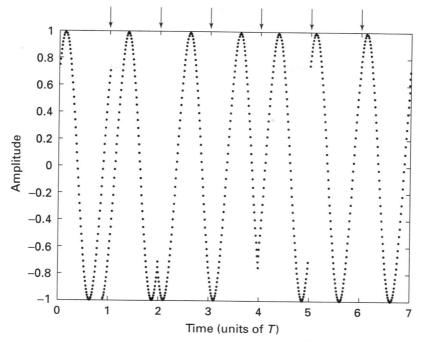

FIGURE 3.41 OQPSK waveform showing (by arrows) locations of phase transitions.

$\pm 3\pi/4$. The schematic of the phase encoding to produce $\pi/4$-QPSK is shown in Figure 3.42. Note that the transmitter configuration is similar to Figure 3.37 for QPSK, except for the alternation between the carriers. The phase constellation associated with $\pi/4$-QPSK is shown in Figure 3.43 along with an explanation of how $\pi/4$-QPSK differs from QPSK.

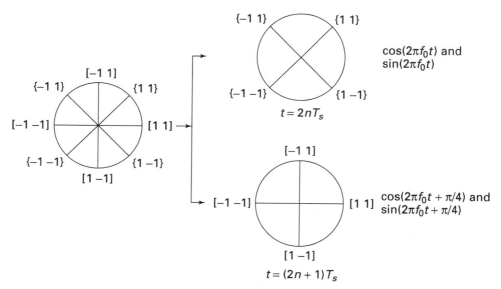

FIGURE 3.42 Phase encoding for $\pi/4$-QPSK. The brackets [ ] and { } correspond to the two respective constellations.

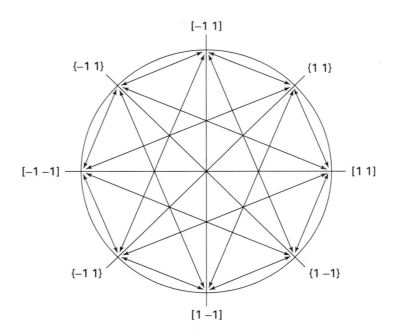

π/4	5π/4	3π/4	π/4
1 1	−1 −1	−1 1	1 1
I Q	I Q	I Q	I Q

FIGURE 3.43 Details of the phase constellation associated with π/4-QPSK. For every alternate symbol, the carrier waves are changed. From (1, 1) to (−1, −1), we go from [1 1] to { −1, −1} or from {1,1} to [−1, −1], resulting in a phase change of 3π/4, as opposed to π/2 in QPSK. In QPSK we can go from [1, 1] to [−1, −1] or from {1 1} to { −1   −1}, resulting in a phase change of ±π. Also, when we go from (1,1) to (1,1) in π/4-QPSK, we go from [1,1] to {1,1}, resulting in a phase change of π/4. The phase changes in π/4-QPSK are limited to ±3π/4 or ±π/4. There are no phase changes of 0, ±π/2, or ±π. All the possible transitions are shown by arrows.

An example of a π/4-QPSK signal is shown in Figure 3.44. The ϕ_n values for the symbols come from Table 3.4. It can be easily seen that the phase transitions occur at intervals of 2T and are either 45° or 135°. Arrows indicate the phase transitions taking place.

Because the phase constellations are alternated, the π/4-QPSK scheme is inherently differential, making it possible to use a differential receiver. It is also possible to encode the data so that a differential approach may be used in transmission, resulting in the π/4-DQPSK format that is used in the U.S. digital as well as in Japanese Digital Cellular systems. The encoding may be done at the bit level or at the symbol level. The principle of differential phase encoding to produce π/4-DQPSK is given below along with a comparison with π/4-QPSK. A typical π/4-DQPSK waveform is shown in Figure 3.45a, showing that the phase transitions are limited to ±3π/4. The representations for π/4-QPSK and π/4-DQPSK are as follows:

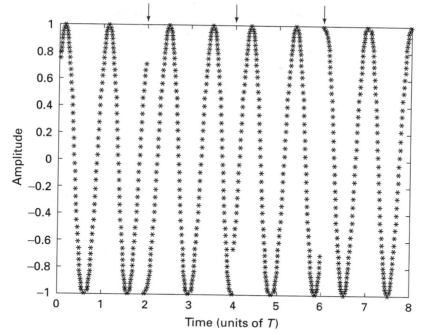

Allowed: { } ⟷ [ ] or [ ] ⟷ { }

Not allowed: [ ] ⟷//⟶ [ ] or { } ⟷//⟶ { }

FIGURE 3.44 A $\pi/4$-QPSK waveform. Arrows indicate the phase transitions.

$\pi/4$-QPSK

nth symbol: $\cos(2\pi f_0 t)\cos(\phi_n) \pm \sin(2\pi f_0 t)\sin(\phi_n)$ $\qquad$ (3.109a)

$(n+1)$th symbol:

$$\cos\left(2\pi f_0 t + \frac{\pi}{4}\right)\cos(\phi_n) \pm \sin\left(2\pi f_0 t + \frac{\pi}{4}\right)\sin(\phi_n) \quad (3.109b)$$

The phase, ϕ_n, takes one of the four possible values, depending on the bit pair. Note the alternation between the two constellations.

$\pi/4$-DQPSK

nth symbol: $\cos\left(2\pi f_0 t + \dfrac{\pi}{4}\right)\cos(\theta_n) \pm \sin\left(2\pi f_0 t + \dfrac{\pi}{4}\right)\sin(\theta_n)$ $\qquad$ (3.110a)

$(n+1)$th symbol:

$$\cos\left(2\pi f_0 t + \frac{\pi}{4}\right)\cos(\theta_{n+1}) \pm \sin\left(2\pi f_0 t + \frac{\pi}{4}\right)\sin(\theta_{n+1}) \quad (3.110b)$$

The phase, θ_n, is given by $\theta_n = \theta_{n-1} + \phi_n$, where ϕ_n is the phase corresponding to the bit pair. The transmitted phase is updated for every symbol by adding the accumulated phase from the previous symbol. For the first symbol, θ is assumed to be zero, making a differential encoding of the phase possible. Note that there is no alternation between the two constellations. The transmitter block diagram is shown in Figure 3.45b.

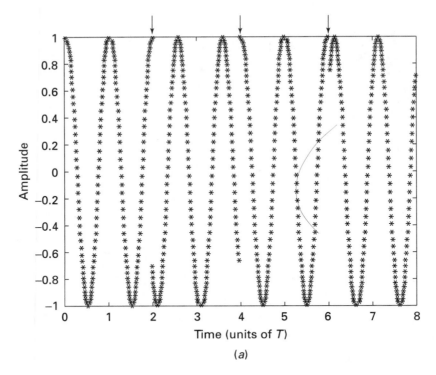

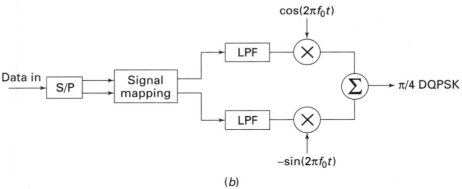

FIGURE 3.45   (a) A $\pi/4$-DQPSK waveform. The arrows indicate phase transitions.
(b) Block diagram of the $\pi/4$-DQPSK transmitter.

$\pi/4$**-QPSK Receiver**   The $\pi/4$-QPSK or $\pi/4$-DQPSK signal can be easily detected using a noncoherent technique, since this eliminates the need for a synchronous local oscillator. There are three different types of noncoherent receivers available (Good 1990, Liu 1991b, Nogu 1986, Fehe 1995, Rapp 1996b, Ono 1991, Tara 1988, Yama 1989, Chen 1991): the baseband differential receiver (Figure 3.46a), the IF differential detector (Figure 3.46b), and the FM discriminator receiver (Figure 3.47).

The details of these receivers are given in a number of papers. The important aspect to note is that the baseband differential receiver uses multiplication by a locally generated carrier; however, there is no need to match the phase of the incoming wave with that of the locally generated carrier. The use of the local oscillator is eliminated

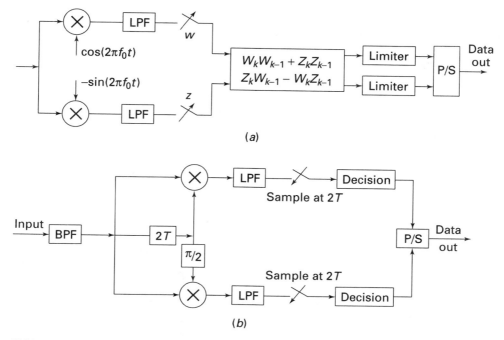

(a)

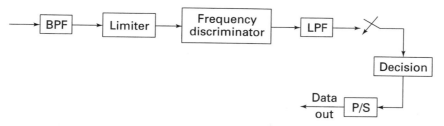

(b)

FIGURE 3.46 *(a)* Baseband differential receiver for $\pi/4$-QPSK. *(b)* IF differential detector for $\pi/4$-QPSK.

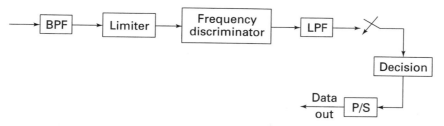

FIGURE 3.47 FM discriminator-based receiver for $\pi/4$-QPSK.

in the IF band differential detector. Making use of the similarity between frequency modulation and phase modulation, the FM discriminator can also recover the data.

EXAMPLE 3.9

For the bit stream in Example 3.8, calculate the symbol phases for transmission in $\pi/4$-DQPSK.

Answer The bits are 0 0 1 0 1 0 1 1 0 1. The symbol phases, respectively, are $-3\pi/4$, $-\pi/4$, $-\pi/4$, $\pi/4$, $3\pi/4$. Transmitted phase for the first symbol $= -3\pi/4 + 0 = -3\pi/4$. Transmitted phase for the second symbol $= -\pi/4 +$ transmitted phase for the first symbol $= -\pi/4 + (-3\pi/4) = -\pi$. Transmitted phase for the third symbol $= -\pi/4 +$ transmitted phase for the second symbol $(= -\pi) = -5\pi/4$. Transmitted phase for the fourth symbol $= \pi/4 +$ transmitted phase for the third symbol $(= -5\pi/4) = -\pi$. Transmitted phase for the fifth symbol $= 3\pi/4 +$ transmitted phase for the fourth symbol $(= -\pi) = -\pi/4$. ∎

Probability of Error in Quaternary PSK The bit error rates for coherently detected QPSK, $\pi/4$-QPSK, and OQPSK are the same and are equal to the probability of error for the BPSK modulation format. However, the symbol error rates for QPSK and BPSK are different (Stei 1964, Skla 1993). It is possible to obtain an expression for the symbol error rate through an examination of the relationship between bits and symbols. Since two bits form a symbol, a symbol error occurs when either one or both that form the symbol are in error. If $p(e)$ is the bit error rate, the symbol error rate, $p_{sy}(e)$, is given by

$$p_{sy}(e) = p(e) + p(e) - p(e)p(e) \approx 2p(e),$$
(3.111)

which is not much different from the bit error rate, considering that typical bit error rates on the order of 10^{-3}. Similar to the case of BPSK, it is possible for us to examine what happens if there is a phase mismatch. We will look only at the case of QPSK. The probability of error for QPSK, $p_{QPSK}(e)$, can be written as

$$p_{QPSK}(e) = \frac{1}{2} \, \text{erfc}\left(\sqrt{\frac{E}{N_0}}\right).$$
(3.112)

However, if we assume that there exists a phase mismatch of $\Delta\phi$ between the recovered carrier and the transmitted carrier, taking note of the quadrature nature of the modulation, we can approximately express the probability of error, $p_{\Delta\phi}(e)$, when a phase match exists as (Coop 1986, Gagl 1988)

$$p_{\Delta\phi}(e) = \frac{1}{4} \, \text{erfc}\left[\sqrt{\frac{E}{N_0}}(\cos\Delta\phi + \sin\Delta\phi)\right]$$

$$+ \frac{1}{4} \, \text{erfc}\left[\sqrt{\frac{E}{N_0}}(\cos\Delta\phi - \sin\Delta\phi)\right].$$
(3.113)

To understand the effects of phase mismatch, the probability of error is plotted as a function of the angular mismatch in Figure 3.48. It shows the increase in error probability as the angular mismatch increases.

The expression for the probability of error for $\pi/4$-DQPSK has been obtained for the three receivers mentioned earlier. It has been shown that the probability of error is the same in all three cases (Guo 1990, Liu 1991a,b) and is given by

$$p_{\pi/4\text{-DQPSK}}(e) = \exp\left(-\frac{2E}{N_0}\right)\left[\sum_{k=0}^{\infty}\left(\sqrt{2}-1\right)^k I_k\left(\frac{\sqrt{2}E}{N_0}\right) - \frac{1}{2}I_0\left(\frac{\sqrt{2}E}{N_0}\right)\right],$$
(3.114)

where $I_k(.)$ is the modified Bessel function of kth order of the first kind. The probability of error for a coherent receiver is compared with the corresponding value for a differential detection receiver in Figure 3.49. At an error rate of 10^{-3}, the differential receiver is about 3 dB worse than the coherent receiver.

Power Spectral Density of QPSK, OQPSK, and $\pi/4$-QPSK The power spectral density of the QPSK family of signals can be easily calculated since the fundamental

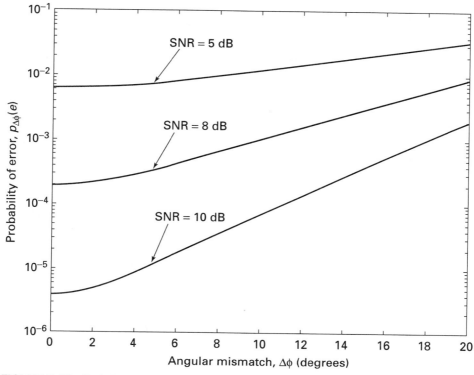

FIGURE 3.48 Probability of error for the case of coherent $\pi/4$-QPSK when phase mismatch is present.

pulse shape is rectangular. Indeed, the power spectra of QPSK, OQPSK, and $\pi/4$-QPSK are the same. The frequency spectrum was shown in Figure 3.26. It can be seen that the required bandwidth of the four-level PSK format is half that of BPSK, leading to an increased bandwidth efficiency of 2 bps/Hz for QPSK over BPSK. Since the bit error rates of BPSK and QPSK are identical and the symbol error rates are extremely close, QPSK offers an attractive means of modulation, with the ability to carry twice the amount of data in the same available bandwidth compared with the BPSK system. Thus the QPSK family of schemes offers excellent bandwidth efficiency in wireless systems, where bandwidth is at a premium.

3.3.4 Constant-Envelope Modulation Techniques

The linear modulation techniques, PSK and different forms of QPSK, do not have a continuous phase, and this can lead to spectral regrowth if the amplifiers are not linear (Ariy 1989, Aghv 1993, Akai 1998). On the other hand, these techniques do possess good bandwidth efficiency and thus permit more users for a given channel bandwidth allocated. If there is an interest in using low-cost amplifiers, which are essentially nonlinear, constant-envelope modulation techniques provide an alternative approach. These modulation methods result in spectra that have less and less power in the sidebands (Amor 1980, Ande 1965, deJa 1978, Chun 1984), so the problems associated with spectral regrowth will be less. The trade-off here is the fact that the main lobes of the spectra of the nonlinearly modulated signals are higher than the ones corresponding to the PSK family. We will examine some of the constant-envelope modulation methods that are employed in wireless communication systems.

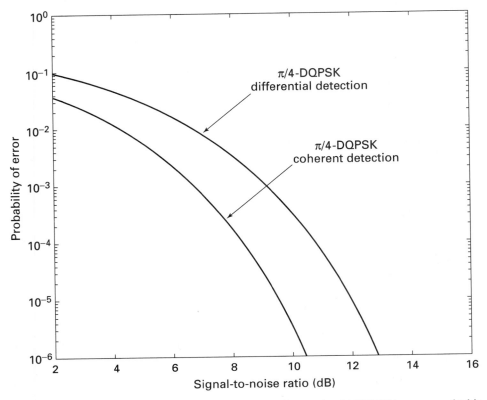

FIGURE 3.49 The probability of error for coherently detected $\pi/4$-DQPSK is compared with the case for a noncoherent detector.

Minimum Shift Keying One of the simplest modulation schemes is the minimum shift keying (MSK). Minimum shift keying can be considered a variation of OQPSK where a half-sinusoidal pulse shape is used *in place* of the rectangular one used in OQPSK (Razb 1972; Pasu 1979; Amor 1976, 1977; Simo 1976; Gron 1976; Hira 1979a). Consider the following data stream $\{b_k\}$, $k = 0, 1, 2$, where $b_k = \pm 1$ at a data rate of $R = 1/T$ bps where T is the bit duration. As indicated earlier, collecting the even bits we get the I stream

$$b_I(t) = b_0, b_2, b_4, b_6, \ldots, \tag{3.115}$$

and collecting the odd bits we get the Q stream

$$b_Q(t) = b_1, b_3, b_5, b_7, \ldots, \tag{3.116}$$

both at a rate of $1/(2T)$ bps. The I stream is delayed by T seconds to produce the OQPSK modulation. The MSK signal, $s_{\text{MSK}}(t)$, can be expressed as

$$s_{\text{MSK}}(t) = b_{Id}(t) \cos(2\pi f_1 t) \cos\left(2\pi f_0 t + \frac{\pi}{4}\right)$$

$$+ b_Q(t) \sin(2\pi f_1 t) \sin\left(2\pi f_0 t + \frac{\pi}{4}\right), \tag{3.117}$$

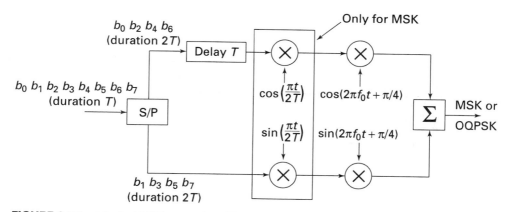

FIGURE 3.50 A typical MSK transmitter. The OQPSK scheme is also indicated.

where b_{Id} is the I stream delayed by T. This delayed I stream is multiplied by $\cos(2\pi f_1 t)$ before being multiplied by the carrier wave. The Q stream is multiplied by $\sin(2\pi f_1 t)$ before being multiplied by the carrier wave. Thus, one can interpret the MSK waveform as a modification of the OQPSK waveform where sinusoidal pulse shaping is used, making it a form of PSK. It is also a form of FSK since the waveform does not exhibit constant frequency; rather, it shows variation in zero crossings, making it similar to frequency shift keying. The minimum value of f_1 to make the pulse shapes orthogonal is equal to $1/4T$, and, the modulation scheme using the minimum value is known as MSK. The MSK transmitter is shown in Figure 3.50.

The data stream is separated and the I stream of duration $2T$ is delayed by T. The I stream is multiplied by $\cos(\pi t/2T)$ and the Q stream is multiplied by $\sin(\pi t/2T)$ before being multiplied by the carrier waves. These steps are illustrated in Figures 3.51 and 3.52. The resulting MSK waveform is shown in Figure 3.53 and does not have a one-to-one correspondence with the data stream. Another version of MSK (type II) can be produced when the pulse shapes are alternatively positive and negative. The different waveforms for QPSK, OQPSK, $\pi/4$-QPSK, and MSK for the same input bit pattern are shown in Figure 3.54. The continuous nature of the MSK waveform and the discontinuous nature of the other three waveforms are clearly seen. The other characteristic that is clear is that MSK looks like a frequency-modulated signal since the zero crossings are not equidistant. The constant envelope of MSK contrasts with the "broken" envelope of the PSK-type waveforms.

Note that MSK signals are a class of signals exhibiting constant envelope modulation and can even be generated using an analog FM modulator, described in the section on GMSK.

Power Spectrum of MSK   The power spectrum of MSK can be calculated in a simple fashion by assuming a sinusoidal pulse shape for the input data. It is compared with the spectra of QPSK and BPSK in Figure 3.26, where it can be seen that the sidelobes of the MSK spectrum decay faster than the sidelobes of QPSK. The first zero crossing of the MSK spectrum is $0.75R$, versus $0.5R$ for QPSK. This means that the spectral efficiency of MSK is lower than that of QPSK. On the other hand, if we

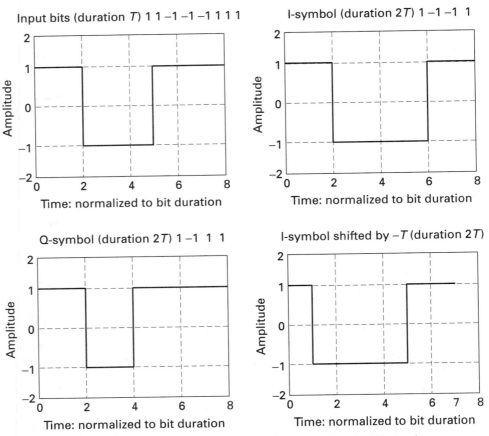

FIGURE 3.51 The bit patterns and symbol patterns for generation of MSK signals.

compare the power containment efficiency, i.e., how fast the power drops off, we see that MSK is more efficient than QPSK.

Gaussian Minimum-Shift Keying (GMSK) Even though problems of spectral regrowth can be reduced through the use of MSK, a significant amount of power remains in the sidebands. One way in which this power can be reduced is through the use of a different form of pulse shaping (Sund 1986, Ande 1965). If one uses a Gaussian-shaped pulse, the amount of power in the sidebands can be reduced at the expense of some intersymbol interference, as seen in Figures 3.55*a* and *b*, where the Fourier transforms of a Gaussian pulse for different pulse widths are shown. Of course, the use of a Gaussian pulse would violate the Nyquist criteria for pulses, and hence it is more appropriate to use an analog modulation approach to describe GMSK (Hira 1979b; Muro 1981, 1985; Moul 1995; Hanz 1994). First we will describe an analog approach to generating MSK using what are known as continuous-phase frequency-shift keying (CPFSK) schemes, and then we will follow it up with the generation of GMSK (Proa 1995, Ande 1999).

Continuous-Phase Frequency-Shift Keying (CPFSK) Before we look at CPFSK modulation, let us briefly review binary frequency-shift keying (BFSK). In

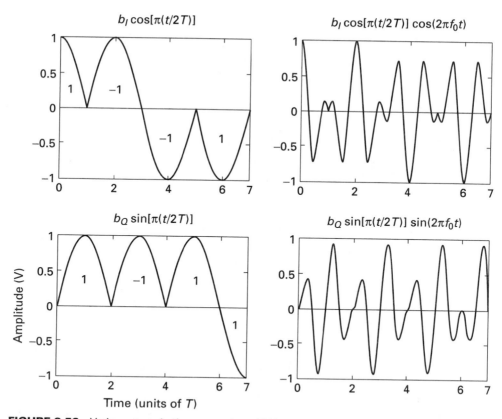

FIGURE 3.52 Various steps in the generation of MSK waveforms.

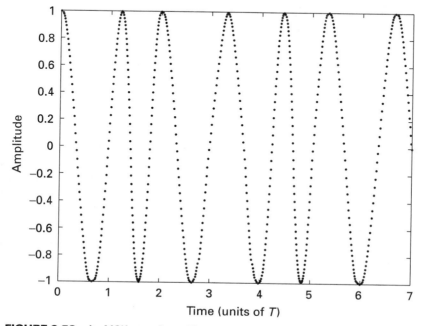

FIGURE 3.53 An MSK waveform. There are no discontinuities.

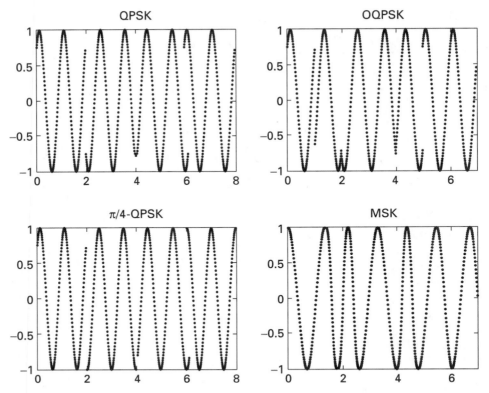

FIGURE 3.54 Four different waveforms are shown for comparison. The different phase transitions can be seen for the three schemes, whereas the MSK has a continuous waveform.

BFSK the carrier frequency is shifted up or down by a fixed value corresponding to a binary 1 or 0. The FSK signal, $s_{FSK}(t)$, is given by

$$
s_{FSK}(t) = \begin{cases} \sqrt{\dfrac{E}{T}} \cos[2\pi(f_0 + \Delta f)t] & 0 \le t \le T \quad 1 \\[3mm] \sqrt{\dfrac{E}{T}} \cos[2\pi(f_0 - \Delta f)t] & 0 \le t \le T \quad 0 \end{cases} \tag{3.118}
$$

Thus, essentially two separate tones are transmitted. The FSK signal given in this equation represents one form of FSK where the signal appears to be discontinuous, as shown in Figure 3.56. This discontinuity can be eliminated through the use of another form of FSK (CPFSK), where the change in frequency from the bit 1 to the bit 0 or vice versa is continuous, making the signal appear continuous.

The FSK signal can be detected coherently or incoherently. The probability of error for coherent detection can be obtained from the results discussed earlier in reference to the constellation. The FSK constellation is given in Figure 3.24a, and the probability of error is given by

$$
p_{FSK}(e) = \frac{1}{2}\,\text{erfc}\left(\sqrt{\frac{E}{2N_0}}\right), \tag{3.119}
$$

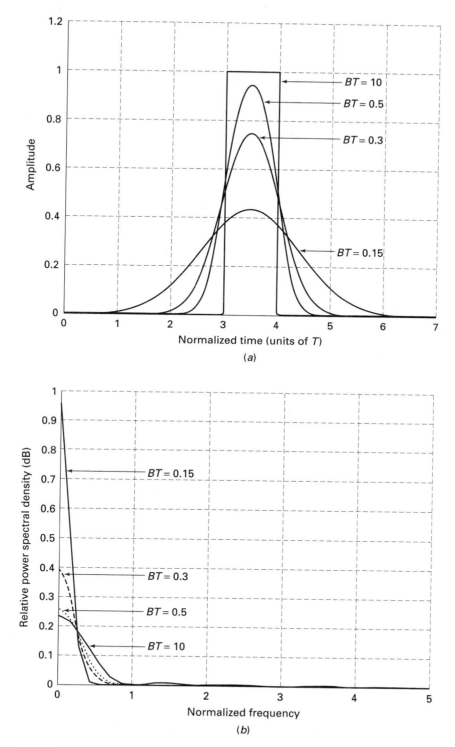

FIGURE 3.55 (a) Filtered pulse shapes. B is the 3 dB bandwidth of the Gaussian low-pass filter, and T is the bit duration. (b) Spectra of filtered pulse shapes.

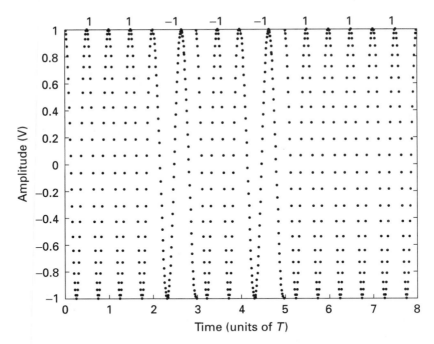

FIGURE 3.56 A frequency shift keying waveform.

which is 3 dB worse than the performance of coherent BPSK. Noncoherent detection results in a probability of error of

$$p_{FSK}(e) = \frac{1}{2} \exp\left(-\frac{E}{2N_0}\right). \tag{3.120}$$

Equation (3.120) can be derived in a fashion similar to the case for DPSK. The coherent receiver is shown in Figure 3.57, and the noncoherent receiver is shown in Figure 3.58.

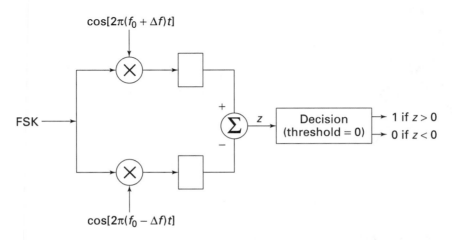

FIGURE 3.57 Block diagram of a coherent demodulator for FSK.

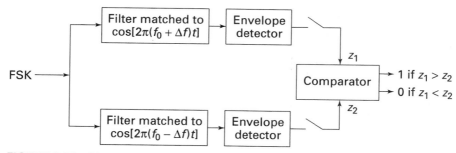

FIGURE 3.58 Block diagram of a noncoherent demodulator for FSK.

As described in the previous paragraphs, the BFSK waveform is likely to be discontinuous. One of the ways in which the FSK signal can be made continuous is through the use of an analog frequency modulator. Such an approach is a digital equivalent of FM, since the input to the modulator will be -1s and 1s, and is therefore referred to as digital frequency modulation (DFM) (Rowe 1975, Ande 1965, Mase 1985b, Sund 1986, Simo 1976). The block diagram of a digital frequency modulator is shown in Figure 3.59a. The input to the modulator, $v(t)$, is given by

$$v(t) = \sum_{n=-\infty}^{\infty} b_n g(t - nT) , \qquad (3.121)$$

where $b_n = \pm 1$ and $g(t)$ is a rectangular pulse-shaping function having an area of $\frac{1}{2}$, given by

$$g(t) = \begin{cases} \dfrac{1}{2T} & 0 \leq t \leq T \\ 0 & \text{otherwise,} \end{cases} \qquad (3.122)$$

as shown in Figure 3.59b.

The output of the FM modulator, $U_{\text{DFM}}(t)$, can be expressed as

$$U_{\text{DFM}}(t) = A_0 \cos\left[2\pi f_0 t + 2\pi h \int_{-\infty}^{t} v(\tau) \, d\tau + \phi_0 \right], \qquad (3.123)$$

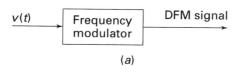

(a)

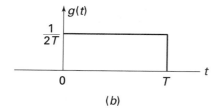

(b)

FIGURE 3.59 (a) Block diagram of a digital frequency modulator. (b) The DFM pulse shape.

where the FM index h is given by

$$h = 2f_d T. \tag{3.124}$$

The quantity ϕ_0 is the initial phase offset, which can be taken to be zero. We can rewrite eq. (3.123) as

$$U_{\text{DFM}}(t) = A_0 \cos\left[2\pi f_0 t + \theta(t)\right], \tag{3.125}$$

where $\theta(t)$ is the instantaneous phase, given by

$$\theta(t) = 2\pi h \int_{-\infty}^{t} v(\tau)\, d\tau. \tag{3.126}$$

Since $\theta(t)$ is a continuous function of t, this form of modulation is often referred to as continuous-phase frequency-shift keying (CPFSK). If $q(t)$ is defined as the phase response given by

$$q(t) = \int_{0}^{t} v(\tau)\, d\tau, \qquad 0 \le t \le T, \tag{3.127}$$

the DFM signal can be rewritten as

$$U_{\text{DFM}}(t) = A_0 \cos\left[2\pi f_0 t + \pi h \sum_{k=-\infty}^{n-1} b_k + b_n 2\pi h q(t')\right], \tag{3.128}$$

$$(n-1)T \le t \le nT,$$

where $t' = t - (n-1)T$. If $h = 0.5$, eq. (3.128) can be rewritten as

$$U_{\text{DFM}}(t) = A_0 \cos\left[2\pi f_0 t + b_n \frac{\pi t'}{2T} + \frac{\pi}{2} \sum_{k=-\infty}^{n-1} b_k\right], \tag{3.129}$$

$$(n-1)T \le t \le nT.$$

For b_n taking a value of ± 1, the two frequencies corresponding to the two bits become $(f_0 - 1/4T)$ and $(f_0 + 1/4T)$. Under these conditions, eq. (3.129) represents MSK. Note that even though this scheme is conceptually simple, the need to maintain the FM index of 0.5 very accurately to produce MSK makes the DFM approach less appealing than the method based on modified OQPSK. Once again, the relationship between MSK and FSK is evident from these equations. This approach to producing the MSK signal is shown in Figure 3.60.

A typical MSK waveform obtained using DFM is shown in Figure 3.61. Note that the waveform differs from the one generated with OQPSK. The MSK waveform generated using DFM bears a one-to-one correspondence with the input bit stream.

FIGURE 3.60   A DFM-based approach to generating a MSK-type waveform.

This form of MSK generated using FSK is often referred to as fast frequency-shift keying (FFSK) (Proa 1995, Couc 1997).

If a Gaussian low-pass filter is added prior to the FM modulator with $h = 0.5$, the modulation scheme is known as Gaussian minimum-shift keying (GMSK). The schematic of a GMSK transmitter is shown in Figure 3.62.

If $h(t)$ is given by

$$h(t) = \frac{\sqrt{\pi}}{\alpha} \exp\left(-\frac{\pi^2 t^2}{\alpha^2}\right),$$ (3.130)

the input to the modulator, $v(t)$, is given by

$$v(t) = \sum_{n=-\infty}^{\infty} b_n h(t) g(t - nT) = \sum_{n=-\infty}^{\infty} b_n c(t - nT),$$ (3.131)

where $c(t)$ is $h(t)g(t)$. The parameter α is related to the 3 dB bandwidth (B) of the low-pass filter and can be expressed as

$$\alpha = \frac{0.5887}{B}.$$ (3.132)

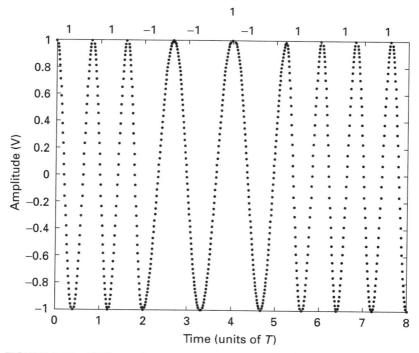

FIGURE 3.61 MSK waveform generated using a digital frequency modulator. The one-to-one correspondence between the waveform and bit pattern is seen here (compare with Figure 3.53).

FIGURE 3.62 A GMSK modulator based on DFM.

The GMSK modulation is characterized by the parameter BT. After some algebra, the pulse shape at the input to the FM modulator can be written as

$$c(t) = \frac{1}{4T}\left[\text{erf}\left(\frac{NT-t}{\sqrt{2}\sigma}\right) - \text{erf}\left(\frac{(N-1)T-t}{\sqrt{2}\sigma}\right)\right]. \tag{3.133}$$

After considerable algebra, the phase response $q_g(t)$ of the GMSK signal can be expressed as

$$q_g(t) = \frac{1}{4T}\left\{T + \left[(N-1)T-t\right]\text{erf}\left[\frac{(N-1)T-t}{\sqrt{2}\sigma}\right] - (NT-t)\,\text{erf}\left[\frac{NT-t}{\sqrt{2}\sigma}\right]\right\}$$

$$+ \frac{2\sigma}{4T\sqrt{2\pi}}\left\{e^{-[(N-1)T-t]^2/2\sigma^2} - e^{-[NT-t]^2/2\sigma^2}\right\}. \tag{3.134}$$

A plot of $c(t)$ is given in Figure 3.63 for various values of BT, showing clearly the broadening of the pulse and the ISI that will certainly result from it. Note that

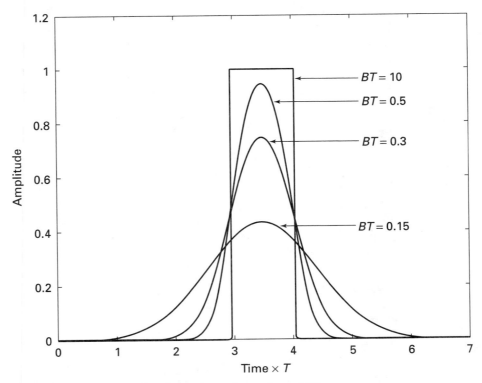

FIGURE 3.63 Output pulse shapes for various values of BT.

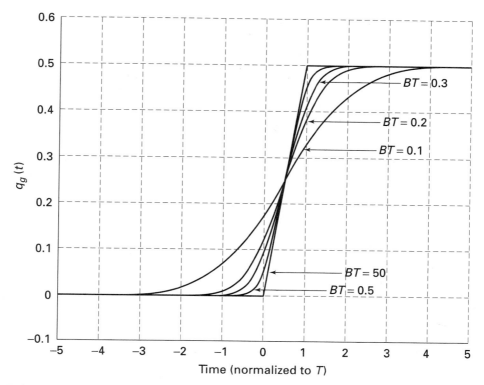

FIGURE 3.64 Phase response of GMSK.

when BT is large, the input pulse to the FM modulator approaches a rectangular shape. As BT becomes smaller, the pulse spills over several bits. This will result in intersymbol interference.

A better way to illustrate this effect is to plot the phase response given in eq. (3.134). The phase response $q_g(t)$ is plotted in Figure 3.64. It is clear that for $BT = 50$ or more, the phase response is almost identical to the phase response of MSK, shown in Figure 3.65. For low values of BT, the response extends over several bits, again demonstrating the presence of intersymbol interference. The GMSK signal can now be expressed as

$$U_{\text{GMSK}}(t) = A_0 \cos\left[2\pi f_0 t + \pi \sum_{n=-\infty}^{\infty} q_g(t - nT) \right]. \tag{3.135}$$

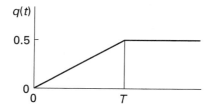

FIGURE 3.65 Phase response of MSK.

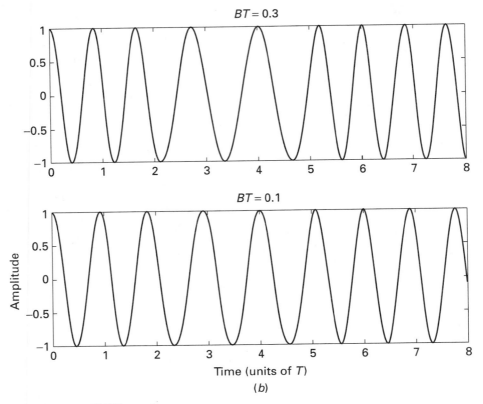

FIGURE 3.66 GMSK waveforms.

The GMSK signal is plotted in Figure 3.66 for various values of BT. As the value of BT decreases, it is difficult to notice any effect of modulation, while as BT increases, the signal approaches the MSK signal generated using a DFM system (shown in Figure 3.61).

It is possible to interpret BT as a form of GMSK modulation index. When $BT \ll 1$, the modulation index is zero (no change in zero crossings). A similar conclusion can be reached by examining the power spectral density of the GMSK signal (Auli 1982a,b,c; Muro 1981). Once again, the advantages of GMSK can be seen clearly in Figure 3.67. The spectrum appears to be much tighter, with much less power in the sidebands (note that sidebands do not exist for low values of BT), providing much better operating conditions where nonlinear amplifiers are used. This, of course, comes at the cost of a significant amount of ISI. The power confinement for GMSK modulation is described in Table 3.5.

TABLE 3.5 Power Confinement in GMSK

	Bandwidth occupied		
BT	*Power = 0.9*	*Power = 0.99*	*Power = 0.999*
0.2	0.52	0.79	0.99
0.25	0.57	0.86	1.09
0.5	0.69	1.04	1.33
Infinite	0.78	1.2	2.76

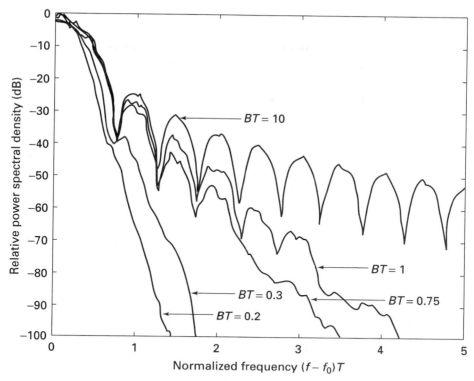

FIGURE 3.67 Power spectra of GMSK signals.

A question often asked is whether there is a quadrature-type transmitter similar to the one used for MSK, or whether QPSK is available for GMSK. One of the problems for the design and implementation of the I-Q transmitter is the need to have signals with shapes that satisfy the Nyquist criteria. Since the Gaussian pulse shape coming out of the LPF does not satisfy the zero crossing or orthogonality criteria, it is not straightforward to use a standard I-Q approach. Still, it is possible to generate GMSK using the approach shown in Figure 3.68 (Dava 1989, Garg 1996, Lee 1997).

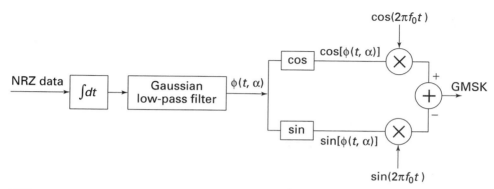

FIGURE 3.68 I-Q approach to generating GMSK signals.

Detection and Reception of MSK and GMSK Since MSK can be generated starting from OQPSK, the bit error performance for MSK is identical to that for BPSK, QPSK, or OQPSK. However, MSK and GMSK can be detected using a one-bit differential detector, a two-bit differential detector, or a frequency discriminator. The block diagrams of these receiver structures are shown, respectively, in Figures 3.69a, b, and c.

To understand the one-bit differential detector, recall the discussion of MSK, where the signal phase is shifted by $+\pi/2$ for a logical 1 and $-\pi/2$ for a logical -1 (or 0). Thus, a one-bit differential detector operates very similarly to the DPSK receiver shown in Figure 3.32. However, GMSK is not MSK, and the limited bandwidth of the premodulation filter in GMSK will degrade the performance because of the increase in ISI brought on by the lower bandwidths. This degradation can be reduced if we use a two-bit delay and perform a correlation as shown in Figure 3.69b. The frequency discriminator uses the property that MSK and GMSK are forms of FSK.

Analytical expressions for the bit error rate are not easily derivable. Numerical results are available in a number of research papers (Abra 1995; Elno 1986a,b; Hiro 1984; Hori 1990; Muro 1981; Simo 1984; Smit 1994; Vars 1993).

The probability of error for coherently detected GMSK is given by

$$p_{\text{GMSK}}(e) = \frac{1}{2} \operatorname{erfc} \sqrt{\frac{\varepsilon E}{N_0}}, \tag{3.136}$$

where ε is a parameter related to the bandwidth of the low-pass filter,

$$\varepsilon = \begin{cases} 0.68 & \text{for GMSK with } BT = 0.25 \\ 0.85 & \text{for GMSK with } BT = \infty/(\text{MSK}). \end{cases} \tag{3.137}$$

A plot of the error probability appears in Figure 3.70. It can be seen that BT approaching ∞ leads to a lower probability of error.

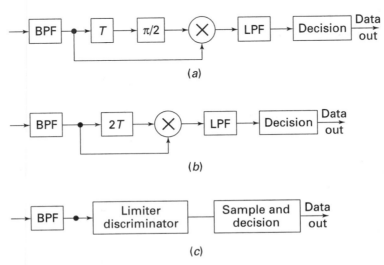

FIGURE 3.69 GMSK demodulators. (a) One-bit differential detector (b) Two-bit differential detector. (c) Frequency discriminator.

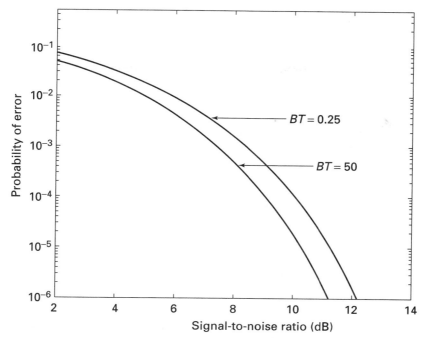

FIGURE 3.70 Probability of error for GMSK.

3.4 GENERAL *M*-ARY MODULATION SCHEMES

QPSK is not the only multilevel modulation format available for wireless communications. We will now look at some other modulation formats and compare their performance with BPSK and QPSK.

3.4.1 *M*-ary Phase Modulation

The PSK modulation scheme can be extended to 8-level PSK, 16-level PSK, and so on. The phase constellation associated with 8-level PSK is shown in Figure 3.71. The *M*-ary PSK signal (Weav 1962, Proa 1995, Skla 1988) can be expressed as

$$S_{\text{MPSK}}(t) = \sqrt{\frac{2E}{T_s}}\, \cos\left[2\pi f_0 t + \frac{2\pi}{M}(m-1)\right],$$

$$0 \leq t \leq T_s; \quad m = 1, 2, \ldots, M, \tag{3.138}$$

where T_s is the symbol period and is given by

$$T_s = [\log_2 M]T. \tag{3.139}$$

As the number of modulation levels increases, the minimum power required to maintain a fixed bit error rate decreases. This can be seen from the decision boundaries in Figure 3.71. With an increase in the number of levels, the energy must be increased to allow the minimum separation between levels to remain at the same value,

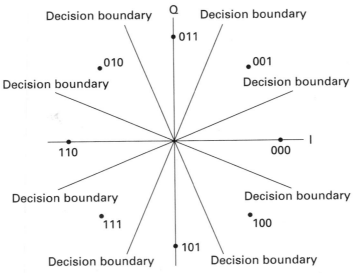

FIGURE 3.71 Phase constellation of 8-level PSK along with the decision boundaries.

since it was shown that the error probability can be expressed in terms of the minimum spacing between the signals in a signal constellation.

The symbol error probability for M-ary PSK is given by (Skla 1988, Hayk 2001)

$$p_{\text{MPSK}}(e) \cong \text{erfc}\left[\sqrt{\frac{E}{N_0}} \sin\left(\frac{\pi}{M}\right)\right], \quad M \geq 4. \qquad (3.140)$$

Values of the minimum signal-noise-ratio (E/N_0) for the different levels of PSK are given in Table 3.6. Even though we do increase our bandwidth efficiency (R/W bps/Hz), beyond $M = 8$, M-ary PSK ceases to be power efficient as the minimum power required to maintain a bit error rate of 10^{-6} goes up. This problem can be solved through the use of a mixture of PSK and ASK modulation formats, resulting in what is referred to as quadrature modulation.

3.4.2 *M*-ary Quadrature Amplitude Modulation (QAM)

In a multilevel PSK format, the in-phase and quadrature components have the same amplitude, enabling the modulated signal to have a constant envelope, except for discontinuities at symbol transitions. This leads to a constellation that is circular in nature. We can remove this restriction of constant amplitude and create a modulation scheme that has a hybrid nature, with variations in both phase and amplitude. This

TABLE 3.6 Power and Spectral Efficiencies for PSK

M	2	4	8	16	32	64
Minimum SNR (dB)	10.5	10.5	14	18.5	23.4	28.5
Spectral efficiency (bps/Hz)	1	2	3	4	5	8

modulation format is referred to as quadrature amplitude modulation. The QAM signal can be expressed as

$$s_{\text{QAM}}(t) = a_i \cos(2\pi f_0 t) + b_i \sin(2\pi f_0 t), \quad 0 \le t \le T, \qquad (3.141)$$

where (a_i, b_i) represent the signal points. For $a_i = \pm 1$, $b_i = \pm 1$, the signal constellation is shown in Figure 3.72, which shows that 4-level QAM is no different from QPSK. Eight-level QAM and 16-level QAM are also shown. The mere fact that there are two parameters, namely, the phase and amplitude information unique for each of the constellation points, makes it possible to detect QAM with less energy compared with *M*-ary PSK. Table 3.7 shows the power efficiency of QAM. Comparing Tables 3.6 and 3.7 indicates that 16-level QAM provides the same spectral efficiency as 16-level PSK, at a much lower power requirement for acceptable performance. For these

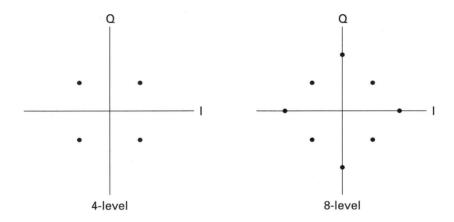

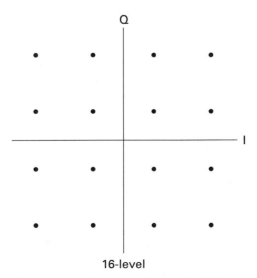

FIGURE 3.72 Various constellations associated with QAM.

TABLE 3.7   Power and Spectral Efficiencies of QAM

M	2	4	8	16	32	64
Minimum SNR (dB)	10.5	10.5	10.5	14.5	17.4	18.8
Spectral efficiency (bps/Hz)	1	2	3	4	5	8

reasons, M-ary QAM is often used for high-data-rate systems. The QAM system has one drawback: Since information is contained in the amplitude as well, the amplifiers must have good linearity.

3.5   COMPARISON OF MODEMS FOR WIRELESS COMMUNICATIONS

Now that we have covered a number of modulation/demodulation schemes used in wireless communications, it is time to make a comparison between the different modulation schemes. It is possible to undertake this comparison in terms of power efficiency, bandwidth efficiency, power containment, linearity, and other factors. Before we look at some of these aspects, let us revisit the concept of bandwidth efficiency with the aid of Shannon's theorem.

3.5.1   Shannon's Theorem

The three most important parameters of a communication system are the bandwidth required for transmission, the minimum power or energy required to transmit efficiently (we define the term *efficiently* very loosely at this time), and the error probability. In a very broad sense, we must be able to transmit a fixed amount of information with the least bandwidth possible, with the lowest power, and with the smallest error probability (certainly less than, say, 10^{-5}). Is it possible that these three characteristics will all be available for a certain communication system and modulation format? Or must we undertake a trade-off between low power and high bandwidth and high power and low bandwidth? Note that we normally need to maintain a certain fixed probability of error to transmit data efficiently from one point to the other. So the only two parameters that we can try to balance are the bandwidth and the power. This trade-off between bandwidth and power (or spectral efficiency and power efficiency) can be understood in terms of the Shannon-Hartley law, or Shannon's theorem (Shan 1949, Skla 1993, Hayk 2001).

The Shannon-Hartley capacity theorem states that the capacity C of a Gaussian channel is given by

$$C = B \log_2 \left[1 + \frac{S}{N} \right] \text{ bps} , \qquad (3.142)$$

where S is the signal power, N is the noise power, and B is the bandwidth. We can rewrite eq. (3.142) by expressing the signal-to-noise ratio in terms of E, the energy contained in a bit of duration T,

$$E = ST, \quad N = N_0 B \qquad (3.143)$$

as

$$\frac{E}{N_0} = \frac{ST}{N_0} = \frac{S}{N_0}\left(\frac{1}{R}\right) = \frac{S}{N}\left(\frac{B}{R}\right), \tag{3.144}$$

where R is the data rate and N_0 is the noise power spectral density. Equation (3.142) now becomes

$$C = B \log_2\left[1 + \left(\frac{E}{N_0}\right)\left(\frac{B}{R}\right)\right] \text{ bps}. \tag{3.145}$$

Since E/N_0 is common to all modulation formats, we can use eq. (3.145) to compare the efficiency of the communication systems. Furthermore, the parameter R/B (bps/Hz) can be identified as the amount of data that can be transmitted in a given bandwidth. These two parameters, E/N_0 and R/B, can be respectively identified as the power efficiency and spectral efficiency; we need low values of E/N_0 and high values of R/B. For the case where the data rate (R) is equal to the capacity (C) of the channel, eq. (3.145) becomes

$$\frac{R}{B} = \log_2\left[1 + \left(\frac{E}{N_0}\right)\left(\frac{B}{R}\right)\right]. \tag{3.146}$$

Rewriting, we get

$$\frac{E}{N_0} = [2^{R/B} - 1]\left(\frac{R}{B}\right)^{-1}. \tag{3.147}$$

The plot of the signal-to-noise ratio versus the spectral efficiency is plotted in Figure 3.73, which is frequently referred to as a bandwidth-efficiency diagram. Regions are identified as $R > C$ (unattainable) and $R < C$. The boundary defines the ideal system, for which $R = C$.

We also observe a few important facets of Shannon's theorem that are very important for the transmission of information.

1. If we have infinite bandwidth available ($B \to \infty$), the signal-to-noise ratio, E/N_0, approaches the Shannon limit (Skla 1988):

$$\left(\frac{E}{N_0}\right)_{\infty} = \lim_{B \to \infty}\left(\frac{E}{N_0}\right) = \log_e(2) = 0.69 = -1.6 \text{ dB}. \tag{3.148}$$

The capacity corresponding to this signal-to-noise ratio is C_{∞}, given by

$$C_{\infty} = \lim_{B \to \infty} C = \frac{S}{N_0} \log_2(e) = 1.44\frac{S}{N_0}. \tag{3.149}$$

In an actual communication system, we can compare the channel capacity obtained at the expense of increased bandwidth to this value in order to determine whether a further increase in bandwidth is worth the cost.

2. The capacity boundary, indicated by $R = C$, gives the regions and thus the parameters necessary to support error-free transmission ($R < C$) and for which error-free transmission is not possible ($R > C$).

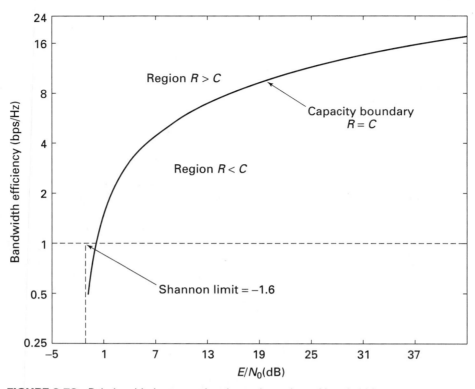

FIGURE 3.73 Relationship between signal-to-noise ratio and bandwidth.

3. *Trade-off behavior* ($R < C$): Assume for a moment that the spectral efficiency, R/B, is fixed. In this case, any increase in E/N_0 (movement to the right) along the E/N_0 axis will reduce the error probability. Now assume that the signal-to-noise ratio, E/N_0, is fixed. Vertical translation along the R/B axis allows one to trade off the error probability against the spectral efficiency. Note that we expect a certain level of performance, and hence the probability of error is *fixed*. We can therefore consider the trade-off between power efficiency and spectral efficiency.

We can now compare the various modulation formats in terms of the relationship between their bandwidth requirements and power requirements. The bandwidth occupied by M-ary PSK can be obtained by observing the power spectra of PSK signals. If we use the criterion of null-to-null bandwidth of the modulated signal, the bandwidth occupied by PSK is

$$B = \frac{2}{T_s},\tag{3.150}$$

where T_s is the symbol duration. For BPSK, $T_s = T$, the bit duration; for QPSK, $T_s = 2T$; for 8-level PSK, $T_s = 3T$; and so on. We can thus express T_s in terms of M as

$$T_s = T \log_2 M.\tag{3.151}$$

Since the data rate R and T are related by

$$R = \frac{1}{T},\tag{3.152}$$

the bandwidth becomes

$$B = 2\frac{R}{\log_2 M}, \tag{3.153}$$

Note that if we use the Nyquist bandwidth criterion instead of the null-to-null criterion,

$$B = \frac{R}{\log_2 M}, \tag{3.154}$$

The bandwidth requirements for QAM are thus identical to those for PSK. Note also that QPSK and 4-level QAM are identical.

The bandwidth required for M-ary FSK can be found in the following fashion. The M-ary FSK signal can be written as (Hayk 2001, Skla 1993)

$$S_{\text{MFSK}}(t) = \sqrt{\frac{2E}{T}} \cos\left[\frac{\pi}{T}(n + k)\right], \quad 0 \le t \le T \tag{3.155}$$

for $k = 1, 2, 3, \ldots, M$. n (an integer) is related to the carrier frequency given by $f_0 = n/2T$. Note that unlike M-ary PSK, the bandwidth of MFSK signal does not go down as M increases. This is because different carrier frequencies separated by $1/2T$ are transmitted for each of the M signals. Consequently, the bandwidth requirement of MFSK modulation goes up as M increases, thereby reducing the bandwidth efficiency. While for M-ary PSK the bandwidth (Nyquist) can be expressed as

$$B_{\text{MPSK}} = \frac{R}{\log_2 M}, \tag{3.156}$$

the bandwidth for M-ary FSK is given by (Skla 1993)

$$B_{\text{MFSK}} = \frac{RM}{\log_2 M}. \tag{3.157}$$

Thus M-ary FSK has limited application in high-volume data transmission. Most data transmission systems use M-ary PSK (up to $M = 8$) or M-ary QAM ($M \ge 8$). The characteristics of MFSK are given in Table 3.8.

It is clear that MFSK has very poor spectral efficiency; however, it does have high power efficiency due to the orthogonal nature of the signals and the fact that there is no need to increase the separation between levels in signal space.

3.5.2 Channel Capacity in Rayleigh Fading

The channel capacity described in Section 3.5.1 needs to be modified when the channel becomes random because of fading. The primary reason for the loss of the deterministic nature of the channel capacity is the randomness of the signal power, which makes the signal-to-noise ratio itself a random variable. This case is considered in Appendix B, Section B.5.

TABLE 3.8 Power and Spectral Efficiencies for FSK

M	2	4	8	16	32
Minimum SNR (dB)	13.5	10.8	9.3	8.2	7.5
Spectral efficiency (bps/Hz)	0.5	0.5	0.37	0.25	0.125

3.6 SUMMARY

The concepts of both analog and digital communication for wireless applications have been presented. The differences between AM and FM were described. Starting with Nyquist's criteria, the digital communication systems were studied in detail.

- Amplitude modulation is a linear modulation technique. The bandwidth required for transmission is twice the message bandwidth and does not depend on the amplitude of the message signal. The bandwidth requirement may be reduced using single-sideband or vestigial-sideband modulation.
- Using a quadrature arrangement, it is possible to transmit two signals with the transmission bandwidth requirement of a single message.
- Frequency modulation is a nonlinear modulation technique. It has a constant envelope; however, the bandwidth depends on the message bandwidth and the amplitude of the message signal. The bandwidth required for FM transmission is equal to or higher than the bandwidth required for AM. In return, the performance of FM is far superior to AM. In fact, we can trade bandwidth for improved performance.
- Compared with analog modulation, digital modulation offers better noise immunity, ease of operation, improved compatibility with fast processors, and other advantages.
- The optimal pulse shapes for digital transmission of data are governed by the Nyquist criteria. Pulse shapes satisfying these criteria will have less intersymbol interference caused by pulse spreading. A raised cosine pulse is the ideal candidate.
- The concept of bandwidth in digital communication can be both confusing and complex. There are several definitions of bandwidth, including absolute bandwidth, 3 dB bandwidth, equivalent noise bandwidth, null-to-null bandwidth, bounded-spectrum bandwidth, and power spectrum bandwidth.
- Two important parameters for comparing digital modulation schemes are power efficiency and bandwidth or spectral efficiency. If a modulation scheme A requires a lower signal-to-noise ratio to maintain acceptable performance in terms of error probability than another scheme AA, scheme A has better power efficiency. If a modulation scheme B can transmit more data in a given bandwidth than another scheme BB, the modulation scheme B has better bandwidth efficiency.
- The performance of a digital communication system is measured in terms of error probability. Error probability or bit error rate depends on the signal-to-noise ratio or, more exactly, on the energy/bit-to-noise ratio, E/N_0. The spectral density of noise assumed to be additive white and Gaussian is $N_0/2$, and E is the bit energy (or energy/bit), given in eq. (3.56).
- The optimal receiver that minimizes the probability of error is a matched filter. The performance of a matched filter is similar to the performance of a correlator.
- The probability of error can be conveniently expressed using signal constellations.
- The need for a coherent demodulator for BPSK can be avoided if the data are differentially encoded. This leads to the DPSK format. The performance of DPSK is approximately 3 dB worse than that of BPSK.
- QPSK uses two bits/symbol. The bandwidth efficiency of QPSK is greater than that of BPSK.

- The bit error rates of QPSK and BPSK are the same. The symbol error rate for QPSK is almost twice the bit error rate.
- The QPSK waveform has sharp phase transitions, which lead to transmission problems in the presence of nonlinear amplifiers. In other words, the inadequate suppression of sidelobes is a problem.
- The phase transition problems can be reduced by using OQPSK, which limits the phase jumps to 0 and $\pm\pi/2$.
- A much better 4-level phase modulation scheme is $\pi/4$-QPSK, where the phase jumps are limited to $\pm\pi/4$ and $\pm 3\pi/4$.
- Taking advantage of the inherent differential nature of $\pi/4$-QPSK, $\pi/4$-DQPSK has been developed to eliminate the need for a coherent receiver. This is the modulation format used in U.S. digital systems as well as Japanese digital systems.
- One disadvantage of linear modulation systems such as QPSK is the existence of significant power in the sidebands. Constant-envelope modulation techniques result in waveforms with very low power in sidebands. These methods have a higher null-to-null bandwidth compared with linear methods such as QPSK.
- One of the nonlinear modulation techniques is the minimum shift keying (MSK). This can be viewed as a form of PSK or FSK. MSK can also be viewed as a form of OQPSK, where a half-sinusoidal pulse shape is used in place of the standard rectangular pulse.
- Even though the problems of sideband power may be reduced somewhat using MSK, a different form of pulse shaping may reduce these problems further. One such scheme is Gaussian minimum-shift keying, or GMSK, which is used extensively in wireless systems in Europe, Australia, and parts of Asia and the United States.
- GMSK can be viewed as a form of continuous-phase frequency shift keying (CPFSK), which is a form of digital frequency modulation where the discontinuous nature of the conventional FSK waveform is removed.
- GMSK can be easily generated using a Gaussian LPF and a conventional FM modulator of index 0.5.
- The Gaussian LPF is characterized by the factor BT, where T is the bit duration and B is the 3 dB bandwidth of the filter. GSM uses $BT = 0.3$.
- Taking advantage of the similarity between GMSK and FM, any standard noncoherent FM demodulator, such as a frequency discriminator, or a one-bit or two-bit differential detector, can be used for GMSK demodulation.
- GMSK schemes are characterized by higher levels of ISI than linear modulation schemes such as QPSK.
- Other M-ary modulation schemes are available for data transmission. The spectral efficiency advantage of multilevel PSK schemes comes at the expense of higher levels of minimum power required to maintain an acceptable bit error rate. Multilevel QAM, on the other hand, has the same spectral efficiency as the corresponding PSK, requiring less power than PSK schemes.
- Bandwidth and spectral efficiencies can be compared with the aid of Shannon's theorem. MFSK has very poor spectral efficiency but has better power efficiency.

PROBLEMS

*** *Asterisks refer to problems better suited for graduate-level students.*

1. Using MATLAB, generate an AM signal. Choose a single-tone message signal. Vary the modulation index and compare the modulated signals in terms of the Fourier transform and bandwidth.

2. Using MATLAB, generate a frequency-modulated signal. Vary the FM index and compare the modulated signals in terms of the Fourier transform and bandwidth. Estimate the transmission bandwidth required by observing the power in the sidebands. What happens to the component at the carrier frequency as the modulation index goes up?

3. Consider a DSB-SC system. If a coherent demodulator is used, calculate the signal-to-noise ratio, assuming that additive white Gaussian noise of power spectral density $\beta/2$ is present. What happens to the signal-to-noise ratio if the modulation scheme is SSB-SC?

4. An amplitude modulator is operating as shown in Figure P3.4, where

$$e_{in}(t) = m(t) + A_0 \cos(2\pi f_0 t)$$

$$e_{out}(t) = a_0 + a_1 e_{in}(t) + a_2 [e_{in}(t)]^2.$$

If the message signal has a bandwidth of W Hz, what is the minimum carrier frequency required for distortionless transmission? How can one *lower* this value?

5. Generate a rectangular pulse and compute the spectrum using MATLAB. Vary the duration of the pulse and calculate the bandwidth required to transmit such a pulse if the BW required is the first zero crossing. Now use the 3 dB bandwidth as the BW criterion.

6. Generate a rectangular pulse and a half-sinusoidal pulse. Pass these signals through a low-pass filter (vary the filter bandwidth if necessary) and observe the distortion in the filtered output. (Hint: Use the functions *butter* and *filtfilt* from MATLAB.)

7. Use the two signals in Problem 6 to create a two-channel AM signal. Plot the two-channel AM signal and plot the power spectrum. (Hint: You must use two separate carrier frequencies.)

8. Use the two signals in Problem 6 and create a QAM signal. Plot the QAM signal and the power spectrum. (Hint: You must use a single carrier frequency, and use sine and cosine carrier waves.) Compare the bandwidth of the QAM with that of the two-channel AM signal.

9. Demodulate the two-channel AM and QAM signals from Problems 7 and 8. Compare the results. Are the demodulated signals similar to the ones created in Problem 6?

10. Generate the various Nyquist pulses shown in Figure 3.14 using MATLAB. You must pay attention to the number of samples required to get a good plot.

11. Generate the spectra of the various Nyquist pulse shapes used in Problem 10.

12. Generate a pulse and a pulse train. Obtain the spectra of these signals.

13. Use MATLAB to obtain the spectra of a PSK signal and a MSK signal.

14. Use MATLAB to compute and plot the rate of decrease of the power in the sidebands (MSK and BPSK).

15. Use a binary on-off keying and show that the outputs of a matched filter and a correlator at $t = T$ are the same. (T is the pulse duration.)

16. Use a RF pulse to show that the outputs of a matched filter and correlator at $t = T$ are the same. (T is the pulse duration.)

17. Using MATLAB, generate two sets of Gaussian random numbers (5000 each), one with a mean of zero and the other with a mean of unity. Choose a standard deviation of 1. If these data streams are the input to a threshold detector, compute the false alarm rate and the miss rate by varying the threshold. Repeat the procedure for standard deviations of 0.7, 0.5, and 0.25. Comment on your results.

18. Repeat the procedure in Problem 17 using two sets of Rayleigh-distributed random numbers. The Rayleigh pdf is

$$f(x) = \frac{x}{b^2} \exp\left(-\frac{x^2}{2b^2}\right) U(x).$$

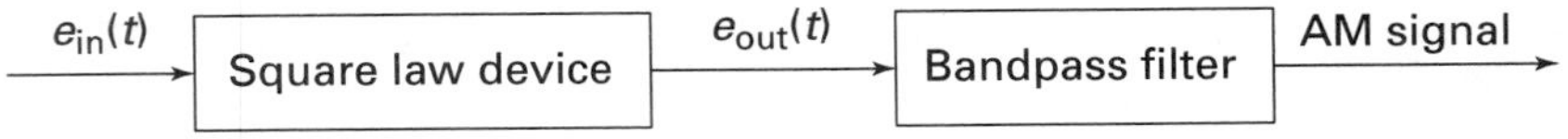

$e_{in}(t)$ → | Square law device | → $e_{out}(t)$ → | Bandpass filter | → AM signal

FIGURE P3.4

Choose a value of $b = 0.5$ for one set and $b = 1$ for the second set. Repeat the calculations for $b = 1, 1.5,$ and 2. Comment on your results by comparing them with those of Problem 17.

19. The curve of the probability of false alarm versus the probability of detection is known as the *receiver operating characteristic* (ROC) curve. Quite often the performance of a receiver is quantitatively described in terms of the area under the ROC curve. Calculate the areas under the various ROC plots for Problems 17 and 18. (Hint: Use the MATLAB command *polyarea*.)

20. Use MATLAB to plot the expressions for the bit error rate given in eqs. (3.83) and (3.105). Compute and compare the minimum powers needed to maintain error rates equal to or lower than 10^{-4}.

21. Derive the equation for the bit error rate, namely, eq. (3.79), of a coherent ASK system. Use a simple correlator receiver (see Figure P3.21). Note that $n(t)$ is white Gaussian noise with a spectral density of $N_0 / 2$.

22. Derive the equation for the bit error rate, namely, eq. (3.83), of a coherent PSK system. Use a simple correlator receiver (see Figure P3.22). Note that $n(t)$ is white Gaussian noise with a spectral density of $N_0 / 2$.

23. Generate an ASK waveform for the data stream

$$1\ 1\ 0\ 0\ 0\ 1\ 0\ 1.$$

24. For the data stream given in Problem 23, generate a BPSK waveform.

25. For the following data stream, generate the transmitted phases in a DPSK scheme. Set the reference bit to be 1.

$$1\ 0\ 0\ 1\ 1\ 1\ 0\ 1\ 0$$

26. For the following bit stream, calculate the phase of the symbol stream for QPSK modulation.

$$1\ 0\ 0\ 1\ 1\ 1\ 0\ 1$$

27. For the OQPSK modulation scheme, compute the phases of the following symbols and the phase differences between the symbols.

$$1\ 0\ 0\ 1\ 1\ 1\ 0\ 1$$

28. For the following bit stream, generate the QPSK waveform using MATLAB. Repeat the simulation with the carrier wave shifted by $\pi / 4$. Now compare the phase change at the symbol transitions.

$$1\ 1\ 0\ 0\ 0\ 1\ 1\ 1$$

29. For the following bit stream, generate the OQPSK waveform using MATLAB. Comment on the phase transitions. At what intervals do they occur? Compare your results with the observations from Problem 28.

$$1\ 1\ 0\ 0\ 0\ 1\ 1\ 1$$

30. For the following bit stream, to be transmitted using $\pi / 4$-QPSK, simulate the transmitted signal using MATLAB. Compare the phase transitions with those observed in Problems 28 and 29.

$$1\ 1\ 0\ 0\ 0\ 1\ 1\ 1$$

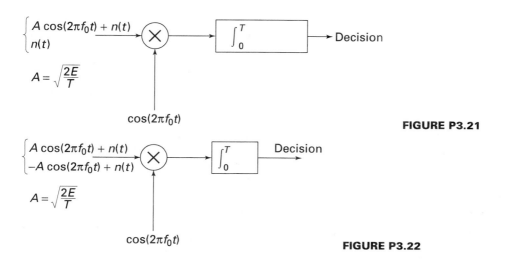

FIGURE P3.21

FIGURE P3.22

31. For the following bit stream, to be transmitted using $\pi/4$-DQPSK, calculate the transmitted phases of the symbols. Simulate the transmitted signal using MATLAB. Compare the phase transitions with those observed in Problems 28–30.

$$1 \ 1 \ 0 \ 0 \ 0 \ 1 \ 1 \ 1$$

32. The following data stream is to be transmitted at a rate of 10^4 samples/s in the QPSK format.

$$1 \ 1 \ 0 \ 0 \ 1 \ 0 \ 1 \ 0 \ 0 \ 0 \ 1 \ 1$$

(a) What is the symbol rate?

(b) Draw the phase constellation for the QPSK system.

(c) What are the transmitted symbol phases?

(d) If the bit stream is to be transmitted by $\pi/4$-DQPSK, what are the transmitted phases?

33. Compare the performance of BPSK and QPSK transmission systems.

(a) If the bit error rate (BER) in the BPSK system is 10^{-4}, calculate the BER and symbol error rate in the QPSK system.

(b) If the minimum bandwidth required for transmission (BPSK) is 60 kHz, what is the minimum bandwidth required for QPSK transmission?

(c) What is the relationship between the spectral efficiencies of the BPSK and QPSK transmission systems?

(d) If you were to use an 8-level PSK system instead of a QPSK transmission system, what would be the minimum bandwidth required?

(e) If you were to use MSK instead of a QPSK transmission system, what would be the resulting transmission bandwidth?

34. For the following bit stream, generate the MSK signal using the I-Q transmitter.

$$1 \ 1 \ 0 \ 0 \ 0 \ 1 \ 1 \ 1$$

35. For the bit stream used in Problem 34, generate the MSK signal using a digital frequency modulation approach. Compare the waveform with the one obtained in Problem 33.

36. For the bit stream in Problem 34, generate a FSK signal.

37. Generate the power spectra of BPSK, QPSK, and MSK waveforms. Comment on the transmission bandwidths.

38. Consider a low-pass filter (Gaussian) of bandwidth B Hz to which the input is a NRZ data stream. The Gaussian response is similar to the one given in eq. (3.106). Obtain and plot the output pulse shapes for

$$BT = 0.5, BT = 1.0, BT = 5, BT = 20.$$

39. Use the pulses and filters of Problem 38 to simulate the eye patterns at the output. Compute the slope of the "eye."

40. Repeat the simulation of Problem 39 by adding white Gaussian noise with a standard deviation of 10% of the amplitude of the NRZ signal. Comment on your results.

41. Generate the plot showing the accumulated phase in GMSK. Comment on the phase characteristics as BT increases.

42. For the data stream in Problem 34, generate the GMSK waveform for $BT = 0.2$, 0.35, and 1. Compare the waveforms with the one generated in Problem 34.

43. Generate the power spectra of GMSK waveforms for the BT values from Problem 42 and compare the spectra with the MSK spectrum obtained.

44. Show that 4-level PSK and 4-level QAM are equivalent.

45. Consider a receiver having a white noise spectral density of 10^{-9} W/Hz. If this receiver is used in a coherent BPSK demodulator, what will be the signal level (amplitude) required to maintain an error rate of 10^{-5} for the following data rates?

(a) 1 kbps

(b) 10 kbps

(c) 100 kbps

(Hint: Note that the bit error rate is given by $\mathrm{erfc}(\sqrt{Z})$, where Z is the energy-to-noise ratio given by $Z = E/N_0$. If the BPSK signal is represented by $A\cos(2\pi f_0 t)$, E is given by $(A^2/2)T$. The noise spectral density, by convention, is $N_0/2$. Use the *erfinv* function from MATLAB.)

46. Repeat the calculations in Problem 45 for the case of a coherent FSK system.

47. Plot eq.(3.86) using MATLAB for the two values of the standard deviation of phase mismatch $\sigma_\phi = 22°, 32°$. What are the error floors?

48. Plot eq. (3.113) for fixed values of $\Delta\phi = 0°, 5°, 10°$.

49. Plot eq. (3.114) using MATLAB. Obtain the excess SNR required in $\pi/4$-DQPSK to have the same BER (10^{-3}) in $\pi/4$-QPSK.

50. Explain the concept of excess power margin in the context of the two related modulation schemes coherent BPSK and DPSK. Generate a MATLAB plot of bit error rate versus excess power margin (dB).

51. Generate a plot of the power efficiency versus spectral efficiency. Mark the points for BPSK, QPSK, and FSK on that plot.

52. Generate a BPSK waveform. Add white noise to it. Demodulate by multiplying with twice the carrier wave. Perform low-pass filtering. Compare the recovered waveform with the transmitted bit stream.***

53. Generate an OOK waveform. Demodulate by multiplying with twice the carrier wave. Perform low-pass filtering. Compare the recovered waveform with the transmitted bit stream.*** Compare the results with those of Problem 52.

54. In Problem 53, instead of using a coherent demodulator, use a noncoherent modulator (square law followed by square root, or just take the absolute values). Perform low-pass filtering and compare the results with those of Problem 53. Now compare the results of Problems 52–54.***

55. In Problems 22 and 52, it was assumed that the local oscillator for BPSK demodulation was perfect. Now repeat Problem 52 by having a local oscillator as $2\cos(2\pi f_0 t + \phi)$, where ϕ is fixed phase offset. Use $\phi = 5°, 30°, 85°$. Comment on your results.***

56. Repeat Problem 55 by treating the phase ϕ as random. Assume the phase to be Gaussian distributed with mean values equal to the phases in Problem 55 and a standard deviation of 10%. Comment on your results.***

57. Implement the coherent BPSK receiver shown in Figure P3.22 in MATLAB. Explore what happens when you increase the additive noise.

58. Repeat Problem 57 by introducing a fixed phase mismatch and explore what happens when you add noise. Compare the input and output bit streams.

59. Repeat Problem 57 by having a random phase mismatch (Problem 56). Compare the input and output bit streams.

60. Implement a coherent QPSK receiver using MATLAB.

61. Using MATLAB, create a DPSK signal.

62. Implement the suboptimal DPSK receiver in MATLAB.

63. Implement the optimal DPSK receiver in MATLAB.

64. Using the Gaussian probability integral, establish the relationship given in eq. (B.7.3) between the error function (erf) and the complementary error function (erfc). Use MATLAB to verify the relationship between these functions.

65. Using the Gaussian probability integral, derive the relationship given in eq. (B.7.4) between erf (u) and erf $(-u)$. Verify your results using MATLAB.

Exercises Involving Noise

66. White Gaussian noise of spectral density $\beta/2$ is passed through a low-pass filter of bandwidth W. What is the noise power at the output of the filter?

67. Verify the results of Problem 66 using MATLAB.

68. White noise of spectral density $\beta/2$ is the input to a coherent demodulator (i.e., it is multiplied by $\cos(2\pi f_0 t)$). What is the noise at the output of the mixer?

69. Verify the results of Problem 68 using MATLAB.

70. Obtain an analytical expression for the noise variance at the output of a correlator. The correlator uses a cosine function.

71. If narrowband (or pass-band) noise $n(t)$ is given in terms of the quadrature representation, i.e.,

$$n(t) = n_c(t)\cos(2\pi f_c t) - n_s(t)\sin(2\pi f_c t),$$

show that the powers of $n(t)$, $n_c(t)$, and $n_s(t)$ are equal.

CHAPTER 4

CELLS AND CELLULAR TRAFFIC

4.0 INTRODUCTION

Mobile systems are designed to operate over a very large geographical area with a limited bandwidth. The service providers use tall antennae and higher transmitted powers to provide coverage over as large an area as possible. However, restrictions exist in different communities for maximum antenna height as well as maximum transmitted power. Once the limiting values of these two parameters have been reached, the only means of providing mobile phone use to an increased number of subscribers is through the use of *cells* (Jake 1974, Lee 1997, MacD 1979, Youn 1979, Oett 1983, Hugh 1985). By dividing a geographical region into smaller-size cells, providers can reuse the channels or frequencies repeatedly and thereby operate with less power. This frequency reuse through the use of cells permits the providers to increase capacity with a limited bandwidth. However, this frequency reuse comes at the expense of co-channel interference (CCI), which is the interference produced by the use of identical channels (frequencies). For example, if a carrier frequency of f_{c1} is used in a cell, as shown in Figure 4.1, it may be used again by cells at a certain radius away from it.

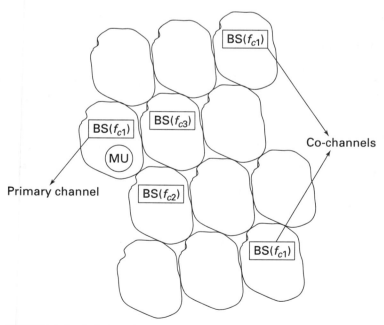

FIGURE 4.1 Splitting of the geographical region into "cells." The frequency f_{c1} is used in multiple cells separated by a fixed distance.

Depending on the distance between the MU and its own base station and the distances between co-channel base stations and the MU, interference from co-channels will be a problem. The unwanted signals generated by other base stations operating at the same frequency band result in co-channel interference. Figure 4.2 shows the power from the primary channel and the total power from co-channels corresponding to the same frequency band. The shaded region shows the relative power of the co-channels. Additional detail on the calculation of CCI is given later in this chapter.

As shown in Figure 4.1, a cell is the smallest geographical area covered by a base station. It can serve a certain number of users depending on the number of channels and the blocking probability. The number of users is the most important factor determining cell size. When the number of users increases, it becomes necessary to reduce the size of the cells so that the channels can be reused more often. Cell splitting, which reduces the size of the cells, also reduces the maximum power that can be transmitted. The reduction in power is necessary to prevent co-channel interference from the channel operating at a distance. A simple way to understand these concepts is to examine the layout of a geographical area covered with cells of symmetric shapes, as shown in Figure 4.3. The ideal shape would be a circle, which makes it possible to define a radial region. However, this will leave a number of "zones" outside the coverage, so the optimal shape of the cell (MacD 1979) is hexagonal, as shown in Figure 4.3*d*. A few other regular shapes, such as triangles and squares, are not well suited since the distance from the center of the cell to different points of the perimeter will be different. A hexagon comes reasonably close to the ideal shape that leaves no gaps and allows only small differences from the center to various points on the perimeter.

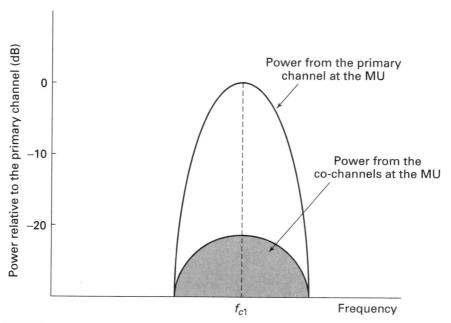

FIGURE 4.2 The power from the primary channel and the total power from the co-channels in the cell subscribing to the primary channel.

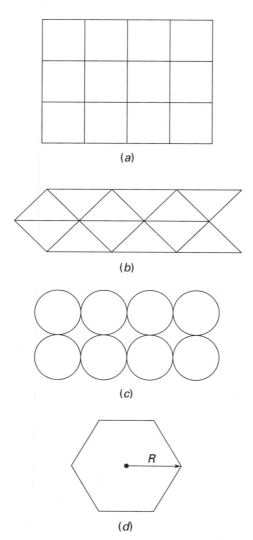

FIGURE 4.3 Different geometrical shapes for cells. (*a*) Square. (*b*) Triangular. (*c*) Circular. (*d*) Hexagonal, where *R* is the "radius" of the hexagon.

4.1 GEOMETRY OF A HEXAGONAL CELL

A geographical region filled with hexagonal cells is shown in Figure 4.4. Note that the *xy*-axes do not match the *uv*-axes of the hexagonal system. The distance, R, from the center of a cell to its vertex is defined as the cell radius. The center-to-center distance between two neighboring cells is therefore $2R \cos(\pi/6)$, or $\sqrt{3}\,R$. The center-to-center distance, D, between two cells with coordinates (u_1, v_1) and (u_2, v_2) is

$$D = \{(u_2 - u_1)^2 \cos^2(\pi/6) + [(v_2 - v_1) + (u_2 - u_1)\sin(\pi/6)]^2\}^{1/2}. \quad (4.1)$$

This equation can be simplified to

$$D = \sqrt{(u_2 - u_1)^2 + (v_2 - v_1)^2 + (u_2 - u_1)(v_2 - v_1)}. \quad (4.2)$$

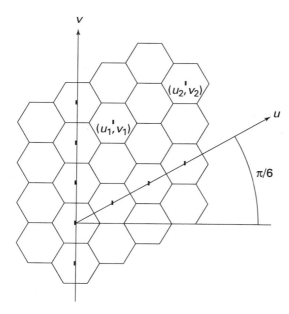

FIGURE 4.4 The concept of a hexagonal cell. The coordinates of the center of a cell are (u,v). The hexagonal coordinates (u,v) do not match the rectangular coordinates (x,y).

In writing eq. (4.2), we have not made use of the information that the radius of the cell is R. Incorporating this information, and assuming that the first cell is centered at the origin $(u = 0, v = 0)$, the equation for the distance between any two cells can be expressed as

$$D = \sqrt{i^2 + j^2 + ij}\sqrt{3}R, \qquad (4.3)$$

where (i,j) represents the center of a cell in the (u,v) coordinate system. For adjoining cells, either i or j can change by 1, but not both. Therefore, the distance between nearby cells is $\sqrt{3}\,R$. Since i and j can take only integer values, it is possible to define the number of cells, N_c, that lie within an area of radius D as

$$N_c = D_R^2, \qquad (4.4a)$$

where

$$D_R = \sqrt{i^2 + j^2 + ij}. \qquad (4.4b)$$

The quantity D_R is the normalized separation between any two cells and depends only on the cell number as counted from the cell at the origin, or the reference cell (Lee 1986, 1991b). The number of cells can thus take values of only 1, 3, 4, 7, 9, 12, 13, etc. Figure 4.5 shows a seven-cell ($N_c = 7$) cluster, and the shaded cells are

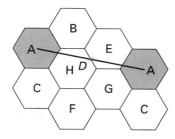

FIGURE 4.5 Frequency reuse pattern of cells. The letters (except D) represent the different channels. D is the distance between cells using the same frequency.

TABLE 4.1   Relationship between i, j, and the Number of Cells in a Cluster

i	j	N_c	$q = D/R$
1	0	1	1.73
1	1	3	3
2	0	4	3.46
2	1	7	4.58
3	0	9	5.2
2	2	12	6

co-channel cells that use the same channel (frequency band A). Table 4.1 shows the relationship between i, j, and the number of cells in a cluster.

4.2   CO-CHANNEL INTERFERENCE (CCI)

If the cell sizes are fixed, the interference from another channel using the same frequency will be determined by the location of the interfering cell from the primary cell. A typical reuse of channels is shown in Figure 4.6. Corresponding letters represent identical channels, and it can be seen that the frequencies can be used and reused over and over again in cells separated by a certain distance. The frequency reuse (Naga 1987; Lee 1986, 1991b, 1993, 1997) is therefore determined by the signal-to-noise (S/N) ratio, or, more specifically, the signal-to-CCI ratio (S/I). The signal-to-noise ratio is given by

$$\frac{S}{N} = \frac{\text{signal power}(S)}{\text{noise power }(N_s) + \text{interfering signal power }(I)}, \tag{4.5}$$

and the signal-to-CCI ratio is given by

$$\frac{S}{I} = \frac{\text{signal power}(S)}{\text{interfering signal power }(I)}.$$

If the noise at the receiver (N_s) is negligible, the signal-to-noise ratio and signal-to-CCI ratio will be equal,

$$\frac{S}{N} = \frac{S}{I}. \tag{4.6}$$

The signal power is proportional to $R^{-\nu}$ while the CCI power is proportional to $D^{-\nu}$, where ν is the loss exponent or loss factor, described in Chapter 2. If we assume that all interfering cells are at the same distance, the signal-to-CCI ratio can be expressed as

$$\frac{S}{I} = \frac{R^{-\nu}}{\sum_{k=1}^{N_I} D_k^{-\nu}} = \frac{R^{-\nu}}{N_I D^{-\nu}}, \tag{4.7}$$

where the number of interfering cells is N_I. For a seven-cell cluster $N_I = 6$. D_k is the distance of the kth interfering cell and is equal to D in the ideal case where all interfering cells are at the same distance. Using eqs. (4.4) and (4.7), S/I becomes

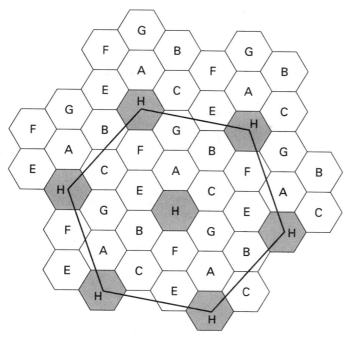

FIGURE 4.6 Expanded view of the cell structure showing a seven-cell reuse pattern.

$$\frac{S}{I} = \frac{1}{6}\left(\frac{D}{R}\right)^{\nu} = \frac{1}{6}(3N_c)^{\nu/2} \qquad (4.8)$$

For a seven-cell cluster, there are six interfering cells and the signal-to-CCI ratio is $(1/6)(21)^{\nu/2}$. For $\nu = 4$, this corresponds to 73.1, or about 18.6 dB. The quantity D/R is a measure of the interference since the farther away the interfering cell is, the smaller the interfering signal power. Because of this, $q = D/R$ is also known as the *frequency reuse factor* or *co-channel interference reduction factor*. Table 4.2 gives values for q along with S/I for the ideal case of equidistant interfering cells.

TABLE 4.2 Relationship between i, j, and S/I

i	j	N	q	S/I (dB)
1	1	3	3.0	16.1
1	2	7	4.6	18.7
2	0	4	3.5	16.8
2	1	7	4.6	18.7
2	2	12	6.0	20.7
3	0	9	5.2	19.6
3	1	13	6.2	21.0
3	2	19	7.5	22.6
3	3	27	9.0	24.0
4	0	16	6.9	21.9
4	1	21	7.9	23.0
4	2	28	9.2	24.2
4	3	37	10.5	25.3

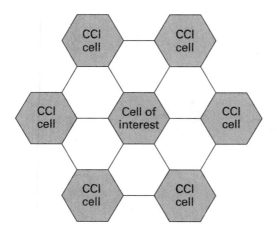

FIGURE 4.7 A three-cell pattern showing six interfering cells $(D = \sqrt{3N_c} = 3)$.

A three-cell reuse pattern is shown in Figure 4.7. Comparison of Figures 4.6 and 4.7 clearly shows that irrespective of the number of cells in a cluster, the number of interfering cells is 6.

4.2.1 Special Cases of Co-Channel Interference

It is possible that the mobile unit operating at the edge of a cell will experience a high level of co-channel interference. In this case, the MU unit is receiving the weakest possible signal from its own base station while it is likely to receive stronger interfering signals from the co-channels, as shown in Figure 4.8.

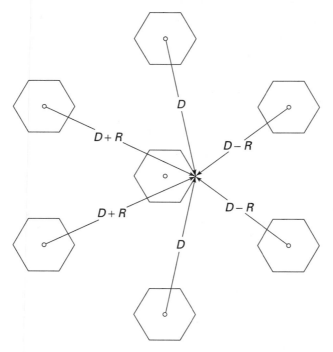

FIGURE 4.8 Geometry for the calculation of co-channel interference.

Based on the geometry shown in Figure 4.8, the signal-to-CCI ratio for a seven-cell reuse pattern can be expressed as (Lee 1986, Garg 1996, Oett 1983, Hamm 1998)

$$\frac{S}{I} = \frac{R^{-\nu}}{2(D-R)^{-\nu} + 2D^{-\nu} + 2(D+R)^{-\nu}}. \tag{4.9}$$

Noting that $q = D/R$ for a seven-cell reuse pattern is equal to 4.6, S/I can be found to be equal to 54.3, or about 17 dB. However, if we use the shortest distance for all the interferers, S/I becomes

$$\frac{S}{I} = \frac{R^{-\nu}}{6(D-R)^{-\nu}} = \frac{1}{6(q-1)^{-\nu}}, \tag{4.10}$$

and will be equal to only about 14 dB. If the acceptable S/I is set at a value of 18 dB, co-channel interference can make the system performance unacceptable. Of course, it is possible to increase S/I by using a cluster of 12 or more cells. But this improvement in performance will be gained at the expense of reduced capacity, since the number of users in a given geographical area will be reduced by increasing the frequency reuse factor. Before we examine ways of maintaining an acceptable S/I, we will consider the case of a single interferer, since this case is more likely to occur than the one where all six interfering cells are active.

EXAMPLE 4.1

In a cellular system, the received power at the mobile unit is –98 dBm. The cell structure is a seven-cell reuse pattern. If the thermal noise power is –120 dBm and the CCI from each interfering station is –121 dBm, calculate the signal-to-CCI ratio and signal-to-noise ratio.

Answer Converting from dBm to mW, we get signal power $P_s = 10^{-9.8}$ mW, noise power $= 10^{-12}$ mW, CCI power $= 10^{-12.1}$ mW. Signal-to-CCI ratio $= 10^{-9.8}/(6 \times 10^{-12.1}) = 33.254$ or $10 \log_{10}(33.254)$ dB $= 15.22$ dB. Signal-to-noise ratio $= 10^{-9.8}/(6 \times 10^{-12.1} + 10^{-12}) = 27.48$ or $10 \log_{10}(27.48)$ dB $= 14.39$ dB. ∎

Special Case of a Single Interfering Cell In many practical conditions, there is a possibility that there is only one interfering cell, as shown in Figure 4.9. Depending on the location of the MU with respect to the interferer, S/I can be very low. The two cases illustrated in Figure 4.9 show the best and worst cases. S/I in this case is given by

$$\frac{S}{I} = \frac{R^{-\nu}}{D^{-\nu}}. \tag{4.11}$$

The S/I values will vary depending on the distance D.

EXAMPLE 4.2

Consider a mobile unit operating at a distance of 6 km from its own base station. A single interfering station is operating at a distance of 15 km from the mobile unit. If the loss factor ν is 3.5, calculate the signal-to-CCI ratio.

Answer Signal power $\propto 6^{-\nu}$; CCI $\propto 15^{-\nu}$. Signal-to-CCI ratio $= (15/6)^{\nu} = (15/6)^{3.5} = 24.7$, or $10 \log_{10}(24.7)$ dB $= 13.92$ dB. ∎

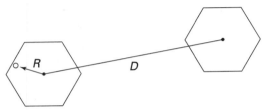

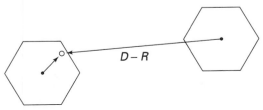

Interference (normal case)

Interference (worst case)

FIGURE 4.9 Special cases of a single interfering cell.

4.3 CCI REDUCTION TECHNIQUES

One of the ways in which CCI can be reduced is through the use of a 12-cell reuse pattern. This may be unacceptable because it reduces the number of users in a given geographical area serviced by a provider. On the other hand, if it is possible to replace the omnidirectional antenna with a directional antenna, some of the interfering signals will never reach the primary mobile unit, thus reducing CCI and improving the signal-to-CCI ratio (Rapp 1996b, Samp 1997, Lee 1986, Wese 1998). This can be accomplished through the use of a 120° sector antenna or a 60° sector antenna.

4.3.1 Directional Antenna Using Three Sectors

In Figure 4.10 each cell is divided into three sectors, with each sector having its own set of frequencies. For a seven-cell cluster, the primary MU will receive interfering signals from only two other cells (instead of six, as with an omnidirectional antenna)

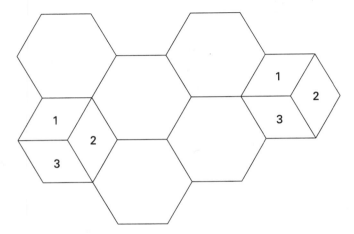

FIGURE 4.10 Concept of a three-sector antenna.

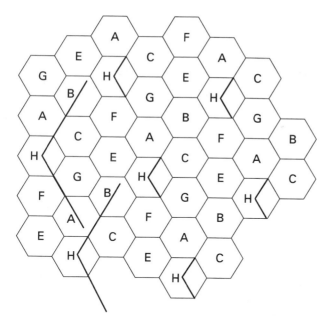

FIGURE 4.11 The interfering cells can be identified by following the radiation pattern of the sector antennae.

because of the directional pattern of the antenna, as seen in Figure 4.11. The S/I ratio can now be expressed in the worst-case scenario (when the MU is at the edge of its own cell as shown in Figure 4.9) as follows:

$$\frac{S}{I} = \frac{R^{-\nu}}{D^{-\nu} + (D + .7R)^{-\nu}}. \tag{4.12}$$

In the best-case scenario, the two interferers will be at a distance of D, and the signal-to-CCI ratio for a 120° sector antenna can now be obtained from the corresponding ratio for the omnidirectional antenna by noting that the omni antenna has six interferers:

$$\left[\frac{S}{I}\right]_{120°} = \left[\frac{S}{I}\right]_{omni} + 10 \log 3 = \left[\frac{S}{I}\right]_{omni} + 4.77 \text{ dB} = 23.4 \text{ dB}. \tag{4.13}$$

Thus, compared with the performance of an omni antenna, a 120° sector antenna provides a signal-to-CCI ratio improvement of 4.77 dB.

4.3.2 Directional Antenna Using Six Sectors

In Figure 4.12 each cell is divided into six sectors and it can be seen that there will be only a single interferer, reducing the CCI and improving the S/I ratio. The improvement in S/I using a 60° sector antenna can be obtained once again by noting that there is only a single interferer, compared with six interferers for an omnidirectional antenna, as shown in Figure 4.13.

$$\left[\frac{S}{I}\right]_{60°} = \left[\frac{S}{I}\right]_{omni} + 10 \log 6 = \left[\frac{S}{I}\right]_{omni} + 7.78 \text{ dB} = 26.4 \text{ dB}, \tag{4.14}$$

indicating a 7.78 dB improvement in the performance of a 60° sector antenna system.

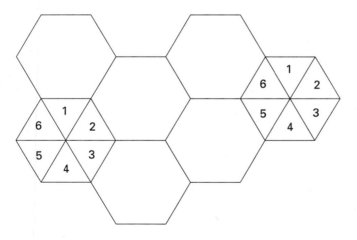

FIGURE 4.12 A 60° sector antenna arrangement.

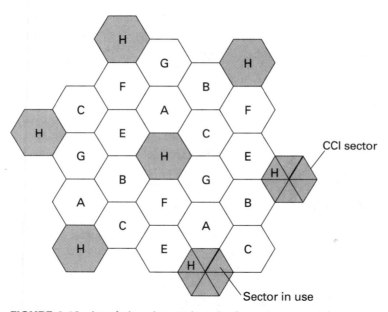

FIGURE 4.13 Interfering element in a six-element antenna scheme.

A Comment on the Use of Sector Antennae There is a major drawback to the sectorized antenna approach to improving the S/I ratio. Sectorized antennae adversely affect the overall capacity of the system (Chan 1992). This can be explained using the concept of *trunking*. We will look at this drawback after we have presented the trunking concepts in Section 4.7.

4.3.3 Geographical Model with Several Tiers of Interferers

The previous sections dealt with the case of a single tier of channel interference at a distance of D. However, as shown in Figure 4.14, other tiers can produce additional interference terms. The first tier is at a distance of D and the second tier is at a

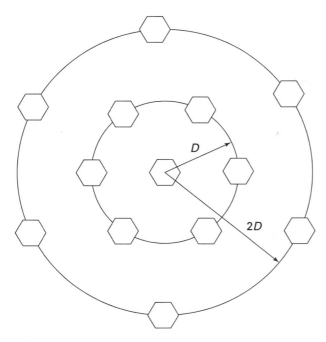

FIGURE 4.14 Geographical model with multiple interferers.

distance of $2D$. In general, tiers exist at distances of kD, $k = 1, 2, \ldots$. The signal-to-CCI ratio in this case can be expressed as

$$\frac{S}{I} = \frac{R^{-\nu}}{6D^{-\nu} + 6(2D)^{-\nu} + 6(3D)^{-\nu}}, \tag{4.15}$$

where the second term in the denominator comes from the six interferers in the second tier, and the third term comes from the six interferers from the third tier. In most cases, these interference terms from the second and third tiers are negligible (Lee 1986, Hamm 1998).

4.4 CELL SPLITTING

One of the ways in which capacity can be increased is through the technique known as cell splitting. In this case, a congested cell is divided into smaller cells. Each smaller cell, a minicell, will have its own transmitting and receiving antennae. Cell splitting thus allows the frequencies/channels to be reused, since the size of the cell has been reduced for a given geographical region.

If each cell size is reduced by half, as shown in Figure 4.15 (where the smaller cells are shown in boldface), the power requirements change (Cox 1982, Lee 1991b, Rapp 1996b). This is necessary to maintain an acceptable signal-to-CCI ratio. It is possible to calculate the reduction in power. If the power required at the cell boundary in an unsplit cell is P_u, we can write this expression as

$$P_u = P_{tu} R^{-\nu}, \tag{4.16}$$

where P_{tu} is the transmitted power. The power received at the new, smaller cell boundary, P_{su}, is

$$P_{su} = P_{st}(R/2)^{-\nu}, \tag{4.17}$$

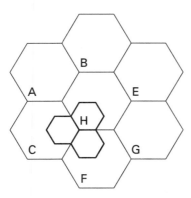

FIGURE 4.15   Cell splitting. The dark borders indicate newly created cells.

where P_{st} is the transmitted power from the antenna of the new cell. To maintain the same CCI performance, P_u must equal P_{su}, and this means that the transmitted power in the smaller cell, P_{st}, is given by

$$P_{st} = \frac{P_{tu}}{16} \tag{4.18}$$

for the case of $\nu = 4$. In other words, the new transmitted power is reduced by about 12 dB.

Note that all cells may not be split. Cell splitting can be accomplished on an as-needed basis. If it is possible to know when there is a likelihood of increased traffic in certain cells, careful planning can be done beforehand to convert some of the cells to minicells or break them down further to microcells.

EXAMPLE 4.3

The transmit power of a base station is 5 W. If the coverage of this region is to be split in half so that minicells (of half the size) can be created to accommodate additional users in the region, what must be the transmit power of the base station (of this minicell) to keep CCI at the same level as that of the unsplit cell? Assume that $\nu = 3.2$.

Answer   To keep the CCI from the interfering stations unchanged, we need to keep the power at the boundary the same, regardless of whether the cell has been split or not. If R is the diameter of the unsplit cell, we must have $5R^{-\nu} = P(R/2)^{-\nu}$, where the left-hand side is the power at the boundary of the unsplit cell and the right-hand side is the power at the boundary of the new cell (half the size). P is the transmit power of the BS of the newly created minicell, $P = 5/(2)^{\nu} = 5/9.186 = 544\,\text{mW}$. ∎

4.5   MICROCELLS, PICOCELLS, AND FIBEROPTIC MOBILE SYSTEMS

One of the drawbacks of cell splitting is the complexity of the multiple hand-offs necessary over small distances. As we decrease the coverage area by going from a macrocell to a minicell, and then to a microcell and a picocell, the complexity of the operation over a given geographical area increases significantly (Lee 1997, Rapp 1996b, Yeun 1996). The number of hand-offs performed by the base station goes up since the geographical region covered becomes split into an increasing number of

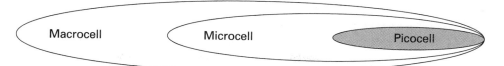

FIGURE 4.16 The areas corresponding to a macrocell, microcell, and picocell.

smaller and smaller cells, as shown in Figure 4.16. Along with this comes the additional workload of switching and control elements of the mobile system. This problem can be solved using a fiberoptic mobile (FOM) system (Chu 1991, Kaji 1996, Shib 1993, Way 1993). In this arrangement, several base stations are linked using optical fibers and are thereby controlled from a single location. The schematic in Figure 4.17 shows this principle (Kosh 1997). The base stations merely act as transmitter/receiver stations, and most of the switching and channel allocation functions are centrally undertaken. The same concept can be applied to systems that employ sector antennae.

The electrooptic (E/O) and optoelectronic (O/E) elements provide the conversion from the electrical to optical and optical to electrical signals, respectively. The low-noise amplifiers (LNA) and high-power amplifiers (HPA) provide the amplification needed. Note that this system is not without problems, most of which arise from the nonlinear behavior of the laser diodes used at the modulator stage.

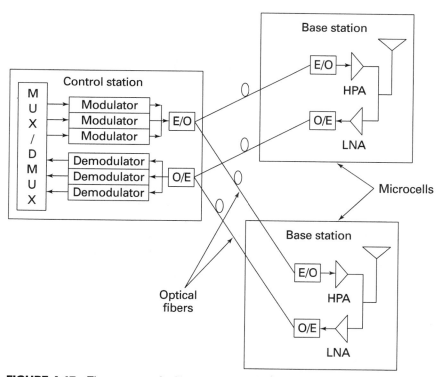

FIGURE 4.17 The concept of a fiberoptic mobile system.

4.6 COVERAGE AREA ESTIMATION

We have examined the concept of a cell and the origins of CCI. We still need to understand how to arrive at the maximum radius, R, of the cell. To estimate the coverage area, we must consider the following:

Transmitter power, P_T (dBm)

Sensitivity of the receiver, or the threshold power, P_{th} (dBm)

Power loss from transmission, L_p (dB)

Under ideal conditions, it is possible to estimate the maximum distance between the transmitter and receiver from the following:

$$L_p = P_T - P_{th}. \tag{4.19}$$

The distance can then be estimated from eq. (2.9) or directly from the Hata model, eq. (2.10). However, this simplistic approach ignores the fact that the signal undergoes long-term fading, which could reduce the received power at any given location by 10 dB in urban areas and a few decibels in rural areas. Since the fading is long term, no improvements can be expected if the MU moves through a short distance. It is therefore necessary to compensate for a fading margin, M, to account for the lognormal fading (Jake 1974, Pale 1991, Pars 1992, Samp 1997, Akai 1998) seen. Equation (4.19) needs to be rewritten as

$$L_p = P_T - P_{th} - M, \tag{4.20}$$

thereby reducing the permitted loss in the transmission and, consequently, reducing the transmission distance. We will now examine ways to calculate the fading margin, M.

Computation of the Fading Margin The probability density function of the received power under long-term fading conditions is given by eq. (2.67), reproduced here:

$$f(p_{LT}) = \frac{1}{\sqrt{2\pi\sigma^2 p_{LT}^2}} \exp\left[-\frac{1}{2\sigma^2}\ln^2\left(\frac{p_{LT}}{p_0}\right)\right], \tag{4.21}$$

where p_{LT} is the power in milliwatts. Whenever the received power falls below the minimum required power, P_{th}, the system goes into outage (Hata 1985, Coul 1998, Pahl 1995). The outage probability, P_{out}, can be defined as the probability that the received signal power fails to reach the threshold (Jake 1974, Hata 1985). If R is the maximum radius of the cell, the outage probability at the boundary, $P_{out}(R)$, can be expressed as

$$P_{out}(R) = \int_0^{P_{th}} f(p_{LT})\, dp_{LT}. \tag{4.22}$$

Using eq. (4.21), the outage probability given in eq. (4.22) becomes

$$P_{out}(R) = \frac{1}{2}\text{erfc}\left[\frac{\ln(P_0(R)/P_{th})}{\sqrt{2}\sigma}\right]. \tag{4.23}$$

The quantity $P_0(R)$ is the median power at the distance R calculated from the Hata model. When the terminal is at a distance r from the transmitter, the received power, $P_0(r)$, can be expressed as

$$P_0(r) = P_0(R)\left[\frac{R}{r}\right]^\nu,$$ (4.24)

where ν is the loss parameter. The outage probability, $P_{out}(r)$, at a distance r can now be expressed as

$$P_{out}(r) = \frac{1}{2}\text{erfc}\left\{\frac{\ln\left[\left(\frac{P_0(R)}{P_{th}}\right)\left(\frac{R}{r}\right)^\nu\right]}{\sqrt{2}\sigma}\right\}.$$ (4.25)

The area outage probability, $P_{aout}(R)$, is given by the integral

$$P_{aout}(R) = \frac{1}{\pi R^2}\int_0^R P_{out}(r)2\pi r\, dr.$$ (4.26)

The integral can be evaluated to

$$P_{aout}(R) = \frac{1}{2}\text{erfc}(Q_1) - \frac{1}{2}e^{(2Q_1Q_2 + Q_2^2)}\text{erfc}(Q_1 + Q_2),$$ (4.27)

where Q_1 and Q_2 are given by

$$Q_1 = \frac{\ln\left[\frac{P_0(R)}{P_{th}}\right]}{\sqrt{2}\sigma}, \qquad Q_2 = \frac{\sqrt{2}\sigma}{\nu}.$$ (4.28)

The fading margin, $M(\text{dB})$, is given by

$$M = 10[\log_{10}P_0(R) - \log_{10}P_{th}]\ \text{dB}.$$ (4.29)

Redefining the fading margin in terms of a scaling factor, m,

$$M = 10\log_{10}\left[\frac{P_0(R)}{P_{th}}\right] = 10\log_{10}(m),$$ (4.30)

the parameter Q_1 can be rewritten without any dependence on the threshold power as

$$Q_1 = \frac{\ln\left(10^{M/10}\right)}{\sqrt{2}\sigma}.$$ (4.31)

It is possible to see the effect of the fading margin (M) on the outage probability. The outage probability at the boundary of the cell, given in eq. (4.23), is shown in

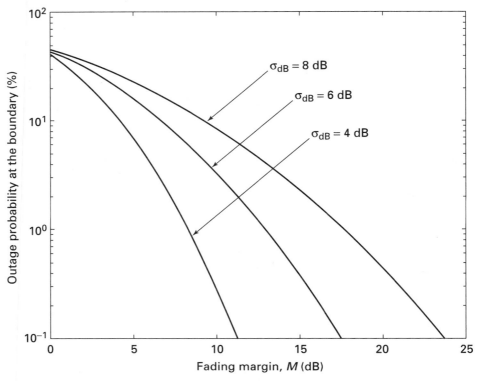

FIGURE 4.18 The outage probability (%) at the boundary of the cell is plotted as a function of fading margin M (dB). The relationship between σ and the standard deviation of fading, σ_{dB}, is given in eq. (2.68).

Figure 4.18. As expected, the outage probability goes up as the standard deviation of lognormal fading goes up, and the power margin required to maintain a fixed outage probability goes up as the fading level (given by σ) increases. Note that there is no explicit dependence on the loss parameter ν.

The area outage probability, given in eq. (4.27), is shown in Figure 4.19. Comparing Figures 4.18 and 4.19 (as well as the two equations for the outage probabilities), the outage probability at the boundary provides us with the worst-case scenario and can therefore be used to calculate the maximum coverage. It is worth mentioning that the area outage probability also depends on the loss parameter ν.

Link Budget Link budget calculations are performed to estimate the maximum coverage that can be obtained. These calculations are based on the minimum power required to maintain acceptable performance, transmitted power, and the power margin needed to mitigate fading. Note that, in general, two margins have to be applied: a power margin, M_1, to account for long-term fading (lognormal) and a second margin, M_2, to account for the short-term fading. This concept is illustrated in Figure 4.20. The distance d_0 is the maximum coverage based on pure attenuation, and with the incorporation of the long-term fading margin, the coverage drops to d_1. If we include the short-term fading margin, the coverage drops to d_2. The schematic in Figure 4.20 reflects the relationship between attenuation, long-term fading, and short-term fading illustrated in Figure 2.5.

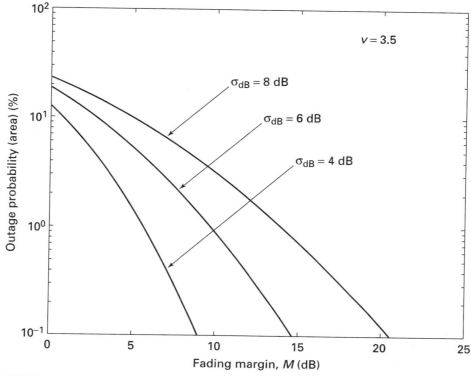

FIGURE 4.19 The area outage probability (%) is plotted as a function of fading margin M (dB). The relationship between σ and the standard deviation of fading σ_{dB} is given in eq. (2.68).

EXAMPLE 4.4

To keep an acceptable level of performance, a service provider is planning to maintain an outage probability of 2%. What is the power margin needed to achieve this goal (assume the worst-case scenario) when the standard deviation of fading has been found to be 5 dB?

Answer The outage probability is given by eq. (4.23), which can be rewritten as

$$P_{\text{out}}(R) = 0.5 \; \text{erfc}\left(\frac{m}{\sigma\sqrt{2}}\right)$$

and the power margin in dB is given by

$$M = 10 \log_{10}(m).$$

The relationship between σ and σ_{dB} is given in eq. (2.68) as

$$\sigma = \left(\frac{1}{10}\right)\ln(10)\sigma_{dB}.$$

We therefore have $\sigma = 1.15$. Thus

$$\text{Outage probability} = 0.02 = 0.5 \; \text{erfc}\left(\frac{m}{1.6282}\right).$$

As we did in Chapter 3, we can solve this using the MATLAB function *erfinv*. We get $m = 2.36$ and $M = 3.73$ dB.

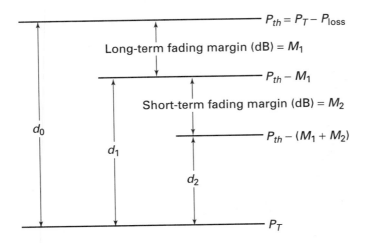

FIGURE 4.20 The margins for fading and the consequent reduction in coverage.

EXAMPLE 4.5

In a cell, the received power measured at a distance of 6 km from the BS is -65 dBm. The threshold level for acceptable performance is -90 dBm. If the power margin is the value obtained from Example 4.4, what is the distance that can be covered? Assume that the path loss exponent is 3.2.

Answer   We have been given the power at 6 km $= -65$ dBm or $10^{-6.5}$ mW. Margin $= 4.3$ dB. Minimum required power $= -90 + 4.3 = -85.7$ dBm or $10^{-8.57}$ mW. Making use of the path loss, we have

$$10^{-6.5} \propto \left(\frac{1}{6}\right)^{3.2}$$

and

$$10^{-8.57} \propto \left(\frac{1}{d}\right)^{3.2},$$

where d is the maximum distance for acceptable performance. Solving, we get $d = 25$ km.   ∎

4.7   TRAFFIC CAPACITY AND TRUNKING

A question commonly asked is, How can we compare the quality of service provided by the various cellular providers? What is the probability of *not* being able to make a call when you want to make one? What is the probability that you will have to wait before getting connected? To answer some of these questions, we need to understand the concepts of telephone trunking and "grade of service" (GOS). Typically there are more users than the number of channels or trunks available. Telephone trunking allows the provider to use a limited bandwidth to accommodate a large number of users. It relies on the principle that not everybody will be using the telephone line at the same time. Thus, each user gets access to a channel as and when the user needs it. When the call is completed and terminated, the channel is freed and made available to other users. There is, of course, a problem. If everybody in the system wishes to make use of their cell phones at the same time, only a limited number of users will be allowed to get connected. This leads to a user being denied access, or being blocked.

"Grade of service" is a measure of the blocking that may take place, or the ability of the user to gain access to the system during the busiest hour.

GOS is determined by the available number of channels and is used to estimate the number of total users that can be supported by the network. To understand the concept of GOS, we first need to define traffic intensity. The unit of traffic intensity is the Erlang (Hong 1986, Jake 1974). If a person is using the phone at a rate of two calls/hour and stays on the phone for an average time of 3 minutes per call, the traffic intensity generated by this user is

$$2 \times \frac{3}{60} = 0.1 \text{ Erl}. \tag{4.32}$$

In other words, this user generates one-tenth of an Erlang. The Erlang is thus a measure of the unit of traffic intensity given by the product of the number of calls/hour and the duration of the call (in hours).

The traffic intensity generated by a user, A_I, is given by

$$A_I = \lambda T_H \text{ Erl}, \tag{4.33}$$

where λ is the average number of calls/hour and T_H is the duration of the calls (in hours). If there are K users in the system supported by a certain provider, the provider must be able to sustain a traffic intensity of A_{tot}, given by

$$A_{tot} = K \times A_I \text{ Erl}. \tag{4.34}$$

The provider certainly must be able to allow the K users to access the system during the peak period, when everybody is trying to access the system, with an acceptable performance criterion or grade of service. To achieve a certain performance criterion or blocking probability, $p(B)$, the provider must be able to offer a certain number of channels or trunks in the network. This will determine the *offered traffic*. The *carried traffic* will be less than the offered traffic (Section B.2, Appendix B). The relationship between carried traffic, offered traffic, blocking probability, and number of trunks or available channels can be obtained from trunking theory.

It has been shown that the calls arriving at the network can be modeled using a Poisson process. The holding time or duration of the calls is exponentially distributed. If the number of channels or trunks available is C, the blocking probability, $p(B)$, can be expressed as

$$p(B) = \frac{\left[\dfrac{A^C}{C!}\right]}{\displaystyle\sum_{k=0}^{C} A^k/k!}, \tag{4.35}$$

where A is the offered traffic, given by

$$A = \frac{\Lambda}{\mu} \text{ Erl}. \tag{4.36}$$

The mean rate of call arrival is Λ, and the mean rate at which calls are terminated is μ. The mean duration of the calls is the inverse of μ. The carried traffic, A_c, can now be expressed as

$$A_c = A[1 - p(B)], \tag{4.37}$$

which takes into account the loss of calls due to blocking (Stee 1999). Equation (4.35) can be solved for various values of C and blocking probabilities to get the carried traffic. These are given in Table 4.3.

The efficiency of the channel usage, η, is given by

$$\eta = \frac{A_c}{C} = \frac{A[1-p(B)]}{C}. \tag{4.38}$$

TABLE 4.3   Erlang B Values

Number of channels	Offered traffic			Number of channels	Offered traffic		
	$p=0.005$	$p=0.02$	$p=0.1$		$p=0.005$	$p=0.020$	$p=0.100$
1	0.005	0.020	0.111	37	24.846	28.254	35.572
2	0.105	0.224	0.595	38	25.689	29.166	36.643
3	0.349	0.602	1.271	39	26.534	30.081	37.715
4	0.701	1.092	2.045	40	27.382	30.997	38.787
5	1.132	1.657	2.881	41	28.232	31.916	39.861
6	1.622	2.276	3.758	42	29.085	32.836	40.936
7	2.158	2.935	4.666	43	29.940	33.758	42.011
8	2.730	3.627	5.597	44	30.797	34.682	43.088
9	3.333	4.345	6.546	45	31.656	35.607	44.165
10	3.961	5.084	7.511	46	32.518	36.534	45.243
11	4.610	5.842	8.487	47	33.381	37.462	46.322
12	5.279	6.615	9.474	48	34.246	38.392	47.401
13	5.964	7.402	10.470	49	35.113	39.323	48.481
14	6.663	8.200	11.474	50	35.982	40.255	49.562
15	7.376	9.010	12.484	51	36.852	41.189	50.644
16	8.100	9.828	13.500	52	37.725	42.124	51.726
17	8.834	10.656	14.522	53	38.598	43.060	52.808
18	9.578	11.491	15.548	54	39.474	43.997	53.891
19	10.331	12.333	16.579	55	40.351	44.936	54.975
20	11.092	13.182	17.613	56	41.229	45.875	56.059
21	11.860	14.036	18.651	57	42.109	46.816	57.144
22	12.635	14.896	19.693	58	42.990	47.758	58.229
23	13.416	15.761	20.737	59	43.873	48.700	59.315
24	14.204	16.631	21.784	60	44.757	49.644	60.401
25	14.997	17.505	22.833	61	45.642	50.589	61.488
26	15.795	18.383	23.885	62	46.528	51.534	62.575
27	16.598	19.265	24.939	63	47.416	52.481	63.663
28	17.406	20.150	25.995	64	48.305	53.428	64.750
29	18.218	21.039	27.053	65	49.195	54.376	65.839
30	19.034	21.932	28.113	66	50.086	55.325	66.927
31	19.854	22.827	29.174	67	50.978	56.275	68.016
32	20.678	23.725	30.237	68	51.872	57.226	69.106
33	21.505	24.626	31.301	69	52.766	58.177	70.196
34	22.336	25.529	32.367	70	53.662	59.129	71.286
35	23.169	26.435	33.434	71	54.558	60.082	72.376
36	24.006	27.343	34.503	72	55.455	61.036	73.467

(Continued)

TABLE 4.3 Erlang B Values *(Continued)*

Number of channels	Offered traffic			Number of channels	Offered traffic		
	p=0.005	p=0.02	p=0.1		p=0.005	p=0.020	p=0.100
73	56.354	61.990	74.558	117	96.599	104.493	122.783
74	57.253	62.945	75.649	118	97.526	105.468	123.883
75	58.153	63.900	76.741	119	98.454	106.444	124.983
76	59.054	64.857	77.833	120	99.382	107.419	126.082
77	59.956	65.814	78.925	121	100.310	108.395	127.182
78	60.859	66.771	80.018	122	101.239	109.371	128.282
79	61.763	67.729	81.110	123	102.168	110.348	129.383
80	62.668	68.688	82.203	124	103.099	111.324	130.483
81	63.573	69.647	83.297	125	104.027	112.302	131.585
82	64.479	70.607	84.390	126	104.962	113.280	132.684
83	65.386	71.568	85.484	127	105.891	114.255	133.784
84	66.294	72.529	86.578	128	106.822	115.234	134.886
85	67.202	73.490	87.672	129	107.753	116.213	135.987
86	68.111	74.453	88.767	130	108.684	117.191	137.087
87	69.021	75.415	89.861	131	109.617	118.167	138.189
88	69.932	76.378	90.956	132	110.550	119.147	139.289
89	70.843	77.342	92.051	133	111.482	120.126	140.390
90	71.755	78.306	93.147	134	112.416	121.106	141.492
91	72.668	79.271	94.242	135	113.348	122.084	142.593
92	73.581	80.236	95.338	136	114.284	123.068	143.696
93	74.495	81.201	96.434	137	115.236	124.052	144.796
94	75.410	82.167	97.530	138	116.160	125.026	145.900
95	76.325	83.134	98.626	139	117.091	126.017	147.003
96	77.241	84.100	99.722	140	118.044	127.005	148.105
97	78.157	85.068	100.819	141	118.984	127.968	149.215
98	79.074	86.035	101.916	142	119.963	128.994	150.306
99	79.992	87.004	103.013	143	120.914	129.994	151.444
100	80.910	87.972	104.110	144	121.837	130.920	152.524
101	81.829	88.941	105.207	145	122.746	131.990	153.614
102	82.748	89.910	106.305	146	123.660	132.899	154.720
103	83.668	90.880	107.402	147	124.588	133.865	155.857
104	84.588	91.850	108.500	148	125.557	134.837	156.911
105	85.509	92.821	109.598	149	126.471	135.871	158.025
106	86.431	93.791	110.696	150	127.397	136.836	159.126
107	87.353	94.763	111.794	151	128.316	137.801	160.216
108	88.275	95.734	112.892	152	129.367	138.835	161.341
109	89.198	96.706	113.991	153	130.274	139.811	162.434
110	90.122	97.678	115.089	154	131.207	140.794	163.587
111	91.046	98.651	116.188	155	132.231	141.785	164.675
112	91.970	99.624	117.287	156	133.161	142.856	165.795
113	92.895	100.597	118.386	157	134.058	143.781	166.851
114	93.820	101.571	119.485	158	134.989	144.832	168.216
115	94.746	102.545	120.584	159	136.030	146.544	169.446
116	95.672	103.519	121.684	160	136.563	147.469	170.895

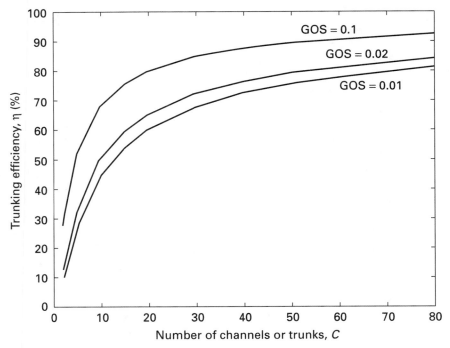

FIGURE 4.21 Trunking efficiency plots.

The efficiency increases as the number of channels, C, increases, as shown in Figure 4.21. This is known as the *trunking effect*.

Let us explore these concepts using a few examples.

EXAMPLE 4.6

If a provider has 50 channels available, how many users can be supported if each user makes an average of four calls/hour, each call lasting an average of 2 minutes? The GOS is 2%.

Answer

$$p(B) = \text{GOS} = 0.02 \,.$$

From Table 4.3 the offered traffic with 50 available channels for a blocking probability of 0.02 is

$$A = 40.255 \text{ Erl} \,.$$

The carried traffic is

$$A_c = A[1 - p(B)] = 40.25 \times 0.98 = 39.445 \text{ Erl} \,.$$

The traffic intensity generated by each user is

$$A_I = \text{calls/hour} \times \text{call duration} = 4 \times 2/60 = 0.1333 \text{ Erl} \,.$$

Therefore, the maximum number of users that can be supported by this provider is

$$\frac{A_c}{A_I} = \frac{39.455}{0.1333} = 296 \,.$$

EXAMPLE 4.7

Two service providers, I and II, are planning to provide cellular service to an urban area. Provider I has 20 cells to cover the whole area, with each cell having 40 channels, and provider II has 30 cells, each with 30 channels. How many users can be supported by the two providers if a GOS of 2% is required? Omnidirectional antennae will be used. Assume that each user makes an average of three calls/hour, each call lasting an average of 3 minutes.

Answer

Provider I:

Number of channels/cell $= 40$

Offered traffic at GOS of 2% $= 30.99$ Erl/cell

Carried traffic $= 30.99 \times 0.98 = 30.4$ Erl/cell

Total traffic carried $= 30.4 \times 20 = 608$ Erl

Traffic intensity/user $= 3 \times 3/60 = 0.15$ Erl

Total number of users $= 608/0.15 = 4054$

Provider II:

Number of channels/cell $= 30$

Offered traffic at GOS of 2% $= 21.99$ Erl/cell

Carried traffic $= 21.99 \times 0.98 = 21.55$ Erl/cell

Total traffic carried $= 21.55 \times 30 = 646.5$ Erl

Traffic intensity/user $= 3 \times 3/60 = 0.15$ Erl

Total number of users $= 646/0.15 = 4310$

4.8 TRUNKING EFFICIENCY OF OMNI VERSUS SECTORIZED ANTENNAE

Now that we have reviewed the concepts of trunking, let us revisit the advantages and disadvantages of the different types of antennae used. We clearly established that the use of 120° sector antennae increases the signal-to-CCI ratio. The use of 60° antennae further increased the signal-to-CCI ratio. We did not examine the trade-off, if any, in going from an omnidirectional antenna to a sector antenna (Chan 1992). Let us compare the performance of these two structures using the traffic capacity that can be attained.

Consider a seven-cell pattern with 56 channels/sector. For an omnidirectional antenna system, this means the availability of all 56 channels in one sector. If we are using a 120° sector antenna system, we have $56/3 = 19$ channels/sector, and for a 60° sector antenna system, we have only $56/6 = 9$ channels/sector.

Omni

Number of channels $= 56$

Offered traffic at GOS of 2% $= 45.87$ Erl

Carried traffic $= 45.87 \times 0.98 = 44.95$ Erl

Trunking efficiency $= 44.95/56 = 80.3\%$

120° Sector

Number of channels/sector = 19

Offered traffic at GOS of 2% = 12.34 Erl/sector

Carried traffic = 12.34 × 0.98 = 12.09 Erl/sector

Carried traffic at this site = 12.09 × 3 = 36.28 Erl

Trunking efficiency = 36.28/56 = 64.8%

60° Sector

Number of channels/sector = 9

Offered traffic at GOS of 2% = 4.35 Erl/sector

Carried traffic = 4.35 × 0.98 = 4.26 Erl/sector

Carried traffic at this site = 4.26 × 3 = 25.58 Erl

Trunking efficiency = 25.58/56 = 45.7%

As we can see, the trunking efficiency of a sectorized antenna system goes down as the number of elements in the sector increases. Even though the signal-to-CCI ratio, S/I, of the antenna goes up as the number of elements goes up,

$$\frac{S}{I} = 18.7 \text{ dB} \quad \text{(omni)}$$

$$\frac{S}{I} = 23.4 \text{ dB} \quad (120° \text{ sector})$$

$$\frac{S}{I} = 26.4 \text{ dB} \quad (60° \text{ sector})$$

the trunking efficiencies are 80.3%, 64.8%, and 45.7%, respectively.

4.9 ADJACENT CHANNEL INTERFERENCE

A discussion of cellular systems would not be complete without a mention of adjacent channel interference (ACI). ACI is caused primarily by inadequate filtering and non-linearity of the amplifiers (Fehe 1995, Samp 1998, Malm 1997, El-Sa 1996, El-Ta 1998). Inadequate filtering arises from the lack of a "brick wall" filter that keeps the spectrum in any channel limited to that channel. The effect of inadequate filtering is shown in Figure 4.22. The energy in the shaded regions corresponding to adjacent channels will cause interference. In most cases, it is sufficient to take under consideration only the interference coming from the two channels on either side of the primary channel. The shaded regions represent the energy contributions to the channel of frequency f_{c2} coming from the two nearby channels, f_{c1} and f_{c3}.

There will be signal intrusions from f_{c4}, f_{c5}, and other channels farther from the channel of interest, f_{c2}. However, the contributions from these channels decrease the farther these channels are from the channel of interest. A comparison of the ACI produced by near neighbors and far neighbors due to inadequate filtering is demonstrated in Figure 4.23.

If we compare ACI (Figure 4.23) with CCI, shown in Figure 4.24, it is clear that ACI is typically attenuated by the receiver filter, while CCI is unaffected by the

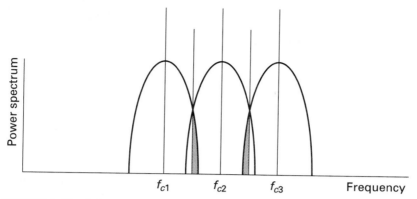

FIGURE 4.22 Adjacent channel interference (shaded regions).

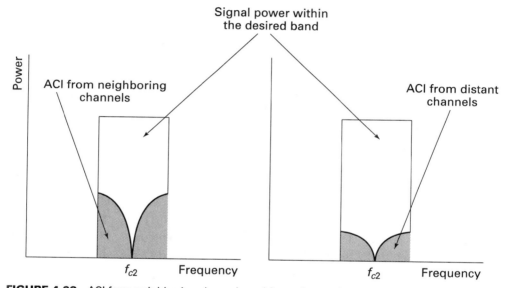

FIGURE 4.23 ACI from neighboring channels and from distant channels.

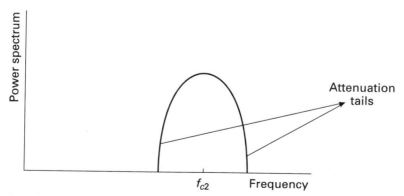

FIGURE 4.24 Characteristics of the receiving filter in the channel of interest (f_{c2}).

receiving filter attenuation characteristics since the ACI components are falling on the "tail ends" of the filter.

Based on Figure 4.23, a measure of ACI (the ACI ratio) can be expressed as (Gole 1994, Samp 1993, Malm 1997)

$$\text{ACI ratio} = \frac{\int_{-\infty}^{\infty} G(f)|H_B(f-\Delta f)|^2 df}{\int_{-\infty}^{\infty} G(f)|H_B(f)|^2 df}, \tag{4.39}$$

where $H_B(f)$ is the transfer function of the bandpass filter and Δf is the channel separation. The power spectral density of the signal is given by $G(f)$.

One of the ways ACI can be reduced is through an increase in the channel separation. It can also be reduced by an appropriate channel allocation scheme and cluster as well as selection of cell sizes. ACI is also a function of its origin; i.e., if the adjacent channel is from the group of frequencies (including the signal) assigned to one base station, the ACI undergoes the same path-dependent attenuation as the signal component, while if the ACI originates from a channel from a different base station, it undergoes a higher attenuation due to the distance compared with the signal from its base station. The latter case is illustrated in Figure 4.25, where distance $d_{\Delta f} > d_f$.

The overall performance of the cellular systems is determined by the total signal-to-interference ratio, $(S/I)_\text{tot}$, given by

$$\left(\frac{S}{I}\right)_\text{tot} = \frac{S}{\text{CCI} + \text{ACI}}. \tag{4.40}$$

It is possible to develop a strategy based on channel separation, cell size, cluster size, and antenna directivity to overcome the problems caused by CCI and ACI (Samp 1993, Malm 1997). The effects of ACI can also be reduced by using advanced signal-processing techniques that employ equalizers (Pete 1992).

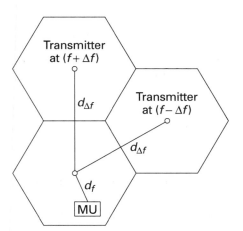

FIGURE 4.25 The MU is transmitting at a frequency of f while the ACI signals are at frequencies of $(f + \Delta f)$ and $(f - \Delta f)$. Note that ACI signals are coming from cells that are far away compared with the signal of interest, which suffers low path-dependent loss since it is closer to its own base station.

EXAMPLE 4.8

In a cellular system operating at a signal power of -85 dBm, CCI of -110 dBm is observed. If an overall signal-to-interference ratio of 20 dB is required, how much ACI can the system tolerate?

Answer

$$\text{Signal power} = 10^{-8.5} \text{ mW}$$

$$\text{CCI} = 10^{-11} \text{ mW}$$

$$\left(\frac{S}{I}\right)_{\text{tot}} = 20 \text{ dB or } 100$$

$$100 = 10^{-8.5} / (\text{ACI} + 10^{-11})$$

$$\text{ACI} = 2.16 \times 10^{-11} \text{ mW or } -106.65 \text{ dBm.} \qquad \blacksquare$$

4.10 SUMMARY

The concept of cells for increasing the capacity of wireless systems was described. The limitation on the performance of cells due to CCI was discussed. Performance improvement can be achieved through the use of sector antennae.

- A hexagonal structure is the optimal cell shape.
- The distance D between co-channel cells is $\sqrt{3N_c}$, where N_c is the number of cells in a pattern given by $\sqrt{i^2 + j^2 + ij}$, where i, j take values 0, 1, 2, 3, ...
- The signal-to-CCI ratio S/I improves as the number of cells in the pattern goes up.
- The co-channel interference reduction factor or frequency reuse factor q is given by D/R, where R is the radius of the cell. Higher values of q result in lower values of interference.
- For any pattern, the number of interfering cells is six.
- S/I is also dependent on the loss factor ν.
- An acceptable value of S/I is about 18 dB or more.
- S/I can be improved by using sector antennae.
- S/I for a 60° sector antenna $> S/I$ for a 120° sector antenna $> S/I$ for an omnidirectional antenna.
- Sector antennae reduce interference since the number of interfering cells goes down as the number of sectors goes up.
- Capacity can be increased through cell splitting.
- When cells are split to increase the coverage, the transmitted power must be reduced to keep the CCI low.
- The complexity of a microcellular system can be reduced through the use of optical fibers.
- The coverage area can be estimated by calculating the attenuation and the power margin needed to take lognormal fading into account.
- GOS is a measure of the ability of a user to gain access to a channel during the busiest period.

- Traffic intensity is expressed in Erlangs. It is given by the product of the number of calls/hour and the duration of the call (hours). For example, if a person makes an average of four calls/hour and the average duration of the call is 3 minutes, the traffic intensity generated is $4(3/60) = 0.2$ Erl.
- If the number of channels or trunks available is C, the blocking probability, $p(B)$, can be expressed as

$$p(B) = \frac{\left[\dfrac{A^C}{C!} \right]}{\displaystyle\sum_{k=0}^{C} A^k / k!} ,$$

where A is the offered traffic in Erl.
- The carried traffic A_c is given by $A[1 - p(B)]$ Erl.
- Even though co-channel interference is reduced through the use of sector antennae, the trunking efficiency is also reduced.
- Additional interference may arise from adjacent channels. ACI must be taken into account to calculate the overall interference.

PROBLEMS

1. In a cellular communication system, the signal power received is −100 dBm. The structure is a seven-cell pattern. If the noise power is −119 dBm and each of the interfering signals is −121 dBm, calculate the overall signal-to-noise ratio. Also calculate the signal-to-CCI ratio.

2. In a cellular communication system, the signal power received is −97 dBm. The structure is a seven-cell pattern. The noise power is −117 dBm and each of the interfering signals is −120 dBm.

(a) Calculate the overall signal-to-noise ratio.

(b) Calculate the signal-to-CCI ratio.

(c) If a 20 dB signal-to-CCI ratio is required, what should be the power of the signal from each of the interfering cells?

3. If the received power at a distance of 2 km is equal to 2 μW, find the received powers at 3 km, 6 km, and 15 km for a path loss exponent ν of 3.8.

4. If a mobile unit is 4 km away from the base station and 10 km away from the CCI site, find the signal-to-CCI ratio. Assume the path loss exponent ν to be 4.0.

5. If a signal-to-CCI ratio has to be 20 dB and the mobile unit is 8 km away from its own base station, how far away should the CCI site be located? Assume the path loss exponent ν to be 3.2.

6. A mobile unit receives a power of −105 dBm. If the predicted loss is 115 dB, what was the transmitted power?

7. For acceptable performance, the signal-to-CCI ratio must be at least 20 dB. What must be the value of D/R? Assume ν to be equal to 3.0.

8. Consider the case of a Rayleigh-faded signal received by a mobile unit (MU). The threshold SNR for acceptable performance is 15 dB. Assume that the average power being received is −96 dBm. The noise power is −115 dBm. Calculate the outage probability. If the acceptable outage is only 2%, what must the noise power be to maintain acceptable performance?

9. Explain the term *Erlang*.

10. If a provider is planning to have 20 voice channels per cell for a grade of service (GOS) of 2%, calculate the offered traffic and carried traffic. If the provider has 10 cells, what is the total carried traffic in Erlangs?

11. A provider has 75 channels in a cell. If the blocking probability is 2%, what is the offered traffic? What is the carried traffic? If each customer uses (holding time) 2 minutes and makes an average of two calls/hour, how many customers can be served?

12. Two service providers, A and B, provide cellular service in an area. Provider A has 100 cells with 20 channels/cell, and B has 35 cells with 54 channels/cell. Find the number of users that can be supported by each provider at 2% blocking if each user averages two calls/hour at an average call duration of 3 minutes.

13. A provider expects to provide coverage for about 700 users/cell. The acceptable GOS is 2%. If the user has an average holding time of 3 minutes/call and makes an average of three calls/hour, how many channels are required?

14. For the case of lognormal fading, given in eq. 2.67, use MATLAB to calculate the outage. Assume that the average power being received is −100 dBm. Standard deviation of fading = 6 dB. Threshold power = −105 dBm. Repeat the calculations for standard deviations of 8, 10, and 12 dB. Generate either lognormal or normal random numbers (at least 5000) to compute the outage probability.

15. Instead of using the random number generator from MATLAB, use the analytical approach to calculate outage probability. Compare your results to the values calculated in Problem 14.

16. Use MATLAB to generate 1000 Rayleigh-distributed random variables such that the average signal power is 0 dBm. If the threshold is −3 dBm, compute the outage probability. Repeat the simulation for a threshold of −5 dBm.

17. Use the analytical approach to calculate the outage for the cases in Problem 16, and compare your results.

18. The transmit power of a base station is 2 W. If the coverage is to be split so that minicells (one-third size) can be created to accommodate additional users in the geographical region, what must be the transmit power (of the minicell) to keep CCI at the same level as that of the unsplit cell? Assume that $\nu = 3.0$.

19. To keep an acceptable level of performance, a service provider is planning to maintain an outage probability of 3%. What should be the threshold power? The standard deviation of fading (lognormal) has been found to be 8 dB. The average received power is −95 dBm.

20. Based on the Hata model for power loss, it has been determined that there is approximately 0.7 dB/km loss in the range of 10–18 km from the transmitter. The long-term fading margin is 6 dB and the short-term fading margin is 4 dB. Calculate the reduction in transmission distance when

(a) Only the long-term fading margin is factored in.

(b) Only the short-term fading margin is factored in.

(c) Both fading margins are taken into account.

21. Use MATLAB to plot the outage probability at the edge of a cell as a function of the power margin.

22. Use MATLAB to plot the area outage probability as a function of the power margin. Comment on the relative variation of the outage probabilities (for this problem and Problem 21) for the two power margin values 6 dB and 10 dB.

23. Use MATLAB to calculate the maximum distance using your calculations from Problem 21 for two outage probabilities, 0.1 and 0.05. Assume that the standard deviation of fading (lognormal) is 6 dB. The path loss exponent is 3.5. The received power at a distance of 4 km from the base station is −50 dBm. The threshold level for performance is −91 dBm.

24. Consider a seven-cell pattern that uses an omnidirectional antenna with 54 channels/sector. Calculate the trunking efficiency if

(a) A 60° sector antenna is used.

(b) A 120° sector antenna is used.

25. Explain the term *adjacent channel interference*. If the signal power received is −90 dBm, the CCI power/channel is −120 dBm, and the ACI is −110 dBm, calculate the overall signal-to-noise ratio. Note that thermal noise has been neglected.

26. Outage probabilities have been calculated for Rayleigh channels assuming that the co-channel power is constant. However, the co-channel signal also undergoes Rayleigh fading, and this must be taken into account. If our requirement for acceptable co-channel interference is that the ratio of the signal power to the co-channel signal power be α dB, derive an expression for the outage probability arising from co-channel interference.

27. Use MATLAB to obtain a plot of outage probability for various values of the "protection ratio" α.

28. Repeat Problem 26 if the desired signal and the CCI are subject to lognormal fading. Assume that the standard deviations of fading are same for both the desired and CCI signals.

29. Use MATLAB to obtain plots of the outage probability as a function of the "protection ratio" α for different values of the standard deviation of fading.

30. Using a random number generator, repeat Problem 26.

31. Repeat Problem 28 using a random number generator.

CHAPTER 5

FADING MITIGATION IN WIRELESS SYSTEMS

5.0 INTRODUCTION

One of the major consequences of fading is the random nature of the wireless signal. Random fluctuations in the received signal power arise primarily from the Rayleigh fading experienced by the signal due to the multipath effect (Pric 1958, Beck 1962, Casa 1990, Stei 1987, Mons 1980, Akki 1994, Ho 1999). Shadowing or lognormal fading (Brau 1991, Hans 1977, Suzu 1977) also contributes to the fluctuations in the received signal. The other consequence of fading is intersymbol interference (ISI) caused by the frequency selectivity (time dispersion) of the channel (Bell 1963a,b). This time dispersion results in the effective channel bandwidth being smaller than the information or message bandwidth (Bell 1963a; Cox 1972, 1975; Bult 1983; Gane 1989a,b; Mela 1986; Rapp 1996b). The problems caused by ISI can be controlled through the use of equalizers (Adac 1988a, Schw 1996, Samp 1997). The equalizer compensates for the various differential delays of the multipath, reducing the effective pulse spread. This leads to an increase in the effective channel bandwidth, reducing or even eliminating the effects of intersymbol interference. Note that the equalizers must be adaptive, since wireless channels are essentially unknown and time varying. We will look at equalizers later in this chapter and concentrate initially on mitigating the effects of fading.

Randomness in the received signal arises from the multipath effect. As seen in Figure 2.5, Rayleigh fading is of a short-term nature while lognormal fading is of a long-term nature. This means that any approaches to mitigate fading must take the short-term (or long-term) nature of fading into account. The randomness of the received signal can be compensated for through diversity techniques that make use of a number receiving antennae instead of a single one. Since Rayleigh fading is of a short-term nature, these receivers can be fairly close; the diversity techniques (Wint 1984, Bren 1959, Gilb 1969, Lee 1971a, Jake 1974, Turk 1990) are commonly referred to as "microscopic." The techniques employed to compensate for long-term fading are referred to as "macroscopic." The terms *microscopic* and *macroscopic* (Bern 1987) are distinguished on the basis of how close the receivers are to each other. We will initially look at the microscopic diversity techniques employed to mitigate the effects of Rayleigh fading.

5.1 EFFECTS OF FADING AND THE CONCEPT OF DIVERSITY

Even though we examined the principle of fading earlier, we need to understand how fading adversely impacts wireless communication systems. Consider the case of signal transmission through a Rayleigh channel. The received signal, $r(t)$, can be expressed as

$$r(t) = A\exp[-j\theta(t)]s(t) + n(t), \qquad (5.1)$$

where $s(t)$ is the transmitted signal; A and $\theta(t)$ are the random gain and random phase, respectively, associated with Rayleigh fading; and $n(t)$ is additive white Gaussian noise (Schw 1996). The effect of fading, therefore, is to create a multiplicative noise term, A, in addition to the additive white Gaussian noise. This multiplicative noise (Schw 1996) will cause problems in bit detection. Rewriting eq. (5.1), the output, $z(T)$, at the sampling instant after the matched filter is

$$z(T) = A\sqrt{E} + n, \qquad (5.2)$$

where E is the bit energy, A is a scaling factor that is Rayleigh distributed, and n is the white Gaussian noise. We can view the first term as the signal and the second term as noise. It is clear that the signal power is now a random variable. Therefore, for a fixed threshold set on the basis of a deterministic signal in the presence of additive white Gaussian noise, we will not be able to correctly identify the bits since the power/energy may go up or down at any given sampling instant. In other words, the signal-to-noise ratio (SNR) is a random variable, and the bit error rates given in Chapter 3 will be conditioned on the random variable A. If we assume that the fading is slow, the average bit error rate in the presence of fading, $p_{fad}(e)$, can be expressed as

$$p_{fad}(e) = \int_0^\infty p(e/a)f(a)\,da, \qquad (5.3)$$

where $p(e/a)$ is the probability of error conditioned on the envelope and $f(a)$ is the Rayleigh density. The error probability under flat fading conditions can be easily evaluated by transforming the random variable A to γ,

$$\gamma = \frac{A^2 E}{N_0}, \qquad (5.4)$$

with a probability density function $p(\gamma)$ given by

$$(\gamma) = \frac{1}{\gamma_0}\exp\left(-\frac{\sqrt{\gamma}}{\gamma_0}\right), \qquad \gamma \geq 0, \qquad (5.5)$$

where γ_0 is the average signal-to-nose ratio under fading conditions, given by

$$\gamma_0 = \frac{E}{N_0}\langle A^2\rangle. \qquad (5.6)$$

The quantity N_0 is the noise power, discussed in Section 3.2. Using eqs. (5.3) and (5.5) and the expressions for error probability from Chapter 3, the error probability in the presence of fading and in the absence of fading can be expressed in terms of SNR, the average signal-to-noise ratio, as

$$\left.\begin{array}{l} p(e) = \dfrac{1}{2}\mathrm{erfc}\left(\sqrt{\gamma_0}\right) \\[1.5em] p_{fad}(e) = \dfrac{1}{2}\left[1 - \sqrt{\dfrac{\gamma_0}{1+\gamma_0}}\right] \end{array}\right\} \quad \text{coherent BPSK} \qquad (5.7a)$$

$$\left. \begin{array}{l} p(e) = \dfrac{1}{2}\exp(-\gamma_0) \\[3mm] p_{fad}(e) = \dfrac{1}{2}\left[\dfrac{1}{1+\gamma_0}\right] \end{array} \right\} \quad \text{DPSK} \qquad (5.7b)$$

$$\left. \begin{array}{l} p(e) = \text{erfc}\sqrt{\varepsilon\gamma_0} \\[3mm] p_{fad}(e) = \dfrac{1}{2}\left[1 - \sqrt{\dfrac{\varepsilon\gamma_0}{1+\varepsilon\gamma_0}}\right] \end{array} \right\} \quad \text{coherent GMSK,} \qquad (5.7c)$$

where ε is a factor determined by the bandwidth of the Gaussian low-pass filter (eq. (3.136) and γ_0 is the average signal-to-noise ratio.

The error probability plots for some modulation formats are shown in Figures 5.1, 5.2, and 5.3. It is clear that for a given SNR (E/N_0), the bit error rate in a faded channel is much higher than the error rate in a fading-free channel. This slower decline of the probability of error versus SNR is a direct consequence of the multiplicative noise. For a fixed bit error rate, this translates into margin of a few dB for the faded channel, as shown in Figure 5.1 for the case of BPSK. For an error probability of 10^{-3}, the fading margin $M \sim 17$ dB. This phenomenon may be explained as follows. The signal-to-noise ratio in a faded channel can be expressed as

$$\left(\frac{S}{N}\right)_{fad} = \frac{S}{f(s)+f(N)}, \qquad (5.8)$$

where the first term in the denominator, $f(s)$, is due to the fading and the second term, $f(N)$, is the system noise. The signal is denoted by S. Note that the first term in the denominator depends on the signal. At low values of the signal, the signal-to-noise ratio in fading, $(S/N)_{fad}$, goes up as the signal power goes up since the signal-dependent noise term is negligible. However, at higher values of signal power, the signal-dependent noise term $f(s)$ starts going up, thus reducing $(S/N)_{fad}$. As the signal-dependent term overtakes the signal-independent noise term, $f(N)$, any further increase in signal power will have less and less impact on the effective $(S/N)_{fad}$ as well as on the probability of error. In other words, in a faded channel, the probability of error cannot be reduced at the same rate as in the nonfaded case by increasing the signal power.

EXAMPLE 5.1

For a coherent BPSK receiver, what is the fading margin required to maintain a bit error rate of 10^{-3}?

Answer From eq. (5.7a), the signal-to-noise ratio γ_0 required to maintain a bit error rate of 10^{-3} for a Gaussian channel (no fading) is obtained by solving

$$p(e) = 10^{-3} = \frac{1}{2}\,\text{erfc}(\sqrt{\gamma_0}).$$

Inverting this equation, we get $\gamma_0 = 6.79$ dB .

For a Rayleigh channel, the signal-to-noise ratio is obtained by solving

$$p_{\text{fad}}(e) = 10^{-3} = \frac{1}{2}\left[1 - \sqrt{\frac{\gamma_0}{1 + \gamma_0}}\right].$$

Inverting this equation, we get $\gamma_0 = 25$ dB. The fading margin is $25 - 6.79 = 18.21$ dB. ∎

One of the ways in which the effects of fading can be reduced is through "diversity" techniques (Jake 1971; Lee 1971a,b; Vaug 1988). Diversity techniques involve the creation of multiple independent versions of the received signal and combining them. Consider, for example, that instead of one receiving antenna, there are M different antennae receiving the same signal. If we assume that the locations of these antennae are such that the signals being received can be considered to be independent of one another, we indeed have created M different versions of the signal. The signal envelopes of these M channels will all be identically distributed Rayleigh-faded signals. This leads to the following scenario. If one examines the signals being received in these M diverse receivers, the likelihood of all of them being severely faded (very weak signals) is very small. For example, if the envelope is Rayleigh distributed, the power will be exponential. Under this condition, the probability that the signal power in any one of the receivers is below the threshold, ζ_T, can be obtained as follows.

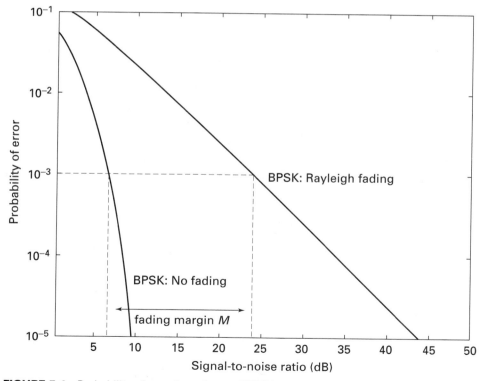

FIGURE 5.1 Probability of error for coherent BPSK in the presence of Rayleigh fading and in the absence of fading.

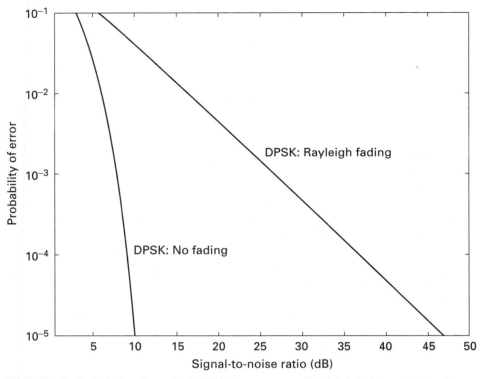

FIGURE 5.2 Probability of error for DPSK in the presence of Rayleigh fading and in the absence of fading.

$$P(\zeta_i \leq \zeta_T) = \int_0^{\zeta_T} p(\zeta_i)\, d\zeta_i = \int_0^{\zeta_T} \frac{1}{\zeta_0} \exp\left(-\frac{\zeta_i}{\zeta_0}\right) d\zeta_i = 1 - \exp\left(-\frac{\zeta_T}{\zeta_0}\right), \quad (5.9)$$

where $p(\zeta_i)$ is the exponential density function of the power ζ_i, and ζ_0 is the average power. The probability, $P_M(\zeta_T)$, that all the receivers simultaneously have signal power less than ζ_T is

$$P_M(\zeta_T) = \left[1 - \exp\left(-\frac{\zeta_T}{\zeta_0}\right)\right]^M. \qquad (5.10)$$

The probability that at least one of the receivers has a signal power above the threshold, ζ_T, is

$$P\{\text{at least one receiver has power} \geq \zeta_T\} = 1 - P_M(\zeta_T). \qquad (5.11)$$

Equation (5.11) is plotted in Figure 5.4 for two values of ζ_T / ζ_0. As the number of diverse channels increases, the probability that at least one receiver has power $> \zeta_T$ also increases. Thus, as M increases, the likelihood of performance degradation becomes lower and lower.

The use of different receivers at various locations is not the only form of diversity available. A number of different ways exist for the engineer to combine the signals from these diverse "branches" to produce the final signal. Descriptions of various diversity techniques and the processing methods used in combining these diverse components are given in Section 5.3.

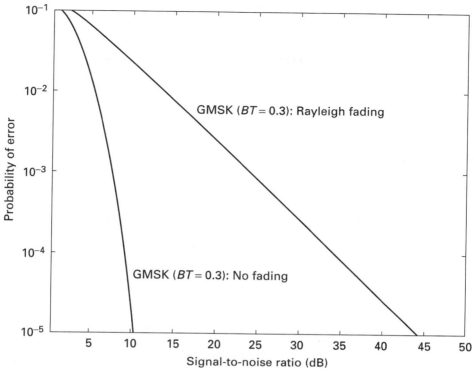

FIGURE 5.3 Probability of error for coherent GMSK in the presence of Rayleigh fading and in the absence of fading.

EXAMPLE 5.2

The average signal-to-noise ratio at the receiver is 14 dB. The receiver goes into outage when the signal-to-noise ratio goes below 10 dB. What is the probability of outage in a Rayleigh channel?

Answer   The outage probability is given by eq. (5.9). Note that 14 dB corresponds to a signal-to-noise ratio in absolute units of 25.1. The threshold signal-to-noise ratio is 10 dB, or 10 in absolute units. Outage $= [1 - \exp(-10/25.1)] = 0.3286$.   ■

EXAMPLE 5.3

In a five-channel diversity receiver scheme, what is the probability of outage if all channels have the same characteristics as in Example 5.2? Assume that the system goes into outage if all channels simultaneously have a signal-to-noise ratio below 10 dB.

Answer   Using eq. (5.10), the outage probability is $[1 - \exp(-10/25.1)]^5 = 0.0038$.   ■

5.2   OTHER SIGNAL DEGRADATION EFFECTS

Rayleigh fading is not the only signal degradation effect imposed by the channel. The relative motion of the MU with respect to the base station introduces random phase changes in the received signal. In addition, as described in Chapter 2, the relative difference between the channel bandwidth and the message bandwidth makes the fading

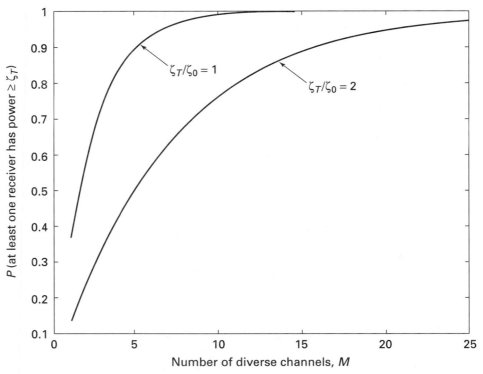

FIGURE 5.4 The number of diverse channels M is plotted against the probability that at least one channel has a power greater than ζ_T.

frequency selective, introducing intersymbol interference. We will briefly review these topics before we explore ways of mitigating the effects of fading.

5.2.1 Effects of Random Frequency Modulation

The preceding discussion is based on the assumption that the only effect of fading is a fluctuation in the power levels due to a Rayleigh-distributed envelope. However, the mobile channels also exhibit fluctuations in the frequency/phase from the motion of the vehicle, or the Doppler effect. In eq. (2.51), the presence of $\cos(\theta_i)$ in the argument of the sine and cosine terms is similar to the frequency or phase modulation seen in eq. (3.20) or (3.24), clearly showing the effect of random FM resulting from the random nature of the phase θ_i. The difficulty with the presence of random fluctuations in the frequency/phase is the fact that any increase in the transmitted power and, hence, in the received power will not help reduce the bit error rates (Adac 1989, 1993; Davi 1971; Jake 1974; Fung 1986; Liu 1991b). Indeed, the error rates reach a low "floor value" as the power increases, and beyond a certain value, any further increase in power has no effect on the bit error rates. Such error floors (Korn 1991) are characteristic of systems suffering random FM, as we have seen in Chapter 3 (Figure 3.30).

The exact error probability equations for all the modulation formats of interest are difficult to express in closed form. However, we will summarize the effects of random FM in the next section along with the effects of frequency-selective fading.

5.2.2 Effects of Frequency-Selective Fading and Co-Channel Interference

As discussed in Chapter 2, frequency-selective fading arises when the coherent bandwidth of the channel is less than the message bandwidth. The error rates vary with the form of modulation and demodulation used. The performance of the modems also depends on the ratio σ_d/T, where σ_d is the rms delay spread (eq. (2.42)) and T is the symbol period. The exact equations governing the error probability, taking frequency- selective fading into account, are again very complex and are beyond the scope of this book. Numerical results are available in a number of research papers.

We will try to explore the effects using the results for the differential detector for MSK. We can express the error probability in terms of the average signal-to noise-ratio (γ_0), average signal-to-CCI ratio (γ_{CCI}), and the Doppler parameter $f_d T$, where f_d is the maximum Doppler shift and T is the symbol period. The average error probability, $p_{av}(e)$, for MSK with differential detection has been shown (Hira 1979b, Akai 1998) to be

$$p_{av}(e) = \frac{1}{2}\left[1 - \frac{J_0(2\pi f_d T)\gamma_0\gamma_{CCI}}{\sqrt{(\gamma_0\gamma_{CCI} + \gamma_0 + \gamma_{CCI})^2 - \left(\frac{\gamma_0}{\pi}\right)^2 J_0^2(2\pi f_d T)}}\right]. \qquad (5.12)$$

The average error probability is plotted in Figure 5.5 for the case where there is no motion (Doppler fading absent, $f_d = 0$). At a low value of signal-to-CCI ratio (10 dB), the average error is quite high and stays high around 0.1. Even when the signal-to-CCI ratio is 60 dB, the effects of error floor are seen, pointing to the fact that any additional increase in the signal power will have no effect on the performance of the system. Note that we still have not included the effects of frequency-selective fading, which will degrade the performance still further.

The average error probability in the absence of CCI (γ_{CCI} = infinite) can be obtained from eq. (5.12) as

$$p_{av}(e)_{\gamma_{CCI} \to \infty} = \frac{1 + \gamma_0[1 - J_0(2\pi f_d T)]}{2(1 + \gamma_0)}. \qquad (5.13)$$

The average error probability under conditions of no CCI is plotted in Figure 5.6. When the speed of the mobile unit goes up (higher values of f_d), the average error rates go up and appear to converge to an "error floor" value, which changes only marginally as the signal-to-noise ratio goes up. The performance of the system mirrors the case shown in Figure 5.5 in terms of the error floor.

The error floor, $p_{fc}(e)$, for the presence of CCI can be obtained from eq. (5.12) by letting the average signal-to-noise ratio γ_0 approach ∞ while $f_d = 0$ (no Doppler fading). It can be shown that the error floor for CCI becomes

$$p_{fc}(e) = \frac{1}{2}\left[1 - \frac{\gamma_{CCI}}{\sqrt{(\gamma_{CCI} + 1)^2 - \left(\frac{1}{\pi}\right)^2}}\right]. \qquad (5.14)$$

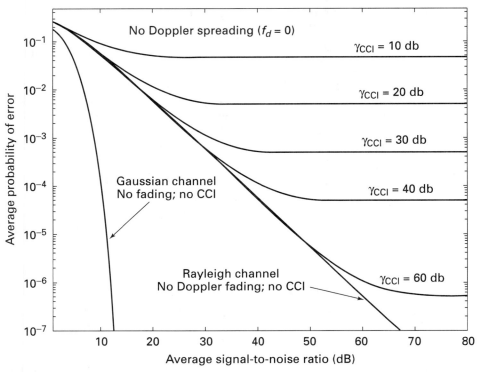

FIGURE 5.5 Error floor behavior in the presence of CCI fading for the case of MSK with a differential detector.

Similarly, the error floor $p_{fd}(e)$ for the case of Doppler fading can be obtained from eq. (5.13) by letting both the average signal-to-noise ratio, γ_0, and the signal-to-CCI ratio, γ_{CCI}, go to ∞. This error floor is

$$p_{fd}(e) = \frac{1}{2}[1 - J_0(2\pi f_d T)] . \tag{5.15}$$

The error floor when CCI is present is shown in Figure 5.7. It is clear that at what may be considered an acceptable value of signal-to-CCI ratio of 20 dB, in the absence of Rayleigh fading, the error is still around 0.004, demonstrating the need to keep the CCI as low as possible under fading conditions.

The error floor under the conditions of a mobile unit in motion is shown in Figure 5.8. The effect of the Doppler shift on the error is clearly seen. The error is about 4×10^{-4} when the Doppler fading term $f_d T = 0.01$.

Note that even though the example of MSK was chosen, the performance of MSK with a differential detector is identical to the case of DPSK. As indicated earlier, we can certainly draw conclusions from these results regarding the general behavior of the error probability in the presence of Doppler fading and CCI. The differences among the various modulation formats will be in the values of the error floor and at what values of the signal-to-noise ratio they are reached. The general behavior of the error rates will be similar.

When the channel is fast and frequency selective, the behavior of the error floor will be similar to that seen in Figures 5.5 and 5.6. Note that the channel is flat if σ_d

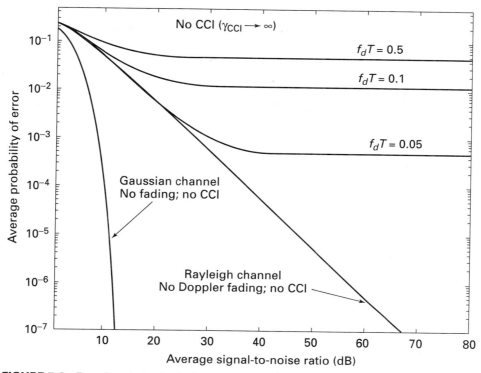

FIGURE 5.6 Error floor behavior in the presence of Doppler fading for the case of MSK with a differential detector.

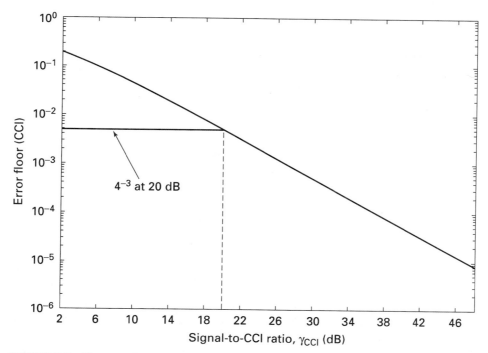

FIGURE 5.7 The error floor plotted as a function of signal-to-CCI ratio.

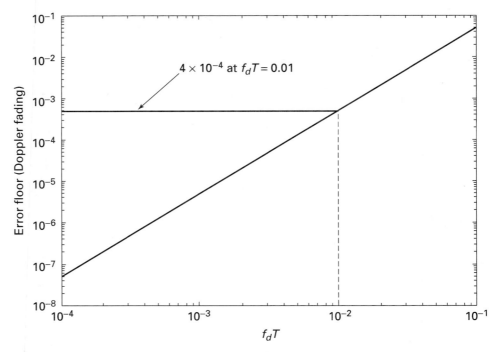

FIGURE 5.8 The error floor plotted as a function of Doppler fading parameter $f_d T$.

is zero. The results are extensively discussed in a number of research papers and books (Fung 1986, Liu 1991b, Fech 1993, Smit 1994, Fehe 1995).

We will now look at ways of overcoming the problems associated with fading using various diversity schemes.

5.3 FORMS OF DIVERSITY

Multiple signals can be created in the following ways (Pier 1960, Gilb 1969, Schw 1996):

Space diversity (antenna locations physically separated)

Angle diversity (different angles of arrival)

Frequency diversity (different frequencies/frequency bands)

Polarization diversity (different polarizations)

Time diversity

Multipath diversity

All these forms of diversity can produce multiple signals depending on a number of factors, such as available space, frequency, angle, and time, except for polarization diversity, where M is limited to 2. The trade-off in the improvement in performance between M and available space, frequency, angle, etc. will be discussed later. The multiple signals can be combined and processed in a number of ways in addition to picking the strongest of the M signals. These methods are

Selection combining

Maximal ratio combining

Equal gain combining.

We will examine the different forms of diversity before we look at the various methods of diversity combining.

5.3.1 Space or Spatial Diversity

Spatial diversity takes advantage of the random nature of propagation in different directions. Consider, for example, the transmitted signal propagating through a multipath scattering environment as shown in Figure 5.9a. Two antenna sites, separated by a distance D, are shown at the receiving location (Schw 1996, Maki 1967, Lee 1971b). It is possible for the signals arriving at the two antenna sites to be independent or at least uncorrelated, depending on the separation between them. This possibility arises from the very nature of multipath fading, specifically the fact that many independent paths exist

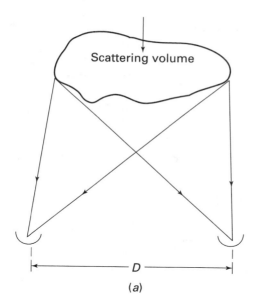

(a)

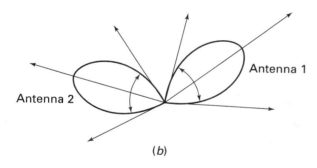

(b)

FIGURE 5.9 (a) Spatial diversity. D is the antenna separation. (b) Angular diversity. Antennae 1 and 2 are two antennae with narrow beam patterns.

at any location, and therefore it is possible for the paths coming to one receiver to be independent of those to the other. It is obvious that if the separation is reduced, we will reach a point where there is a significant overlap between the groups of paths arriving at the two sites. One can now visualize locating a number of such antennae at the receiving site, each antenna separated from its neighbor by the minimum distance necessary to make the signals arriving at the particular antenna independent of (or at least uncorrelated with) the signals arriving at the other antenna sites. Independence of the signals from the different diversity branches is essential for the improvement in performance from multiple observations expressed in eq. (5.11). Note that if the radio frequency components are uncorrelated and Gaussian, they are also independent. Thus, in the case of Rayleigh fading, the requirement of zero correlation (uncorrelated) is sufficient since the Rayleigh density function results from two Gaussian random variables (Appendix A).

5.3.2 Angle Diversity

Angular diversity makes use of antennae having directional properties. Each antenna responds to a received signal propagating in a specific direction or angle, and this faded signal will be uncorrelated with (preferably independent from) other signal components being received by the other directional antennae (Peri 1998, 1999). A schematic of this arrangement is shown in Figure 5.9*b*. A mobile unit is shown with two antennae having narrow beam widths, positioned in different angular directions.

In spatial diversity, the two antennae are separated horizontally and pointed in the same direction as shown in Figure 5.9*a*; in angular diversity, the antennae are co-located, collecting signals coming from different angular directions. Whereas space diversity is implemented at the base station, angular diversity may be implemented at the base station or at the mobile unit (Lee 1997, Kuch 1999, Peri 1999). Improvement in performance is realized when the signals from the two antennae are uncorrelated.

5.3.3 Frequency Diversity

Rather than using several antennae separated by a specific distance, we can transmit the same information at different carrier frequencies. The resulting diversity is referred to as frequency diversity. The carrier frequencies must be separated enough so that the signals corresponding to the different carrier frequencies are at least uncorrelated, and preferably independent. Under these conditions, the signals at these carrier frequencies will result in independently fading signals that can be combined using any of the techniques described later in this chapter. The limitation of the technique is the availability of bandwidth to allow a number of different frequencies and the ability of the receiver to pick up these diverse signals without the need for multiple receivers tuned to different frequencies. In some implementations of cdma2000/IMT 2000, frequency diversity is used to reduce the effects of fading (Knis 1998a,b; Rao 1999). This has been made possible through the allocation of a significant amount of bandwidth for third-generation (3G) wireless communications.

5.3.4 Polarization Diversity

Polarization diversity makes use of the inherent property of most scattering media to depolarize the propagating beam even when the transmitted signal may be vertically (or horizontally) polarized. The depolarized signal arriving at the antenna can be split into two orthogonal polarizations (vertical and horizontal), producing two independently fading signals (Bren 1959, Vaug 1990, Atan 2000). The only limitation of this technique is the inability to generate more than two diverse signals, whereas it is possible to have many diverse signals in spatial, frequency, or angular diversity.

5.3.5 Time Diversity

If we can transmit the same data stream repeatedly at intervals that exceed the coherence time of the channel, these multiple information streams will be subjected to independent fading. These independently faded components can then be combined using any diversity combining techniques. Time diversity techniques have one major advantage over the space and frequency diversity systems in not requiring multiple antennae and needing only a single antenna at the receiver. Time diversity systems, however, do require storage of the received data streams for processing (Turk 1990, Stei 1987). Time diversity (similar to in-band diversity, since we are using the same frequency characteristics) might require some form of transmit power division penalty since one transmitter is used to transmit the data repeatedly. One can consider the case of error control coding or sending more data than necessary (redundant data transmission) as a form of time diversity. Error control coding is discussed in Section B.3, Appendix B.

5.3.6 Multipath Diversity

Instead of transmitting at different times, it is possible to take advantage of the existing multipath signals present in fading wireless channels (Lehn 1987). If the time duration of the pulses is very small, then the multipath signals corresponding to the different paths will be nonoverlapping, and, therefore, uncorrelated. One such case is shown in Figure 5.10.

The three pulses, 1, 2, and 3, are time-delayed versions of the same pulse and are resolvable. Such pulses can be combined like any other diverse signals by taking into account the amount of delay that exists between the multiple paths, as shown in Figure 5.11. For most data transmission systems, the multipath components are

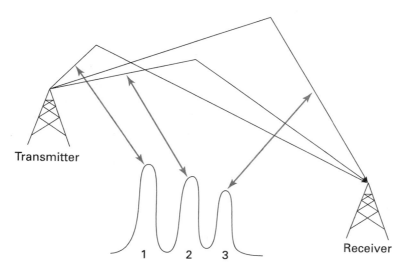

Transmitter

Receiver

FIGURE 5.10
Multipath diversity.

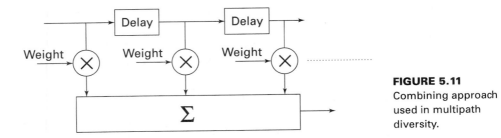

FIGURE 5.11

Combining approach used in multipath diversity.

unresolvable and overlap, making them useless for multipath diversity approaches. On the other hand, in wideband systems employing code division multiple access, since the pulse widths used are extremely small, it is possible to "pick" each delayed version of the same pulse for the purposes of employing multipath diversity. A RAKE receiver collects these time-shifted (or delayed) versions of the transmitted signals for processing. We will look at RAKE receivers in Chapter 6 after we discuss the details of code division multiple-access techniques.

5.4 DIVERSITY COMBINING METHODS

The multiple versions of the signals created using the various diversity techniques must be combined in some fashion to provide improved performance in terms of lowering the minimum signal-to-noise ratio required to maintain an acceptable level of bit error rate. There are three primary means of combining the signals; selection combining, maximal ratio combining, and equal gain combining (Bren 1959; Alhu 1985; Beau 1991; Adac 1989, 1991a,b; Anna 1998, 1999).

5.4.1 Selection Combining

Selection combining is the simplest form of diversity combining techniques. Its effectiveness is based on the simple principle that not all signals coming out of the diversity branches will have low values at the same time. Under these conditions, it is possible to look for the branch having the highest signal-to-noise ratio and use that particular branch as the primary received signal. This is illustrated in Figure 5.12. Five sets (1000 each) of exponentially distributed random numbers corresponding to the power of a Rayleigh-faded signal were generated. Two of these sets (only 40 samples) are shown. Also shown is the largest of the five sets for every sample number. It demonstrates that it is possible to ensure that the signal power will not always be small if we choose the largest of a number of outputs. The processor principle is shown in Figure 5.13.

Consider M independent Rayleigh-fading components produced by any diversity scheme. Since the envelope is Rayleigh distributed, the signal power and therefore the signal-to-noise ratio, γ_n, will be exponentially distributed. If γ_0 is the average signal-to-noise ratio of any one of the diversity components, the pdf, $f(\gamma_n)$, of the instantaneous signal-to-noise ratio, γ_n, of any one of the M components can be expressed as

$$f(\gamma_n) = \frac{1}{\gamma_0}\exp\left(-\frac{\gamma_n}{\gamma_0}\right)U(\gamma_n), \quad n = 1, 2, ..., M. \tag{5.16}$$

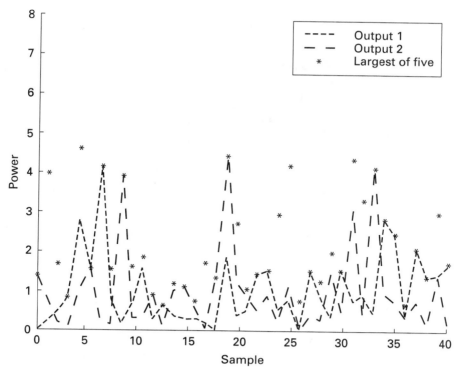

FIGURE 5.12 Principle of selection diversity.

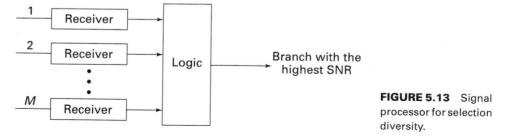

FIGURE 5.13 Signal processor for selection diversity.

The probability that the signal-to-noise ratio in any one of the diversity components will be less than any specific value, γ, is given by

$$P[\gamma_n \le \gamma] = \int_0^\gamma f(\gamma_n)\, d\gamma_n = 1 - \exp\left(-\frac{\gamma}{\gamma_0}\right), \quad n = 1, 2, 3, \dots, M. \quad (5.17)$$

Since all of the M components or branches are independent, the probability that all of them would have a SNR less than γ is

$$P_r(\gamma_1, \gamma_2, \gamma_3, \dots, \gamma_M \le \gamma) = \left[1 - \exp\left(-\frac{\gamma}{\gamma_0}\right)\right]^M. \quad (5.18)$$

The probability, $P_M(\gamma)$, that at least one branch achieves a signal-to-noise ratio greater than γ will now be

$$P_M(\gamma) = 1 - P_r(\gamma_1, \gamma_2, \gamma_3, \dots, \gamma_M \le \gamma) = 1 - \left[1 - \exp\left(\frac{\gamma}{\gamma_0}\right)\right]^M. \quad (5.19)$$

This is the probability associated with the selection combining algorithm (Adac 1991a, Fehe 1995, Rapp 1996b, Garg 1996). It provides a measure of the degree of improvement we can expect if we are able to examine the various diversity components and pick the one having the signal-to-noise ratio exceeding the value that has been set. To compute the actual improvement we can expect out of this selection combining, we must calculate the pdf of the output of the selection combining algorithm, namely, the pdf associated with the probability given in eq. (5.19). The pdf of the output of the "selection combiner," $f(\gamma)$, is obtained by differentiating eq. (5.16) with respect to γ, leading to

$$f(\gamma) = \frac{d}{d\gamma}\left\{1 - \left[1 - \exp\left(-\frac{\gamma}{\gamma_0}\right)\right]^M\right\}$$

$$= \frac{M}{\gamma_0}\left[1 - \exp\left(-\frac{\gamma}{\gamma_0}\right)\right]^{M-1} \cdot \exp\left(-\frac{\gamma}{\gamma_0}\right). \quad (5.20)$$

Equation (5.20) may also be obtained directly from the results on "order statistics" (Papo 1991). The probability density function is plotted in Figure 5.14 for

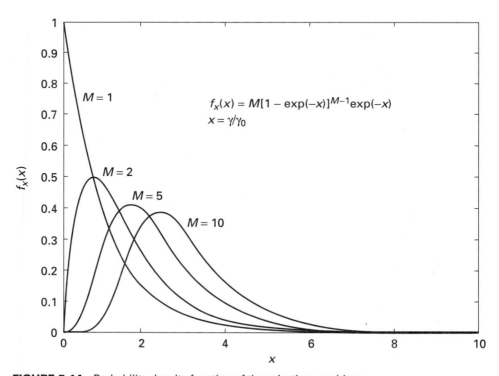

FIGURE 5.14　Probability density function of the selection combiner.

different values of M. For $M = 1$, the pdf is identical to the exponential density function associated with the signal-to-noise ratio of a single branch. It is seen that the peak of the density function moves to the right as M increases. This shift to the right will make the pdf attain more and more symmetry, thus making it closer and closer to being a Gaussian density function. When that happens, the fading channel can be considered to be no different from an ideal channel, i.e., a Gaussian channel. This transformation from Rayleigh to Gaussian is the primary reason for the expected reduction in bit error rates after compounding. We can also draw an analogy with Rician statistics (Chapter 2) and the improvement brought on by the presence of a steady component as we look at this density function as its peak continues to move to the right.

The average signal to noise ratio, γ_{se}, of the output of the selection combiner will be (Jake 1971, 1974)

$$\gamma_{se} = \int_0^\infty \gamma f(\gamma) \, d\gamma = \gamma_0 \sum_{n=1}^M \frac{1}{n}. \tag{5.21}$$

The improvement in signal-to-noise ratio obtained through selection combining is

$$\frac{\gamma_{se}}{\gamma_0} = \sum_{n=1}^M \frac{1}{n}. \tag{5.22}$$

The actual improvement in performance will depend on the modulation/demodulation format used.

Practical Considerations   The implementation of selection combining starts with the examination of the received signal components from the diversity branches, with the one that has the strongest signal being selected. This is a little cumbersome since the process of examination for the strongest signal has to be undertaken continuously. This problem can be overcome if we resort to the "scanning selection combining" method. In this approach, at the beginning of the process of selection combining, the branch with the strongest signal is chosen. This particular branch is used until such time that the SNR of this branch goes below the threshold value. As the SNR value in that branch goes down, a new selection is made, and the process is continued.

EXAMPLE 5.4

Consider a two-channel selection combiner. The outage occurs when the signal-to-noise ratio goes below one-fourth of the average. Show that the outage probability with a two-channel selection combiner is smaller than the outage probability with no selection diversity.

Answer   The pdf of the signal-to-noise ratio of a two-channel selection combiner can be written as

$$f(\gamma) = \frac{2}{\gamma_0} \left[1 - \exp\left(-\frac{\gamma}{\gamma_0} \right) \right] \exp\left(-\frac{\gamma}{\gamma_0} \right) U(\gamma).$$

The outage probability for a two-channel selection combiner is

$$\int_0^{\gamma_0/4} f(\gamma)\, d\gamma = 2[0.5 + 0.5 \times \exp(-0.5) - \exp(-0.25)] = 0.0489\,.$$

The outage probability in the absence of diversity is

$$\int_0^{\gamma_0/4} \frac{1}{\gamma_0} \exp\!\left(-\frac{\gamma}{\gamma_0}\right) d\gamma = 1 - \exp(-0.25) = 0.2212\,.$$

The reduction in outage probability because of diversity is clearly seen in the two probabilities obtained. ■

5.4.2 Maximal Ratio Combining

The selection combining approach to the processing of diversity branches, though simple, does not take full advantage of the availability of multiple signals. Maximal ratio combining (Wint 1984, Schw 1996, Akai 1998) creates a new signal that is a linear combination of all the multiple signals with appropriate weighting. The concept is shown in Figure 5.15. The outputs of the different diversity branches are added in phase with a gain factor such that the processed signal, r_{MR}, can be expressed as

$$r_{MR} = \sum_{n=1}^{M} g_n a_n + \sum_{n=1}^{M} g_n n_n\,, \tag{5.23}$$

where g_n is the weighting factor for the nth diversity branch and a_n is a multiplicative component that represents the Rayleigh fading. We can rewrite eq. (5.23) as

$$r_{MR} = r_M + N_M\,, \tag{5.24}$$

where r_M and N_M, respectively, are the signal and noise terms, given by

$$r_M = \sum_{n=1}^{M} g_n a_n, \quad N_M = \sum_{n=1}^{M} g_n n_n\,. \tag{5.25}$$

The signal-to-noise ratio of the maximal ratio combiner, γ_{MR}, is

$$\gamma_{MR} = \frac{r_M^2}{2N_{MP}}\,, \tag{5.26}$$

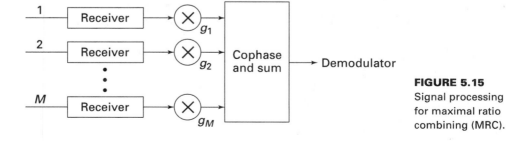

FIGURE 5.15
Signal processing for maximal ratio combining (MRC).

where N_{MP} is the total noise power. This value is obtained by assuming that the noises in all the channels are identical and independent, and therefore

$$N_{MP} = N \sum_{n=1}^{M} g_n^2,$$ (5.27)

where N is the average noise power in any channel. It can be shown that the signal-to-noise ratio of the maximal ratio combiner will be maximum when the gain factor, g_n, is chosen to be equal to

$$g_n = \frac{a_n}{N}.$$ (5.28)

Rewriting eq. (5.26) for the signal-to-noise ratio of the maximal ratio combiner, γ_{MR} becomes

$$\gamma_{MR} = \frac{1}{2N} \frac{\left(\sum_{n=1}^{M} \left(a_n^2/N\right)\right)^2}{\sum_{n=1}^{M} (a_n/N)^2} = \sum_{n=1}^{M} \gamma_n,$$ (5.29)

where γ_n is the instantaneous signal-noise-ratio given by

$$\gamma_n = \frac{a_n^2}{2N}.$$ (5.30)

The SNR of the maximal ratio combiner is the sum of the signal-to-noise ratios of all the diversity branches. Since all of them are independent and identically distributed, the signal-to-noise ratio is

$$\gamma_{MR} = \sum_{n=1}^{M} \gamma_n = M\gamma_0,$$ (5.31)

where γ_0 is the signal-to-noise ratio in any one of the branches. It is thus seen that the maximal ratio combiner results in a higher signal-to-noise ratio than the selection combiner.

Once again, the probability density function, $f(\gamma)$, of the signal-to-noise ratio γ at the output of the maximal ratio combiner can be expressed as (Schw 1996)

$$f(\gamma) = \frac{1}{(M-1)!} \frac{\gamma^{M-1}}{\gamma_0^M} \exp\left(-\frac{\gamma}{\gamma_0}\right).$$ (5.32)

The density function is plotted in Figure 5.16. Once again, for $M = 1$, the pdf is identical to the density function of the signal-to-noise ratio of a single branch (no diversity). We see that as M increases, the peaks of the density functions move to the right. Just as in the case of selection diversity, the density functions of the compounded signals become more Gaussian-like. The channels after fading thus become Gaussian channels, leading to a reduction in bit error rates and outage probabilities.

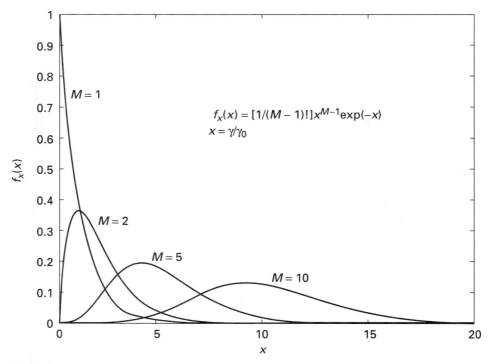

FIGURE 5.16   Probability density function for the maximal ratio combiner.

EXAMPLE 5.5

Consider a two-channel MRC system. The outage occurs when the signal-to-noise ratio goes below one-fourth of the average signal-to-noise ratio. Show that the outage probability with a two-channel selection combiner is smaller than the outage probability with no selection diversity.

Answer   The density function for a two-channel MRC receiver is

$$f(\gamma) = \frac{\gamma}{\gamma_0^2}\exp\left(-\frac{\gamma}{\gamma_0}\right)U(\gamma) \ .$$

The outage probability is obtained as

$$\int_0^{\gamma_0/4} \frac{\gamma}{\gamma_0^2}\exp\left(-\frac{\gamma}{\gamma_0}\right) d\gamma = 1 - \frac{5}{4}\exp(-0.25) = 0.0265$$

The outage probability in the absence of compounding is 0.2212 (from Example 5.4). We once again see the reduction in the outage probabilty as a result of diversity. We also see that the outage with the MRC system is less than the outage for a selection combiner.   ■

Practical Considerations   The maximal ratio combiner does outperform the selection combiner. However, this improvement is realized at the cost of increased complexity. Considerable signal processing is necessary to achieve the correct weighting factors. This implementation problem naturally leads to the case where the weighting factors can be made equal, as described next.

5.4.3 Equal Gain Combining

The equal gain combiner is a maximal ratio combiner in which all the weights are equal. All the weights are set to unity, and the processing described in the previous section is undertaken. The signal-to-noise ratio for the equal gain combiner, γ_{EC}, can be shown to be equal to

$$\gamma_{EC} = \gamma_0 \left[1 + \frac{\pi}{4}(M-1) \right], \tag{5.33}$$

which lies between the SNRs of the maximal ratio combiner and the selection combiner. In fact, as M increases, the difference between equal gain combining and maximal ratio combining becomes much smaller.

Once again, the approximate probability density function, $f(\gamma)$, of the signal-to-noise ratio (energy) γ at the output of the equal gain combiner can be expressed as (for $\gamma \ll \gamma_0$),

$$f(\gamma) = \frac{2^{M-1} M^M}{(2M-1)!} \frac{\gamma^{M-1}}{\gamma_0^M}. \tag{5.34}$$

The performance of the three signal-processing schemes is compared in Figure 5.17, with γ_{av} equal to the signal-to-noise ratio after combining. The performance comparison is also given in tabular form in Table 5.1.

EXAMPLE 5.6

Consider a two-channel equal gain diversity system. Outage occurs when the signal-to-noise ratio goes below one-fourth of the average SNR. Show that the outage probability with a two-channel equal gain combiner is smaller than the outage probability with no diversity.

Answer The density function of the signal-to-noise ratio of the equal gain combiner output is given by eq. (5.34),

$$f(\gamma) = \frac{3}{2} \left(\frac{\gamma}{\gamma_0^2} \right) \text{(approx.)}.$$

The outage probability is

$$\int_0^{\gamma_0/4} f(\gamma) \, d\gamma = 0.0469.$$

The outage in the absence of compounding is 0.2212 (from Example 5.4). We once again see the reduction in outage probability because of diversity. We also see that the outage with the equal gain system is less than that for the case of a selection combiner. However, the outage probability for an equal gain combiner is higher than the outage probability for a MRC system. ∎

The performance of the equal gain combiner is midway between the performance levels of selection combining and maximal ratio combining, the latter providing the best performance. The selection combiner is much easier to implement than the maximal ratio combiner.

TABLE 5.1 Signal-to-Noise Ratio Improvement through Diversity

Number of branches, M	Signal-to-noise ratio improvement (dB)		
	Selection combiner $\sum_{K=1}^{M} \frac{1}{K}$	Maximalratio combiner M	Equal gain combiner $1 + (M-1)\frac{\pi}{4}$
1	0	0.00	0.00
2	1.761	3.01	2.52
3	2.632	4.77	4.10
4	3.187	6.02	5.26
5	3.585	6.99	6.17
6	3.892	7.78	6.92

Effects of Correlated Signals   All of the analysis presented so far has assumed that the multiple signals generated by any one of the diversity techniques are all uncorrelated. Because of incorrect positioning of the receiving antennae (spatial diversity) or insufficient separation between the frequency bands (frequency diversity), the signals from the different branches can be correlated. For correlation in the range of a few percentage points, the degradation in performance will not be significant. Even when the correlation is high (>50% for example), the performance of the diversity-based system will still be better than for nondiversity systems, even though the im-

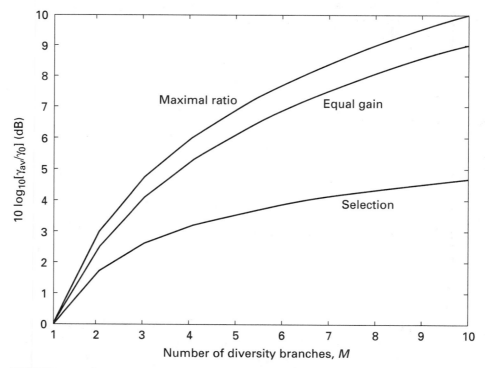

FIGURE 5.17 Comparison of the three combining schemes.

provement will be much less than if the signals were truly uncorrelated. In fact, the improvement (signal-to-noise ratio) in performance due to diversity goes down by a factor of $\sqrt{1 - \rho^2}$, where ρ is the correlation coefficient (Schw 1996, Akai 1998).

5.5 PERFORMANCE IMPROVEMENT FROM DIVERSITY IN TERMS OF REDUCED BIT ERROR RATE

The improvement in performance realized through the use of diversity techniques can now be calculated. We will simplify the analysis by expressing the equation for the probability of error in terms of a general expression such as

$$p(e) = \frac{1}{2}\text{erfc}(\sqrt{b\gamma}) \tag{5.35}$$

for coherent modulation techniques, with $b = 1$ for BPSK, $b = 0.5$ for FSK, and $b = 0.85$ or 0.68 for GMSK. The signal-to-noise ratio (energy) is γ. Similarly, the probability of error for a noncoherent receiver can be expressed as

$$p(e) = \frac{1}{2}\exp(-b\gamma) . \tag{5.36}$$

The error probability, $P_{\text{av}}(e)$, at the output of the diversity combiner is given by

$$p_{\text{av}}(e) = \int_0^\infty p(e)f(\gamma)\, d\gamma , \tag{5.37}$$

where $f(\gamma)$ is the pdf of the signal-to-noise ratio γ for the specific diversity combining algorithm. The equations governing the bit error rate at the output of the diversity receiver are complex (Schw 1996; Proa 1995; Akai 1998; Samp 1997; Eng 1996; Wint 1984; Bigl 1998; Biya 1995; Bour 1993; Chen 1991, 1995). However, approximate expressions are available. For example, at high signal-to-noise ratios, the bit error rate at the output of a selection combiner system for BPSK can be expressed as

$$P_{\text{avsc}}(e) = \int_0^\infty \frac{1}{2}\text{erfc}\sqrt{\gamma}\, \frac{M}{\gamma_0}\left[1 - \exp\left(-\frac{\gamma}{\gamma_0}\right)\right]^{M-1}\exp\left(-\frac{\gamma}{\gamma_0}\right)d\gamma . \tag{5.38}$$

This equation can be simplified (Schw 1996, Samp 1997, Proa 1995) to

$$P_{\text{avsc}}(e) = \frac{1}{2}M\sum_{k=0}^{M-1}(-1)^k C_k^{M-1}\frac{1}{k+1}\left(1 - \frac{1}{\sqrt{1 + \frac{k+1}{\gamma_0}}}\right) . \tag{5.39}$$

Similarly, the average error probability for a maximal ratio combiner can be expressed as

$$P_{\text{avmc}}(e) = \int_0^\infty \frac{1}{2}\text{erfc}\left(\sqrt{\gamma}\right)\frac{1}{(M-1)!}\frac{\gamma^{M-1}}{\gamma_0^M}\exp\left(-\frac{\gamma}{\gamma_0}\right)d\gamma . \tag{5.40}$$

This equation can be simplified to

$$P_{\text{avmc}}(e) = 2^{-M}\left(1 - \sqrt{\frac{\gamma_0}{1+\gamma_0}}\right)^M \sum_{k=0}^{M-1} C_k^{(M+k-1)}\left(\frac{1}{2}\right)^k\left(1 + \sqrt{\frac{\gamma_0}{1+\gamma_0}}\right)^k. \quad (5.41)$$

For noncoherent receivers, the average error probability for the cases of selection combining and maximal ratio combining can be expressed as

$$P_{\text{avsn}}(e) = \int_0^\infty \frac{1}{2}\exp(-\gamma)\frac{M}{\gamma_0}\left[1 - \exp\left(-\frac{\gamma}{\gamma_0}\right)\right]^{M-1}\exp\left(-\frac{\gamma}{\gamma_0}\right)d\gamma, \quad (5.42)$$

which simplifies to

$$P_{\text{avsn}}(e) = \frac{1}{2}M\sum_{k=0}^{M-1}(-1)^k C_k^{M-1}\frac{1}{k+1}\left(1 - \frac{1}{\gamma_0/(k+1)}\right), \quad (5.43)$$

and

$$P_{\text{avmn}}(e) = \int_0^\infty \frac{1}{2}\exp(-\gamma)\frac{1}{(M-1)!}\frac{\gamma^{M-1}}{\gamma_0^M}\exp\left(-\frac{\gamma}{\gamma_0}\right)d\gamma, \quad (5.44)$$

which simplifies to

$$P_{\text{avmn}}(e) = \frac{1}{2}\left[\frac{(M-1)!}{(1+\gamma_0)^M}\right]. \quad (5.45)$$

The bit error rate curves for the different cases are shown in Figures 5.18–5.21.

Comparison of Figures 5.18 and 5.19 shows the usefulness of diversity. As M increases, the performance approaches that of an ideal channel, i.e., a Gaussian one. This convergence to Gaussian behavior was discussed in Sections 5.4.1 and 5.4.2 in connection with the shape of the density functions after compounding. It is also obvious from these figures that MRC diversity is better than selection combining in terms of how quickly Gaussian channel conditions are realized. For example, with the selection combiner, a bit error rate of 10^{-5} is reached at a signal-to-noise ratio of about 13 dB. The same bit error rate is reached at a lower SNR value with the MRC system. The performance of the DPSK system (Figures 5.20 and 5.21) for the two methods of diversity combining is similar to that for the BPSK system.

These results do not include the effects of Doppler fading, CCI, or frequency-selective fading. Indeed, the analytical results of the analysis of GMSK and $\pi/4$-DQPSK systems are not simple to derive. Numerical results are available in a number of publications (Guo 1990; Kaas 1998; Liu 1991a,b; Ng 1993, 1994). We will, however, look at a simple case of a DPSK system employing two-branch diversity using the MRC scheme in the presence of Doppler fading (Jake 1974). The average error probability has been shown (Jake 1974) to be

$$P_{\text{avm}}(e) = \left[\frac{2(\gamma_0+1)+\gamma_0 J_0(2\pi f_d T)}{1+\gamma_0}\right]\left(\frac{1+\gamma_0[1-J_0(2\pi f_d T)]}{2(1+\gamma_0)}\right)^2. \quad (5.46)$$

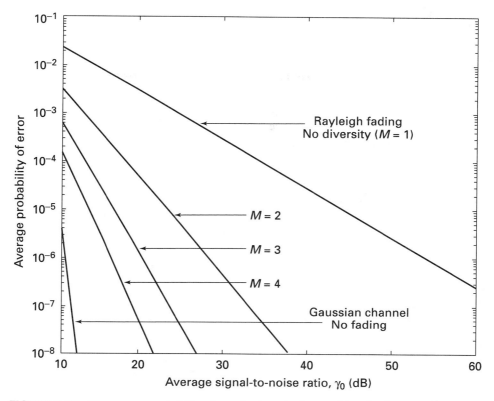

FIGURE 5.18 The average probability of error for the selection combiner for the case of coherent BPSK.

The results in Figure 5.22 show that even for a low value of $f_d T$ ($= 0.05$), the error floor exists (10^{-2}) and diversity techniques can bring the error floor down to 10^{-3}. Diversity techniques also lead to a steeper drop in error rates as the signal-to-noise ratio values go up. The error floors seen here are typical of communications systems operating in a random FM environment, as we saw in Section 5.2.

The lowering of the error floor through the use of diversity techniques can be seen clearly if we plot the error floor values as a function of the parameter $f_d T$. These results are shown in Figure 5.23. The steep drop in error floor seen is evidence of the importance of diversity techniques in mobile communication systems.

5.6 MACROSCOPIC DIVERSITY

The diversity techniques described so far are sufficient to mitigate the effects of short-term fading, i.e., Rayleigh fading. The fading in this case occurs over very short distances; in the case of lognormal fading or shadowing, the fading is very slow and occurs over distances of many wavelengths. For the case of lognormal fading, the use of multiple receivers separated by short distances will not produce any difference between the multiple signals. To compensate for lognormal fading, the diversity technique must be implemented on a site-by-site basis, with the sites separated by many wavelengths. This form of diversity, where receivers are located at

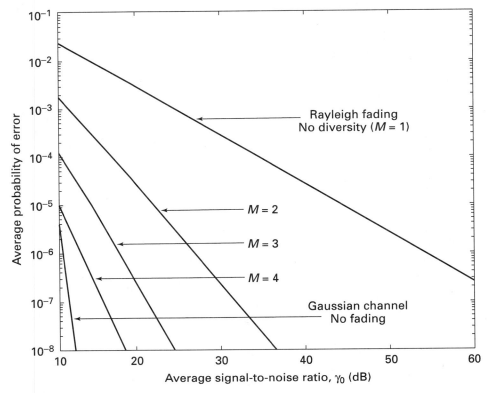

FIGURE 5.19 The average error probability for MRC diversity for the coherent BPSK receiver.

distances orders of magnitude larger than those in the case of microscopic diversity, is known as macroscopic diversity. Macroscopic diversity is also sometimes referred to as site diversity (or multiple-base-station diversity). One can visualize macroscopic diversity as a means by one site of collecting the signal from another site when the primary site is obstructed by a shadowing region. An example of this form of diversity exists in cellular systems: the hand-over between base stations can be treated as a form of macroscopic diversity.

The different combining techniques described in connection with microscopic diversity are also applicable to macroscopic diversity. A detailed description of macroscopic diversity techniques is given in Section C.3, Appendix C.

5.7 EQUALIZATION/FREQUENCY-SELECTIVE FADING

So far we have considered the attempts made to overcome problems associated with flat fading. If the channel is frequency selective, i.e., the coherent bandwidth happens to be less than the information bandwidth, pulse spreading takes place, leading to intersymbol interference, as shown in Figure B.6.2 (see Appendix B). ISI is a serious problem in high-speed data transmission. We could certainly reduce the data rate so as to reduce the pulse spreading, and, thus, reduce the effects of ISI. But this

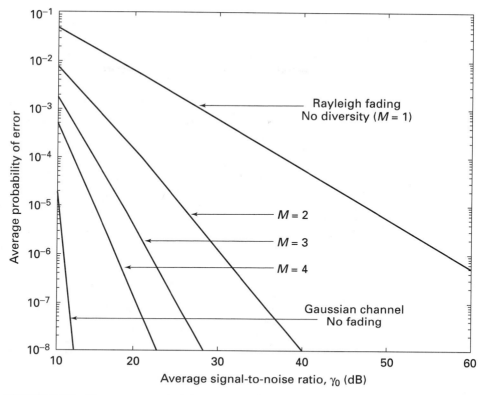

FIGURE 5.20 The average probability of error for selection diversity for the case of DPSK.

is not an option, since the goal is to transmit data at higher rates. One of the ways in which the frequency selectivity of the channel can be reduced or eliminated is through the process known as equalization (Mons 1980, Skla 1988, Proa 1995, Hayk 2001).

The transfer function of the channel, $H_c(f)$, can be expressed as

$$H_c(f) = A(f)\exp[\theta(f)], \tag{5.47}$$

where $A(f)$ is the envelope and $\theta(f)$ is the phase response. The envelope delay, $\tau(f)$, is expressed as

$$\tau(f) = -\frac{1}{2\pi}\frac{d\theta(f)}{df}. \tag{5.48}$$

The channel is distortionless if

$$A(f) = A_0 \tag{5.49}$$

and if

$$\tau(f) = \tau_0, \tag{5.50}$$

where A_0 and τ_0 are constants in the frequency range of interest. The latter condition means that the phase $\theta(f)$ must be linear. If $A(f)$ is not constant, we have envelope or amplitude distortion, and if $\tau(f)$ is not constant, we have delay distortion. The consequence of distortion is ISI.

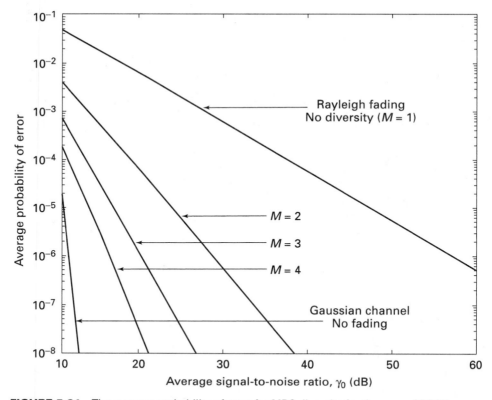

FIGURE 5.21 The average probability of error for MRC diversity for the case of DPSK.

To understand equalization in purely qualitative terms, let us look at the frequency response of a frequency-selective channel, shown in Figure 5.24. The ideal response is almost flat over the complete spectrum. The phase response is also linear for the ideal channel. If the channel impulse response could be altered to look like the ideal response shown in Figure 5.24, i.e., if it is possible to equalize the response at all frequencies, the pulse spreading would be eliminated. Note that pulse distortion also occurs if the phase response of the channel is nonlinear. An equalizer must correct the magnitude as well as the phase response of the system.

A simple conceptual block diagram of an equalizer is shown in Figure 5.25. The transmitting filter includes the modulator. The receiving filter includes the receiver. As explained in the previous paragraph, the ultimate goal of the equalizer is to ensure that the overall impulse response, when the channel and equalizer together are taken into account, is a delta function. This is realized if

$$H_{\text{eq}}(f)H_c^*(-f) = 1 . \tag{5.51}$$

In other words, the equalizer is nothing but an inverse filter, as shown in Figure 5.26.

Even though the concept of equalization appears simple, there are practical problems associated with its implementation owing to the fact that the channel is time varying. This means that an equalizer must be able to adjust or retune to keep track with the changes in the channel characteristics. Thus the idea of a simple filter

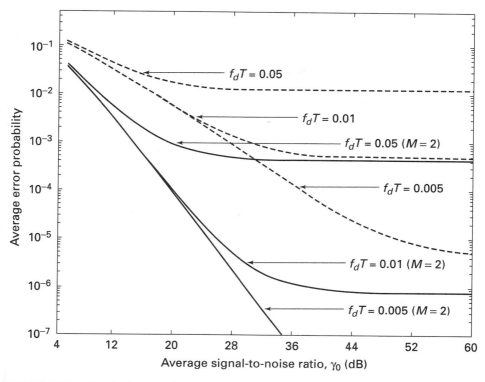

FIGURE 5.22 Results for a two-branch diversity receiver using the MRC scheme for a DPSK system ($M = 2$). Results are also shown for the case of no diversity.

with the required response after the demodulator/receiver filter is inadequate. Therefore, equalizers must be designed to be adaptive. We will briefly review some of the equalization techniques commonly used in wireless communications (Tayl 1998).

5.7.1 Linear Transversal Equalizer

A block diagram depicting a simple linear equalizer, more commonly known as a linear transversal equalizer, is shown in Figure 5.27. It consists of a number of elements, successively introducing a delay of τ. The maximum time delay τ is T. For the case where the delay is the symbol duration (T), the equalizer is known as a symbol-spaced equalizer.

Each delayed version of the input signal is weighted appropriately. The number of taps required is determined by the severity of ISI. If the ISI spreads a given symbol over many symbol periods, the number of required taps increases. It is possible to calculate the weights denoted by c_0, c_1, c_2, ... to ensure that the ISI is zero. The ISI approaches zero when the number of taps is infinite. This equalizer may be considered to be a zero-forcing equalizer (ZF algorithm), since in an ideal case, the inverse filter given in eq. (5.51) is expected to force the ISI to be zero at the sampling instants.

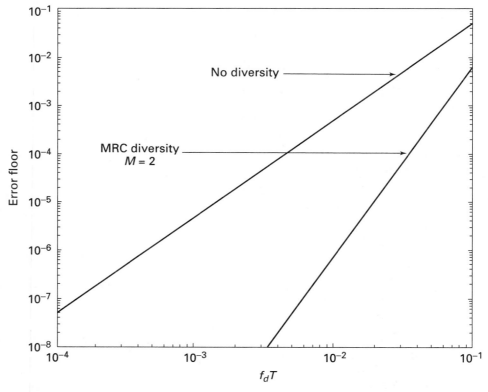

FIGURE 5.23 The error floor for a two-channel MRC diversity system for DPSK compared with the case of no diversity.

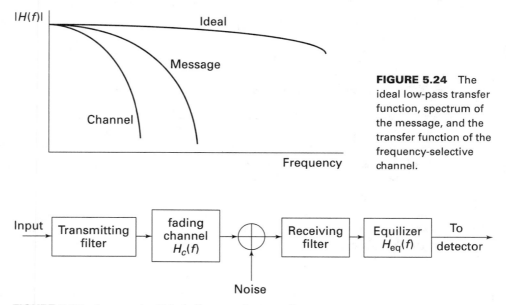

FIGURE 5.24 The ideal low-pass transfer function, spectrum of the message, and the transfer function of the frequency-selective channel.

FIGURE 5.25 A conceptual block diagram of an equalizer.

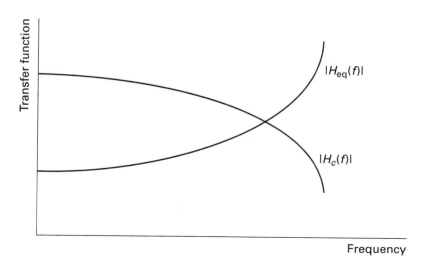

FIGURE 5.26 The overall low-pass transfer function of the channel, $H_c(f)$, in the absence of an equalizer and the transfer function of an equalizer, $H_{eq}(f)$.

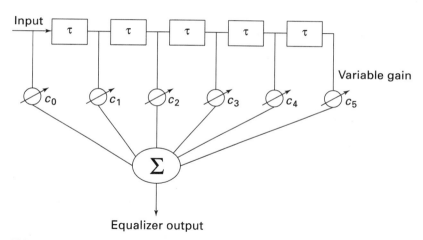

FIGURE 5.27 Linear transversal equalizer.

EXAMPLE 5.7

Consider a mobile channel with a channel response $h_c(t)$ given by

$$h_c(t) = a\delta(t - t_0) + b\delta(t - t_1),$$

where a and b are constants and t_0 and t_1 are time delays. If a delay line with three taps is used as an equalizer, calculate (1) the transfer function of the channel and (2) the transfer function of the "equalizer" and the relationship between a and b given that $b \ll a$ and $t_1 > t_0$. Assume that there is no noise.

Answer Taking the Fourier transform of $h_c(t)$, we obtain the transfer function, $H_c(f)$, of the channel:

$$H_c(f) = a\exp(-j2\pi ft_0) + b\exp(-j2\pi ft_1).$$

The equalizer transfer function, $H_{eq}(f)$, must be such that

$$H_{eq}(f)H_c^*(-f) = 1 \ .$$

(eq. (5.51)). For a three-tap equalizer that uses delays of T,

$$H_{eq}(f) = c_0 + c_1 \exp(-j2\pi fT) + c_2 \exp(-j2\pi f2T)$$

$$= c_0 \left[1 + \frac{c_1}{c_0} \exp(-j2\pi fT) + \frac{c_2}{c_0} \exp(-j2\pi f2T) \right].$$

Using the equation for the "inverse filter,"

$$H_{eq}(f) = \frac{\exp(-j2\pi f\Lambda)}{H_c^*(-f)}$$

$$= \frac{\exp(-j2\pi f\Lambda)}{a \exp(-j2\pi ft_0) + b \exp(-j2\pi ft_1)} = \frac{(1/a)\exp\left[-j2\pi f(\Lambda - t_0)\right]}{1 + (b/a)\exp\left[-j2\pi f(t_1 - t_0)\right]}.$$

Since $b \ll 1$, we can rewrite this equation as

$$H_{eq}(f) = \left(\frac{1}{a}\right)\exp\left[-j2\pi f(\Lambda - t_0)\right] \times \left\{ 1 - \frac{b}{a}\exp\left[-j2\pi f(t_1 - t_0)\right] + \left(\frac{b}{a}\right)^2 \exp\left[-j4\pi f(t_1 - t_0)\right] \right\}.$$

If we assume that $\Lambda \approx t_0$, comparing the two equations we get

$$c_0 = \frac{1}{a}$$

$$\frac{c_1}{c_0} = -\frac{b}{a}$$

$$\frac{c_2}{c_0} = \left(\frac{b}{a}\right)^2$$

and

$$t_1 - t_0 = T \ .$$

The coefficients of the taps can now easily be found. ∎

In contrast to a symbol-spaced equalizer, the time delay between adjacent taps may be made less than T. This is known as a fractionally spaced equalizer. Quite often the delay is chosen to be $T/2$. In this case, the impulse response of the equalizer is

$$H_{eqf}(f) = \sum_{n=-N}^{N} C_n \exp\left(-j2\pi fn\frac{T}{2}\right), \tag{5.52}$$

where there are $2N + 1$ taps. By choosing taps spaced by delays less than T, the chances of aliasing are eliminated, and hence, ISI is likely to be compensated for more fully. The number of taps is chosen such that it exceeds the number of symbols spanned by ISI, as shown in Figure 5.28.

Note that this type of equalizer is typically operated by sending a training data set to calculate the weights. However, such an approach is not adaptive. It is possible to make the equalizer adaptive by continuously updating the weights. This can be

Transmitted symbols

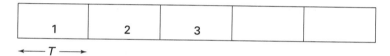

Received symbols

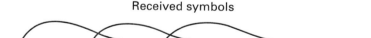

FIGURE 5.28 Transmitted symbols and received symbols, showing the "overflowing" nature of ISI.

accomplished by observing the equalized bits on a block-by-block basis and comparing them to some known or expected signal. The weights are then adjusted to minimize the mean square error (MMSE algorithm).

5.7.2 Nonlinear Equalizer

It is possible that the distortion introduced by the channel is too severe to be compensated for through the use of a linear transversal equalizer. To understand the problem with linear equalizers, let us examine the impulse responses of three channels shown in Figure 5.29. The first one can be considered to correspond to low-ISI channels such as telephone lines, while the other two correspond to the wireless channels. Their transfer functions can be approximated as shown in Figures 5.30*a*, *b*, and *c*, respectively.

The existence of a null (Figure 5.30*c*) and the sharp dropoff (Figure 5.30*b*) show the presence of significant amounts of ISI, and the linear equalizer may not be able to compensate for it; the frequency response in Figure 5.30*a* points to a low amount of ISI, which the linear equalizer should be able to handle. Another drawback of the linear equalizer may be amplification of noise present around the spectral null, since a linear equalizer will introduce a large gain to compensate for the spectral null.

Severe cases of ISI can be handled using nonlinear equalizers (Tayl 1998). Non-linear techniques include decision feedback equalization (DFE), maximum likelihood symbol detection (MLSD), and maximum likelihood sequence estimation (MLSE). We will briefly examine two of these techniques.

Decision Feedback Equalizer (DFE) To combat severe distortion brought on by frequency-selective fading, DFE uses a feedback mechanism to eliminate the ISI caused by previously detected symbols interfering with the current symbols being detected.

A block diagram of the DFE is shown in Figure 5.31. The input information is passed through a linear transversal filter like the one discussed earlier. This filter is identified as the *feedforward* filter and is normally a fractionally spaced equalizer. The output of this filter is applied to a second filter, the *feedback* filter, which is a symbol-spaced equalizer with adjustable tap gains.

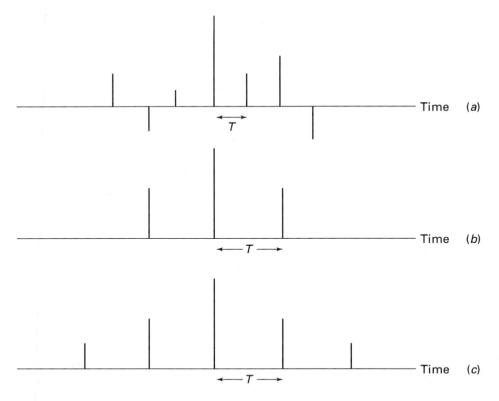

FIGURE 5.29 Impulse responses of three channels.

The input to the feedback filter is the previously detected symbols. The feedback filter subtracts the portion of ISI produced by these symbols. The equalizer is nonlinear because of the existence of the decision device in the feedback loop.

DFE can compensate for moderate to severe amounts of ISI.

Maximum Likelihood Sequence Estimation (MLSE) Even though the decision feedback equalizer is better than the linear transversal equalizer in its ability to mitigate ISI caused by fading, it is not the optimal processor for minimizing the probability of error in the detection of the transmitted symbols from the received symbol sequences. The MLSE equalizer estimates the channel impulse response and uses the Viterbi algorithm (Vite 1971, Forn 1973) to generate estimates of the transmitted symbols. A block diagram of the MLSE equalizer is shown in Figure 5.32. It consists of a channel estimator and the MLSE algorithm (Viterbi algorithm). The channel estimator is typically an adaptive one and estimates the channel response. The MLSE receiver performs a correlation of the received signal with all possible transmitted symbol sequences and chooses the optimal one based on the maximum likelihood procedure. The Viterbi algorithm is a computationally efficient means to accomplish this task.

The MLSE technique is more efficient than DFE; however, MLSE is computationally very exhaustive.

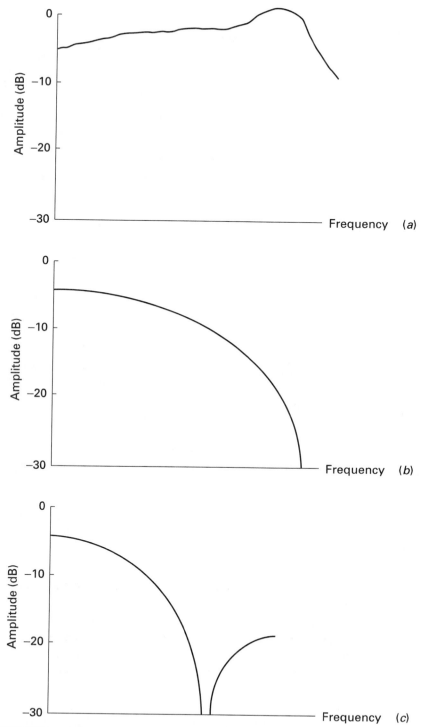

FIGURE 5.30 Transfer functions corresponding to the three impulse responses shown in Figure 5.29.

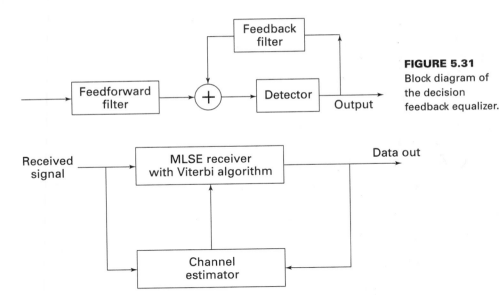

FIGURE 5.31
Block diagram of the decision feedback equalizer.

FIGURE 5.32 Block diagram of a MLSE equalizer.

5.8 SUMMARY

The effects of fading on data transmission have been studied in this chapter. The random nature of the received power due to fading increases the probability of error. In frequency-selective fading, the probability of error reaches a "floor" beyond which any increase in signal-to-noise ratio does not seem to lower the bit error rate. Similar effects are observed in the presence of random frequency modulation induced by relative motion of the transmitter with regard to the receiver.

The effects of flat fading may be reduced using diversity techniques. In diversity, multiple versions of the signal are created and then combined using some algorithm to result in a new signal that has an acceptable bit error rate. Other fading mitigation approaches use a multifrequency transmission technique known as orthogonal frequency division multiplexing (OFDM). A brief discussion of OFDM is given in Appendix B, Section B.1.

- The additional signal-to-noise ratio necessary to keep the bit error rate in fading channels the same as in unfaded channels (Gaussian channels) is referred to as the fading margin.
- Fading margins may be reduced using diversity techniques.
- Diverse signals may be created using the following methods: spatial diversity, frequency diversity, polarization diversity, and time diversity.
- These diverse signals must be uncorrelated to result in improved performance.
- Diverse signals may be combined using selection diversity, maximal ratio combining, and equal gain diversity.
- The maximal ratio combining algorithm has the best performance, followed by equal gain diversity and selection diversity.
- Some modulation schemes show better fading mitigation than others.

- Performance improvement goes up with an increase in the number of diverse signals.
- RAKE diversity operates on the basis of combining delayed versions of the signals arriving through a multipath. For this form of diversity to work, the duration of the pulse must be very short so that multiple versions of the signals coming through the various paths are resolvable.
- Macroscopic diversity is implemented to mitigate the effects of long-term fading. In this case, the multiple receivers must be far apart.
- Frequency-selective fading can be mitigated through the use of equalizers.

PROBLEMS

*** *Asterisks refer to problems better suited for graduate-level students.*

1. The instantaneous signal-to-noise ratio in a Rayleigh channel is given by A^2/N_0, where N_0 is the noise power. The average signal-to-noise ratio expected at the receiver is $\langle A^2 \rangle / N_0 = 3$ dB. A is Rayleigh distributed. To mitigate the effects of Rayleigh fading, the service provider is planning to use a six-channel diversity receiver with MRC. If outage occurs when the instantaneous SNR goes below 0 dB, calculate the outage probability with and without use of the diversity system.

2. In Problem 1, if the diversity is instead based on selection combining, calculate the outage probability.

3. If selection combining is to be carried out on the basis of envelopes instead of powers, derive an expression for the pdf of the selection diversity of a system with M diverse channels.

4. Plot the density functions of the selection combiner of Problem 3 for $M = 1, 3$, and 6.

5. Derive the expression for the selection combiner using order statistics.

6. Using MATLAB, generate 10 sets of Rayleigh-distributed random variables. Plot the samples and show that selection diversity will improve the performance. If the performance is defined in terms of $U = $ mean/std. dev., compute U with and without diversity.

7. Using MATLAB, plot the bit error rates for the Rayleigh-faded and unfaded cases for coherent BPSK. For error rates equal to 10^{-3} and 10^{-4}, calculate the power margin.

8. Consider a transmission system where the received signal $r(t)$ is expressed as

$$r(t) = As(t) + n(t),$$

where $s(t)$ is the transmitted signal and $n(t)$ is white Gaussian noise of power spectral density $N_0/2$. The parameter A is a scaling factor that is a random variable (similar to fading) having the following pdf:

$$f(a) = 0.5, \quad 0 \le a \le 2.$$

Calculate the average probability of error if a matched filter is used at the receiver. Assume that the modulation is BPSK.

9. The bit error rate for BPSK in the presence of phase mismatch is given by $\frac{1}{2}\text{erfc}(\sqrt{z}\cos\phi)$ (see Chapter 3). Proceed in a manner similar to Problem 8, considering the phase to be random (analogous to fading). If the phase is uniform in the range $(-20°, 20°)$, compare the bit error rates when ϕ is zero and when ϕ is random.

10. Evaluate the average error probability for a BPSK receiver for the case of a Rician-faded channel. Use MATLAB and plot the average BER for values of K ranging from -10 to 20 dB in steps of 5 dB. Comment on your results.

11. Evaluate the average error probability for a DPSK receiver for the case of a Rician-faded channel. Use MATLAB and plot the average BER for values of K ranging from -10 to 20 dB in steps of 5 dB. Comment on your results.

12. In Problem 8, if you are asked to use a diversity receiver to mitigate fading present in the signal, calculate the performance improvement arising from selection diversity.

13. Repeat Problem 8 if the pdf of a is given by

$$f(a) = 0.1\delta(a) + 0.5\delta(a-1) + 0.4\delta(a-2).$$

14. In Problem 13, if you are asked to use a diversity receiver ($M = 2$) to mitigate fading in the signal, calculate the performance improvement arising from selection diversity. (Proceed as in Problem 12.)

15. In Problem 13, if you are asked to use a diversity receiver ($M = 2$) to mitigate fading in the signal, calculate the performance improvement arising from equal gain diversity. (Proceed as in Problem 12.)

16. Derive the expression for the bit error rate for a DPSK system in the presence of Nakagami fading.

17. Plot the error rate for a BPSK system in the presence of Nakagami fading. Compare the results to those of Problem 10.

18. Derive the expression for the bit error rate in Rayleigh fading for BPSK.

19. Derive the expression for the bit error rate in Rayleigh fading for DPSK.

20. Repeat Problem 6 using a set of exponentially distributed random variables.

21. Compare the outages after diversity combining, using MRC for $M = 3$ and 4. The average signal-to-noise ratio of a single channel is 5 dB. The threshold signal-to-noise ratio required is 0 dB.

22. Repeat Problem 21 if the selection diversity technique is used.

23. Use MATLAB to compare outages for MRC and selection diversity. Use an average SNR of 3 dB and a threshold of 0 dB. Consider M of 1 through 8.

24. Explain the concept of excess power margin in the context of the modulation scheme, and coherent BPSK in the absence of fading and in the presence of fading. Generate a MATLAB plot of bit error rate versus excess power margin (dB).

25. Explain the concept of excess power margin in the context of the modulation scheme, and DPSK in the absence of fading and in the presence of fading. Generate a MATLAB plot of bit error rate versus excess power margin (dB).

26. Explain why the performance of digital communication systems in Rician fading is better than in Rayleigh fading.

27. Examine the relationship between power efficiency and spectral efficiency in the presence of Rayleigh fading using MATLAB (see Figure B.5.1 in Appendix B).***

28. Examine the relationship between efficiency and spectral efficiency in Rayleigh fading when diversity is used to mitigate fading (see Figure B.5.1 in Appendix B).***

29. Examine the relationship between power efficiency and spectral efficiency in the presence of Nakagami fading. Note that the signal-to-noise ratio z in Nakagami fading has the distribution given by

$$f(z) = \left(\frac{m}{z_0}\right)^m \frac{z^{m-1}}{\Gamma(m)} \exp\left(-m\frac{z}{z_0}\right) U(z),$$

where z_0 is the average signal-to-noise ratio and m is the Nakagami parameter. Use $m = 1, 1.5, 2, 5, 50$. Comment on the results and see what happens when $m = 50$.***

30. Based on the results of Problem 29, comment on the relationship between power efficiency and spectral efficiency when fading is Rician.***

31. Use the results of Problems 16 and 17 to show that diversity will improve the performance of the communication system. (Hint: Use the results showing that the sum of two Nakagami random variables, each with a parameter m, is another Nakagami random variable, with a parameter $2m$.)***

32. Generate a BPSK waveform subject to Rayleigh fading and noise. Demodulate the waveform in a manner similar to that used in the problem in Chapter 3. Compare the recovered data stream in conjunction with fading and no fading.***

33. Generate a number (M) of BPSK waveforms subject to fading and noise. Demodulate them. Combine the detected signals by adding them (a form of equal gain diversity) and then averaging the resultant waveform. For $M = 3, 5, 10, 15$, examine the results and compare them with those of Problem 32.***

CHAPTER 6

MULTIPLE-ACCESS TECHNIQUES

6.0 INTRODUCTION

Because of the limited amount of bandwidth available, it is necessary to explore methods to allow multiple users to share the available spectrum simultaneously. Three major approaches exist for accomplishing "sharing" of the resources (Coop 1979, Yue 1983, Pahl 1995, Jake 1974, Samp 1997, Stee 1994, Dixo 1994). These are FDMA, TDMA, and CDMA. In frequency division multiple access (FDMA), the available frequency band is divided among different users. In time division multiple access (TDMA), a fixed amount of spectrum is allocated to a number of users, who use the spectrum only for a very short period and share the time on a reserved basis. On the other hand, in code division multiple access (CDMA), the whole bandwidth is shared all the time by all the users, who use different codes (Vite 1979, 1985; Taub 1986; Shap 1994; Scho 1977, 1982). The three forms of multiple access are represented in Figure 6.1.

It is also important that users be allowed to transmit information to the base station and receive information from the base station simultaneously. In other words, systems must also have a full-duplex provision. This may be achieved in the frequency domain or in the time domain. In frequency division duplex (FDD) systems, two distinct bands of frequencies are provided for every user. These bands are separated by a guard band, as shown in Figure 6.2a. The forward band will provide the traffic flow from the base station to the mobile unit while the reverse band will provide the traffic flow from the mobile unit to the base station (see details on frequency bands in Section 6.4). For all users and whichever frequency bands they are using, the frequency split between the forward and reverse channels is always the same. Since the transmission and reception are undertaken using the same antenna, a duplexer (Figure 6.2b) is needed to separate the two signals. The carrier frequencies of the uplink and downlink should be separated by an amount sufficient to allow the use of low-cost methods to separate the two signals.

In time domain duplex (TDD) systems, time rather than frequency is used to separate the forward and reverse channels, as shown in Figure 6.3. Since the time split between the forward and reverse channels is very small, almost continuous transmission is possible. While a duplexer is required in FDD systems to separate the frequency bands for the forward and reverse channels, no such equipment is required in TDD systems.

The multiple access schemes are used in conjunction with the duplex configurations to provide simultaneous transmission of uplink and downlink information. We therefore have FDMA/FDD, TDMA/FDD, TDMA/TDD, etc.

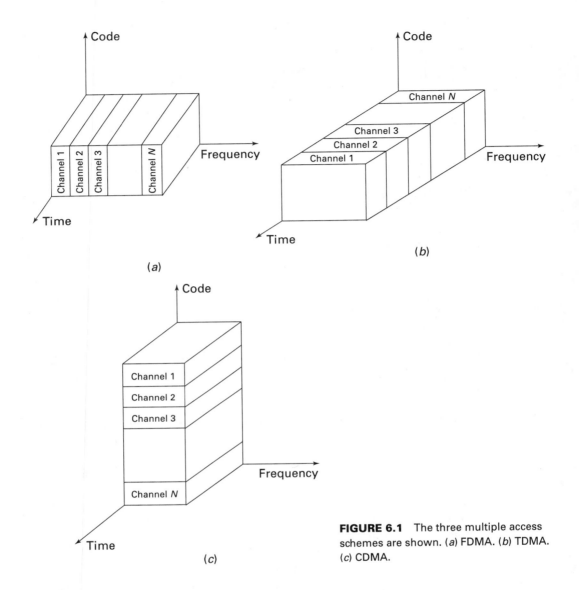

FIGURE 6.1 The three multiple access schemes are shown. (*a*) FDMA. (*b*) TDMA. (*c*) CDMA.

6.1 FREQUENCY DIVISION MULTIPLE ACCESS

FDMA is one of the simplest schemes used to provide multiple access. It easily separates different users by having each user operate at a different carrier frequency. The multiple users can therefore be isolated using bandpass filters. Frequency division multiple access is the mechanism used in analog cellular systems.

The principle of FDMA is shown in Figure 6.4. The available bandwidth W is divided into N nonoverlapping bands, each of width W_{ch}. A small guardband is provided so that interference from adjacent channels will be reduced in the event of any instability in the carrier frequencies of the neighboring channels. When a user makes a call request to the base station, the BS assigns one of the unused channels, which then becomes the exclusive "property" of that particular user, and nobody

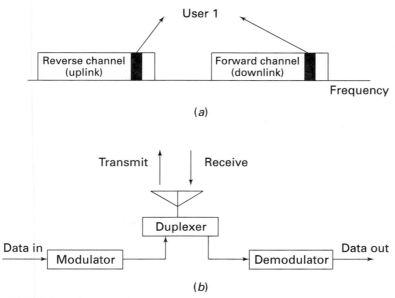

FIGURE 6.2 Concept of frequency domain duplex (FDD).

FIGURE 6.3 Concept of time domain duplex (TDD).

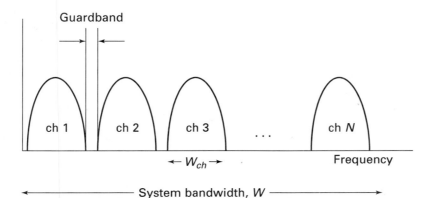

FIGURE 6.4 Principle of frequency division multiple access.

else will be assigned that channel. When the user terminates the call, the frequency may be reassigned to another user. If during the call, the caller moves into another cell, the caller will be assigned an unused channel in the new cell. If frequency domain duplex (FDD) is used, the available band is divided in two; one half is used for the forward channel, and the other half is used for the reverse channel. The caller has one frequency for the forward channel and another frequency for the reverse channel.

The major advantage of the FDMA system is its hardware simplicity, since discrimination between users is managed using simple bandpass filters. No timing information or synchronization is required. Since the bandwidth assigned each user is relatively small, the problems of frequency-selective fading are essentially nonexistent and the fading is purely flat. FDMA systems also have a number of major disadvantages. Let us briefly review them.

If a FDMA channel is not in use, it sits idle and cannot be used to enhance the capacity of the system. This is to say that the idle channel cannot be assigned to another cell unless some form of dynamic channel assignment is possible, in which unused channels may be assigned to the other cells that need more channels. Since the multiple access schemes rely heavily on bandpass filters, these filters must have excellent cutoff characteristics. The major problem in FDMA systems is the cross-talk arising from adjacent channel interference produced by nonlinear effects. The many channels that compose the FDMA system use the same antenna and, therefore, the associated power amplifiers. Since amplifiers have some level of nonlinearity, intermodulation products will result.

Consider a simple example of a three-channel case where the composite signal, $c(t)$, at the receiver can be expressed as

$$c(t) = a_1(t)\cos(2\pi f_1 t) + a_2(t)\cos(2\pi f_2 t) + a_3(t)\cos(2\pi f_3 t), \qquad (6.1)$$

where f_1, f_2, f_3 are the carrier frequencies and $a_1(t), a_2(t), a_3(t)$ are information-bearing signals. The output of a nonlinear amplifier, $c_{out}(t)$, will be

$$c_{out}(t) = b_0 + b_1[c(t)] + b_2[c(t)]^2 + b_3[c(t)]^3 + \cdots \qquad (6.2)$$

where the b_i are the scaling factors. Depending on the ratio of the carrier frequencies, the nonlinear terms can result in terms of the type

$$f_1 = 2f_2 - f_3 \qquad (6.3)$$

or any other combination such that signals from other channels will appear in the same frequency window of the signal being received, leading to interchannel interference. An appropriate frequency planning system can reduce the cross-talk induced by intermodulation.

A problem with the FDMA system is its inability to be used in variable-rate transmission, which is becoming common in digital systems. Variable-rate transmission makes it necessary to employ a number of modems at the terminal. This eliminates FDMA as the choice for combined voice and data transmission.

Another drawback of the FDMA system is its inherent need for transmitters and receivers with high Q values to ensure excellent channel selectivity. Monitoring this may also be difficult.

6.2 TIME DIVISION MULTIPLE ACCESS (TDMA)

The TDMA technique enables users to access the whole bandwidth, which is allocated on a time basis (Skla 1988, Naka 1990, Samp 1997). Each user/channel occupies the whole bandwidth for a fraction of the time, called a slot. But if the user continues to have access or reservation to the bandwidth on a periodic or rotational basis, it is possible for the user to carry on the conversation or transmission of information on a nearly continual basis. If there are N users, we have a frame of N time slots, as shown in Figures 6.5 and 6.6.

A conceptual setup of the TDMA scheme is shown in Figure 6.6. The uplink (to the base station) consists of transmission from each user at specific time intervals. The

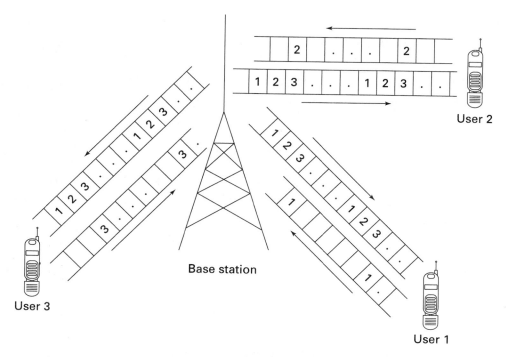

FIGURE 6.5 A number of terminals communicating with a base station in the TDMA scheme.

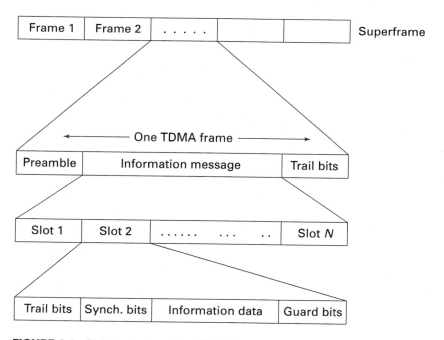

FIGURE 6.6 Details of a complete TDMA frame.

downlink consists of signals corresponding to all the users in the frame. As we can see from Figure 6.6, the uplink consists of signals only at the assigned time slots. A number of frames can be combined to produce a superframe. Each frame is made up of a preamble, an information message, and trail bits. In Figure 6.6, the preamble consists of the address and synchronization information that both the base station and the subscribers can use for identification. If time domain duplex (TDMA/TDD) is used, half the time slots in the frame will correspond to the uplink channels and the other half will correspond to the downlink channels. On the other hand, if frequency domain duplex (TDMA/FDD) is used, the downlink channels and uplink channels will be separate frames, at different carrier frequencies. As shown in Figure 6.5, several time slots of delay are introduced between the uplink and downlink time slots for a particular user.

One of the important issues in TDMA systems is the synchronization. If the time slot synchronization is lost, the channels may collide with each other. To maintain synchronization, the base station periodically sends a frame timing signal, which the MUs use for synchronization. In fact, we need synchronization of the slot, frame, and superframe. While susceptibility of FDMA to fading is not severe, TDMA systems suffer from frequency-selective fading. Processing to overcome the effects of fading must be undertaken. On the positive side, TDMA makes it relatively easy to measure the power levels, and hence to perform hand-off procedures.

6.3 CODE DIVISION MULTIPLE ACCESS

Most of the systems described in the previous chapters achieve multiple access using a frequency division technique, a time division multiple access technique, or a combination of the two. These techniques are used in environments where there is a premium on the available bandwidth. On the other hand, if a large chunk of bandwidth is available, it is possible to utilize this large bandwidth for a single user (Coop 1979; Pick 1982; Shap 1994; Yue 1983; Vite 1979, 1985; Wang 1993). This, however, is an extremely inefficient way of utilizing the available bandwidth. Highly efficient use of the bandwidth can be accomplished if a large number of users can occupy the same bandwidth at all times and if each user is assigned a different set of "bits" or codes. In other words, it is possible to share the bandwidths through code division multiple access or CDMA, where each user is assigned a pseudorandom or pseudonoise (PN) binary-valued sequence or code (MacW 1976).

This distinguishing feature of the CDMA system, namely, the wider bandwidth, involves nothing but spreading of the spectrum of the transmitted signal through the use of much narrower pulses (Scho 1977, 1982; Skla 1988; Taub 1986; Schi 1994; Proa 1995; Wu 1995). Thus the spectrum of the transmitted signal is wider than the spectrum associated with the data rate. To illustrate, the spectrum of the signal associated with the data rate is shown in Figure 6.7 along with the time domain pulse. Now, if we were to transmit a set (K) of extremely narrow pulses during the period T, with randomly chosen values of amplitude (+ or -1), each of duration $T_c = T/K$, the spectrum of the transmitted signal would become wider. The duration of these narrow pulses, or chips, as they are commonly referred to, is T_c. This spreading of the spectrum, or the spread-spectrum approach, is the key to implementation of the CDMA technique.

The basic principle of the CDMA technique is illustrated in Figure 6.8. The input bit stream and the chip sequence are shown. The number of chips/bit is K. Once the data sequence is multiplied by the chip sequence, the spectrum of the information is spread.

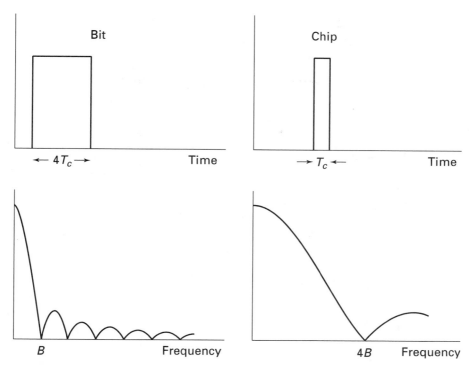

FIGURE 6.7 A bit of duration $T = 4T_c$ and a chip of duration T_c are shown along with their spectra. In this figure, $K = 4$.

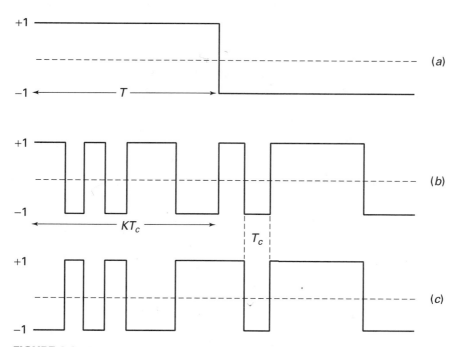

FIGURE 6.8 (*a*) Input data stream. (*b*) Chip sequence. (*c*) Modulo-2 addition of the bit stream and the chip sequence.

The PN sequence is unique to each user and is almost orthogonal to the sequences of other users. Thus, the interference from other users in the same band will be much less. The number of orthogonal codes, however, is limited. As the number of users increases, the codes become less and less orthogonal and the interference increases. This allows a soft capacity limit for allowing new users into the system. As more and more users enter the system, the quality of the channel goes down, but nobody needs to be turned away. This also makes the frequency planning and design of cells relatively easy, since all users are using the same frequency band. This is in sharp contrast to systems based on TDMA/FDMA, where there is a hard capacity limit and a need for strict frequency planning and cell design.

In addition to the advantages of soft capacity and the relative ease of frequency planning, the use of very short pulses provides an easy means to combat the effects of multipath fading. In non-CDMA systems, the multipath effects lead to broadening of the pulses owing to the nonresolvable paths. In CDMA, the pulses are very narrow, and therefore the multipath fading produces nonoverlapping, resolvable pulses at the receiver, corresponding to the distinct paths. This resolvable multipath scenario is akin to multipath diversity since each of these resolvable paths corresponds to a different branch of the diversity system. These nonoverlapping pulses can be combined to combat the effects of fading using a RAKE receiver, which is nothing more than a means of combining the multipath pulses with appropriate delays and weights, as we will see later in this chapter.

The spread-spectrum system may use either of the two modulation schemes PSK or FSK. The PN sequence in conjunction with PSK generates phases of 0 and π pseudorandomly in accordance with the code, at a rate that is an integral multiple of the data rate. This approach results in direct-sequence systems (DS-CDMA). On the other hand, if the PN sequence is used in conjunction with M-ary FSK to select the frequency of the transmitted signal pseudorandomly (Skla 1988, Pick 1982), the result is a frequency-hopped direct-sequence system (FH-CDMA).

6.3.1 Description of a PN Code Generator

The pseudonoise (PN) sequence is a periodic binary sequence that appears noiselike. It can be generated using a feedback shift register, as shown in Figure 6.9. The setup consists of a number of flip-flops (two-state memory stages) and a logic element, typically a modulo-2 adder, interconnected to form a feedback circuit. A single timing clock regulates all of the flip-flops.

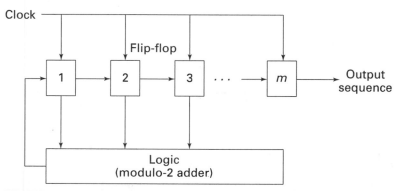

FIGURE 6.9 Block diagram of a PN code generator.

6.3.2 Properties of Pseudonoise or Pseudorandom Sequences

A PN sequence is a collection of positive and negative 1s, in almost equal numbers, occurring randomly. Using the feedback shift register of m stages shown in Figure 6.9, we expect to have $2^m - 1$ sequences of numbers, also known as a maximal-length sequence. The sequence repeats itself, and thus has a period of $2^m - 1$.

Some of the properties of the maximal-length sequences are as follows.

- In each period of a maximal-length sequence, the number of $+1$s is exactly one more than the number of -1s. This is known as the *balance property*.
- The autocorrelation function of a maximal-length sequence is binary valued and periodic.

Even though the sequence is deterministic and, therefore, not completely random, it appears like white noise to an unauthorized/unknown user. The autocorrelation of the code can therefore be identified with the autocorrelation of band-limited white noise. A typical code sequence and its autocorrelation are shown in Figure 6.10. Figure 6.10*a* shows the autocorrelation, $R(\tau)$, of white noise, and Figure 6.10*b* shows the autocorrelation, $R_{PN}(\tau)$, of the PN code. We can see that the code decorrelates with shifted versions of the code, as evidenced by the fast dropoff of the autocorrelation. Note that when the code length becomes infinite, the autocorrelation of the PN code and that of the band-limited white noise become equal.

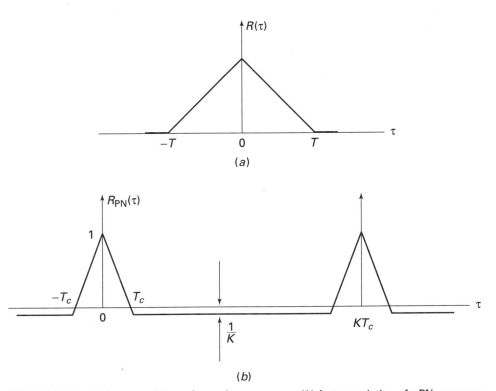

FIGURE 6.10 (*a*) Autocorrelation of a random sequence. (*b*) Autocorrelation of a PN sequence of length K. The chip duration is T_c.

This decorrelation property of the code makes it possible to use time diversity to combine delayed versions of the signals when a multipath exists. We will examine this aspect in Section 6.3.4.

6.3.3 Direct-Sequence Spread-Spectrum Modulation

The direct-sequence spread-spectrum modulated signal (Skla 1988; Taub 1986; Pick 1982, 1991; Pado 1994) can be generated using the setup shown in the block diagram of Figure 6.11. The bit duration is T and the chip duration is T_c, where a "chip" is identified as a single pulse of the PN waveform. The ratio of T to T_c gives the number of chips in a bit. The encoded data are added to the PN code chips in a modulo-2 fashion before being modulated using a BPSK scheme.

The transmitted signal, $s_{DSSS}(t)$, can be expressed as

$$s_{DSSS}(t) = \sqrt{\frac{2E}{T}}m(t)p(t)\cos(2\pi f_0 t + \theta),\tag{6.4}$$

where $m(t)$ is the encoded data, $p(t)$ is the PN chip sequence, and θ is the phase at $t = 0$. Note that $m(t)$ consists of data symbols (± 1) of duration T while $p(t)$ consists of chips (± 1) with a chip duration of T_c, with $T_c \ll T$, with the provision that $T = KT_c$, where K is an integer. The very short duration of the chips also means that the bandwidth of the DSSS signal is K times the bandwidth of conventional BPSK, where the symbol duration is T. This increase in bandwidth is shown in Figure 6.7. We will come back to the issue of increased bandwidth when we discuss interference suppression in DS-CDMA systems.

The receiver for the DSSS system (CDMA-R) is shown in Figure 6.12. The received signal is multiplied by the PN code and filtered. This, in effect, results in a correlator. The output is passed through a BPSK demodulator to recover the original data.

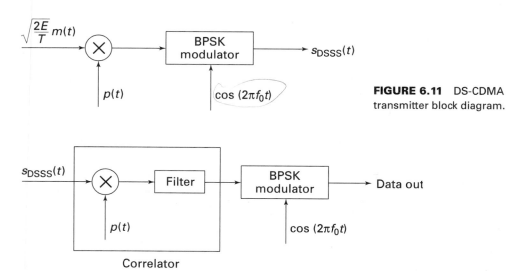

FIGURE 6.11 DS-CDMA transmitter block diagram.

Correlator

FIGURE 6.12 Block diagram of a DS-CDMA receiver.

The question arises about the expected probability of error in this form of modulation/demodulation. The multiplication by the PN code $p(t)$ at the transmitter, and later at the receiver stage, results in no change in the signal level since the product is always unity. Thus, the overall performance of the DSSS system in an ideal case should have no effect from the presence of the PN code as far as thermal noise is concerned. The error probability is once again given by the error probability for the BPSK receiver, eq. (3.83):

$$p_{\text{DSSS}}(e) = p_{\text{BPSK}}(e) = \frac{1}{2}\text{erfc}\left(\sqrt{\frac{E}{N_0}}\right). \tag{6.5}$$

Even though the performance of the DSSS system is no better than that of a pure BPSK system in the presence of thermal noise, the DSSS system has an exceptional ability to suppress in-band interference. This can be understood by treating the interference as akin to noise. Consider the case of a single-tone interference. The input to the receiver, $c_{\text{in}}(t)$, can be written as

$$c_{\text{in}}(t) = m(t)p(t)s(t) + A_{\text{int}}s(t), \tag{6.6}$$

where the second term is the interfering term, with A_{int} being a scaling factor. In eq. (6.6),

$s(t)$ is a BPSK carrier signal.

$m(t)$ is the bipolar data.

$p(t)$ is the PN code.

For A_{int} equal to 1, the powers of the signal of interest and the interferer are the same. If A_{int} is larger than 1, the interfering signal will be stronger, and this will lead to problems. The issue of this "near/far problem" will be discussed later in this chapter. The spectra of the signal and interferer are shown in Figure 6.13.

Note that the interfering term has a much smaller bandwidth since it has not been multiplied by the PN code, $p(t)$, while the signal spectrum is broad. Input $c_{\text{in}}(t)$ is first multiplied by the code $p(t)$ and then applied to a BPSK demodulator. The input to the modulator can be expressed as

$$c_{\text{in}}(t) = \text{signal} + \text{noise}, \tag{6.7}$$

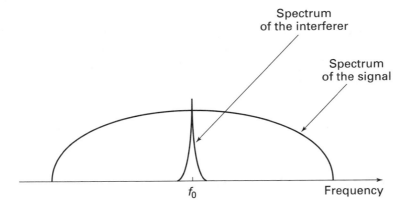

Spectrum of the interferer

Spectrum of the signal

f_0

Frequency

FIGURE 6.13
Spectrum of the signal along with the spectrum of the interferer during transmission.

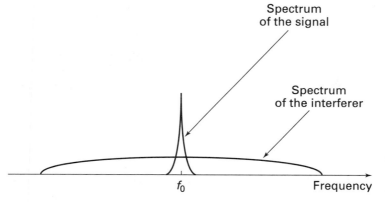

FIGURE 6.14
Spectrum of the signal along with the spectrum of the interferer after correlation.

where the signal term is identical to the signal component at the output of the BPSK receiver (since $p(t) \times p(t) = 1$) and the noise is given by

$$\text{Noise} = A_{int}p(t)n_{int}(t). \tag{6.8}$$

The value $n_{int}(t)$ is the interfering signal that has the same power as the primary signal. Note that the spectrum of the *signal* now is narrower, while the spectrum of the *noise* is broader because of the presence of the spreading code $p(t)$. The bandwidth of the *noise* is now K times the bandwidth of the *signal* component. This is shown in Figure 6.14.

We can now calculate the error probability by treating the second term in eq. (6.6) as noise (neglecting thermal noise):

$$p(e) = \frac{1}{2}\text{erfc}\sqrt{\frac{E}{N_0}} = \frac{1}{2}\text{erfc}\sqrt{\frac{P_s T}{P_{int}T_c}} = \frac{1}{2}\text{erfc}\sqrt{\frac{K}{A_{int}^2}}, \tag{6.9}$$

where P_s is the signal power and P_{int} is the interfering signal power. As the value of K increases, the length of the code increases and the error from interference decreases. We can therefore regard K as the processing gain from the use of the spread-spectrum technique. For $A_{int} = 1$ (equal power of the interfering signal), the error probability becomes

$$p(e) = \frac{1}{2}\text{erfc}\sqrt{K}. \tag{6.10}$$

We can now consider the case of DS-CDMA. The block diagram of the receiver is shown in Figure 6.15. The length of the code is once again KT_c, and we consider the case where there are k users in a given frequency band. We also assume that the k codes are almost uncorrelated with one another. The composite received signal, $c_k(t)$, at any one of the receivers will be

$$c_k(t) = \sum_{i=1}^{k} m_i(t)p_i(t)s(t), \tag{6.11}$$

where it has been assumed that all the k signals, identified by $m_i(t)$, and the corresponding PN codes, $p_i(t)$, have the same energy. The carrier signal, $s(t)$, is $\cos(2\pi f_0 t)$.

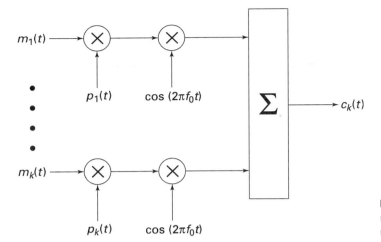

FIGURE 6.15 DS-CDMA receiver showing all the multiple channels.

Consider the case of a receiver of the first channel ($k = 1$). The output, $c_{k1}(t)$, of the multiplier with $p_1(t)$ will be

$$c_{k1}(t) = \sum_{i=1}^{k} m_i(t)p_i(t)p_1(t)s(t). \qquad (6.12)$$

Once again, this signal will be the input to a BPSK demodulator, as shown in Figure 6.16. The input, $c_{k1\text{in}}(t)$, to the demodulator can now be expressed as

$$c_{k1\text{in}}(t) = \text{signal} + \text{noise} \qquad (6.13)$$

where the signal comes from the first term of the summation in eq. (6.12) and the noise represents all the other ($k - 1$) terms, given by

$$\text{Noise} = \sum_{i=2}^{k} m_i(t)p_1(t)p_i(t). \qquad (6.14)$$

Assuming that all the codes are nearly uncorrelated, we can see that there will be $k - 1$ interfering components, each with the same power as the signal, and, hence, the error probability from the interfering signals becomes

$$p(e) = \frac{1}{2}\text{erfc}\left(\sqrt{\frac{K}{k-1}}\right). \qquad (6.15)$$

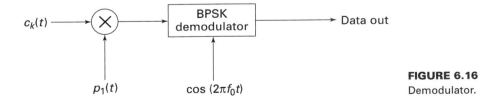

FIGURE 6.16
Demodulator.

Note, however, that it is possible to have the powers of the interfering signals be higher than the power of the signal of interest being received, creating significant variation in the performance.

EXAMPLE 6.1

In a DS-CDMA cell, there are 24 equal power channels that share a common frequency band. The signal is being transmitted using a BPSK format. The data rate is 9600 bps. A coherent receiver is used for recovering the data. Assuming the receiver noise to be negligible, calculate the chip rate needed to maintain a bit error rate of 10^{-3}.

Answer Assuming there is no thermal noise, the bit error rate is given by eq. (6.15), where K is the processing gain and k is the number of channels. BER $= 10^{-3} = 0.5$ erfc($\sqrt{z}$), where $z = K/(K-1)$. Using the MATLAB function *erfinv(.)*, we can solve for $z = [erfinv(1 - 2*BER)]^2$.

We get $z = 4.77$. We are given $k = 24$; $K = 23 \times 4.77 = 109.82$. Since $K =$ (chip rate)/ (data rate), chip rate $= 109.82 \times 9600 = 1.05$ Mchips/s. ∎

EXAMPLE 6.2

A DS-CDMA system is expected to have a processing gain of 30 dB. The expected data rate is 9600 bps. What should the chip rate be? If BPSK modulation will be employed, what is the bandwidth required for transmission using a null-to-null criterion?

Answer Processing gain $= 30$ dB $= 10\log_{10}(K)$; $K = 1000$; data rate $= 9600$ bps

Chip rate $= K \times$ data rate $= 9,600,000 = 9.6$ Mchips/s.

BW $= 2 \times 9.6$ MHz $= 19.2$ MHz. ∎

6.3.4 RAKE Receiver in DS-CDMA Systems

The multipath effects present in wireless systems adversely impact the TDMA-based systems since the multiple pulses arriving at the receiver are nonresolvable (see Chapter 2). These pulses thus overlap, leading to broadening of the pulse, resulting in frequency-selective fading as discussed in Chapter 2. However, the chip duration in DS-CDMA systems is very narrow, and the delays between multiple paths may be larger than the chip duration. Under this condition, the delayed versions of the chips are resolvable, providing time diversity (Skla 1993; Chan 1994a; Pick 1982, 1991; Kohn 1995; Proa 1995; Pani 1996). The fact that the PN sequences have very low correlation makes it possible to separate these delayed versions and separately perform the correlation. The block diagram in Figure 6.17 shows the concept of a RAKE receiver implementation for DS-CDMA systems.

The different correlators (see Chapter 3 for a discussion of correlators) can be synchronized to various paths with different delays and programmed to capture the strongest signals coming from multipath components. Note that the signal arriving at any given time is synchronized with one of the correlators and, as a result, will have negligible correlation with the other two correlators by virtue of the low correlation of the code. The outputs from the correlators are appropriately weighted and combined for decisionmaking and recreation of the data bits.

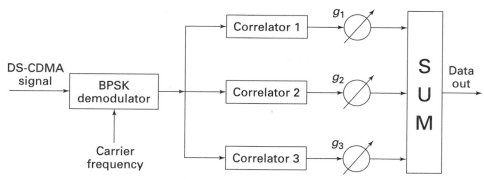

FIGURE 6.17 Conceptual block diagram of a RAKE receiver. Three multiple paths are arriving and being processed in three separate correlators. Each output is weighted by a factor g_i, $i = 1, 2, 3$.

EXAMPLE 6.3

A RAKE receiver is being designed to take advantage of the multipath effects in the channel. If the minimum delay difference is 300 m, what is the minimum chip rate necessary to successfully resolve the multipath components and operate the RAKE receiver?

Answer   If the chip rate is not sufficient, the multiple paths corresponding to the chips will not be resolvable, and the requirement of separable pulses will not be met. This means that the chip duration must be smaller than

$$\tau = \frac{\text{delay distance}}{\text{speed of the e.m. wave}} = \frac{300}{3 \times 10^8} = 1\,\mu s$$

The chip rate must be greater than $1/\tau = 1$ Mchip/s.   ◼

6.3.5 Frequency-Hopping Spread-Spectrum Technique

While the direct-sequence spread-spectrum technique uses phase modulation (and may also be viewed as amplitude modulation), the frequency-hopping spread-spectrum technique uses frequency modulation. In fact, frequency hopping involves the hopping of the carrier frequency in a random fashion. The set of possible frequencies used is referred to as a *hopset* (Skla 1988, Taub 1986, Kohn 1995). The starting point of a FH-SS technique is either a BFSK signal or a MFSK signal. If one considers a conventional BFSK signal, the carrier frequency is alternated between two *fixed* frequencies. In a FH/BFSK system, the data symbol modulates a carrier whose frequencies are *pseudo-randomly* determined. This statement can be understood with the aid of the block diagram of a FH/BFSK modulator shown in Figure 6.18.

The output of a conventional BFSK modulator and the output from a digital frequency synthesizer are applied to a mixer. The bandpass filter selects the sum frequency coming out of the product modulator. The successive k-bit segments of the PN sequence drive the frequency synthesizer, enabling the carrier frequency to hop over 2^k distinct values. For a given hop, the bandwidth of the FH/BFSK signal is the same as for conventional BFSK. Over many hops (2^k), the bandwidth occupied by the transmitted signal will be orders of magnitude higher than for conventional BFSK. Indeed, it is possible to have a spectral spreading in FH/BFSK far exceeding that

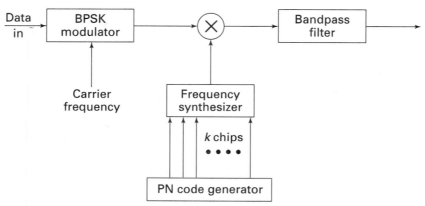

FIGURE 6.18 FH-CDMA transmitter.

observed in direct-sequence techniques. Consequently, the processing gain in FH/BFSK systems can be superior to that in DS-SS systems.

The rate at which the frequency hops determines whether the system is a slow-hopping or a fast-hopping system. We have *slow frequency hopping* (SFH) if the symbol rate, R_s, is higher than the hopping rate, R_h. This means that several symbols will be transmitted on each frequency hop. If, on the other hand, R_h is higher than R_s, we have *fast frequency hopping* (FFH), where the carrier frequency will change several times during the transmission of a single symbol. The term *chip* is defined differently in the frequency hopping context. In the DS systems, a "chip" refers to the shortest duration. In FH systems, a "chip" refers to the shortest uninterrupted waveform in the system. The chip rate, R_c, for a FH system is

$$R_c = \max[R_h, R_s]. \tag{6.16}$$

Figure 6.19 demonstrates the difference between FFH and SFH. In Figure 6.19*a*, we see that the symbol rate is 20 symbols/s while the frequency hopping rate is 40 hops/s. The chip rate here is the hop rate, since it is the maximum of the two (1 chip = 1 hop). This is FFH. In Figure 6.19*b*, the data rate is still 20 symbols/s but the hopping rate is reduced to 6.66 hops/s. In this case, changes in the waveform are due to modulation, and therefore, the chip rate is the symbol rate (1 chip = 1 symbol) since the data symbol is shorter than the hop duration. This is an example of SFH.

A typical demodulator setup for frequency-hopping spread-spectrum is shown in Figure 6.20. Since the demodulator is essentially a noncoherent BFSK receiver, the performance of FH spread-spectrum systems can be evaluated. The probability of error, $p(e)$, associated with a noncoherent BFSK receiver is given by

$$p(e) = \frac{1}{2}\exp\left(-\frac{E}{2N_0}\right). \tag{6.17}$$

6.3.6 Comparison of DS and FH Systems

The bandwidth of DS systems is related to the PN sequence clock rate or chip rate. The bandwidth of FH systems depends on the tuning range of frequencies and can, with ease, be hopped over a wide bandwidth. The synchronization/timing becomes very crucial in DS systems because of the extremely small chip duration. The timing is less critical in FH systems since hop rates range up to several thousands/s compared with Mbit/s rates

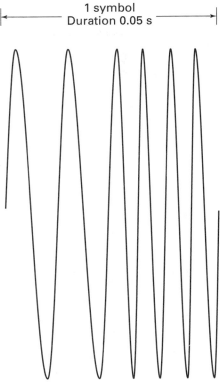

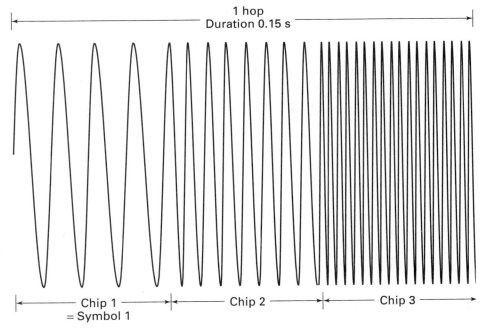

FIGURE 6.19 (*a*) Fast frequency hopping. There are two hops (chip 1 and chip 2) for each symbol. (*b*) Slow frequency hopping. There are three symbols in a hop.

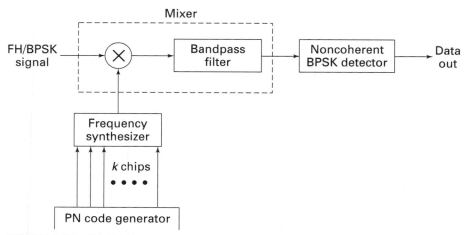

FIGURE 6.20 FH-CDMA demodulator.

in DS systems. The spectrum of the DS system always appears very wide, as it should, while the spectrum of the FH systems is narrow; but the FH spectrum center frequency is changed several times. DS and FH systems are ideal candidates for multipath diversity. The DS system uses extremely short chips, resulting in time diversity. The FH system has an inherent frequency diversity present in fast hopping, with many frequencies used over one data period. While near/far problems are more likely to occur in DS systems because of sharing of the same frequency/bandwidth, they are less likely to occur in FH systems because of the different frequencies.

6.4 OVERVIEW OF WIRELESS SYSTEMS AND STANDARDS

We will now briefly compare the characteristics and features of the major wireless communication systems: AMPS, North American Digital Systems, IS 54 and IS 95, Japanese Digital Cellular Systems, and the Pan-European GSM system.

6.4.1 Advanced Mobile Phone Systems (AMPS)

The first-generation mobile communication systems based on AMPS use frequency modulation for the transmission of signals. For the reverse link (uplink) from the MU to the base station, the frequency band of 824–849 MHz is used. For the forward link (downlink) from the base station to the MU, the frequency band of 869–894 MHz is used. Between the uplink and downlink communication in simplex mode, a separation of 45 MHz exists between radio channels. This large separation permits the use of low-cost duplexers. The maximum frequency deviation of the FM modulator is ±12 kHz. The control channel transmissions and blank-and-burst data streams are transmitted in FSK mode at 10 kbps with a maximum frequency deviation of ±8 kHz. Each base station transmits control data in FSK mode on the forward control channel (FCC) at all times, allowing the MU to lock onto the strongest FCC wherever it is. The base station reverse control

TABLE 6.1 Characteristics of AMPS

Parameter	AMPS Specification
Multiple access	FDMA
Duplex technique	FDD
Channel bandwidth	30 kHz
No. voice channels/rf channel	1
Uplink (reverse channel) band	824–849 MHz
Downlink (forward channel) band	869–894 MHz
Modulation technique (voice)	FM
Modulation technique (control channel)	FSK
Peak frequency deviation	±12 kHz (FM); ±8 kHz (FSK)
Data rate on control channel	10 kbps
Spectral efficiency	0.33 bps/Hz
Total number of voice channels	832

channel continuously monitors the transmission from the MU. The wideband FSK data is used in a blank-and-burst mode to initiate hand-offs to adjoining cells, change the MU transmit power, and, if necessary, initiate a hand-off to another channel in an "intracell hand-off mode" in places where sector antennae are used. Note that these "blank-and-burst" events are of extremely short duration (~100 ms), and any interruptions caused by them are not discernible. The typical parameters associated with AMPS are shown in Table 6.1.

To limit the peak frequency deviation to a preset value, the voice signal is passed through a compander and associated electronics, as shown in the block diagram in Figure 6.21. Since human speech has a large dynamic range, it is necessary to limit the amplitude range by compressing it in a nonlinear fashion, that is, by a 1 dB increase at the output of the compander for every 2 dB of increase in input. This confines the signal energy to the allocated bandwidth of 30 kHz. Note that at the receiver, an inverse process recovers the original speech characteristics.

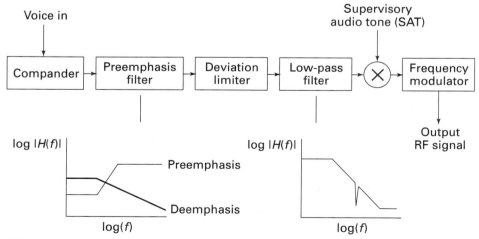

FIGURE 6.21 Block diagram of the transmitter associated with an AMPS system.

The preemphasis filter improves the performance of the FM and has the transfer function shown in the figure. At the receiver, a deemphasis filter performs the inverse operation to preserve the signal characteristics. The low-pass filter, with a notch at about 6 kHz, serves two purposes: It limits the spectrum of the signal going into the modulator and provides a place to insert the supervisory audio tone (SAT) signal at around 6 kHz. The SAT signal allows the BS and MU to confirm the existence of a proper link. In addition to SAT, the signals from the BS contain a 10 kHz sinewave supervisory tone (ST), which is used to indicate the "call termination" process. This tone thus acts like an on-off hook feature of a plain old telephone system (POTS).

The control data channel uses Manchester code binary frequency shift keying at a rate of 10 kbps.

Hand-off Procedure The hand-off process from one base station to another is controlled by the MTSO (mobile telephone switching office). The MTSO measures the signal strengths from the current BS and the neighboring ones. Initiation of the hand-off is based on two thresholds. For a current user, if the power is around −100 dBm, a hand-off is initiated. A threshold of −90 dBm is set for a new call.

6.4.2 United States Digital Cellular (IS 54)

The AMPS system has a limited ability to support the need for increased capacity. Digital systems offer higher capacity and the possibility of advanced features such as voice mail and paging. The U.S. digital system was designed to operate using the same frequency range, channel bandwidth, and frequency reuse plans as AMPS.

Some of the characteristics of IS 54 are given in Table 6.2. The radio-frequency (RF) channel bandwidth is 30 kHz, the same as for AMPS; however, in IS 54, three users share the bandwidth, thereby using only 10 kHz/voice channel. The gross bit rate is 48 kbps, giving each user 16.2 kbps. This individual rate is split as follows: speech = 7.95 kbps, error correction = 5.05 kbps, control channel = 0.6 kbps. This leaves 2.6 kbps as the overhead for such things as guard time, ramp-up, and synchronization, resulting in an overhead rate of 2.6/16.2 or 16%.

The modulation scheme used is $\pi/4$-DQPSK, with a symbol rate of 24.3 kbps and a symbol duration of 41.1523 μs. Raised cosine pulse shaping is used to reduce the effect

TABLE 6.2 Characteristics of U.S. Digital Systems (IS 54)

Parameter	Specification
Multiple access	TDMA/FDMA
Duplex	FDD
Channel bandwidth	30 kHz
Uplink (reverse channel) band	824–849 MHz
Downlink (forward channel) band	869–894 MHz
Modulation technique (voice)	$\pi/4$-DQPSK
Forward/reverse channel data rate	48.6 kbps
Spectral efficiency	1.62 bps/Hz
Number of users/channel	3 at 7.95 kbps/user
Equalizer	Unspecified

of intersymbol interference. The specific demodulation method varies from manufacturer to manufacturer.

Hand-off Procedure The hand-off procedure employed in the U.S. digital system is different from the technique used in AMPS, where MTSO controls the sequence of events. In mobile-assisted hand-off (MAHO), the responsibility for the hand-off rests with the mobile unit. In this scheme, the MU measures the quality of the signal received at the mobile unit from all the base stations and continuously communicates with its base station. The MU initiates a hand-off when it determines that the power from another base station is higher than the power from its own base station by a certain amount. Thus the process is quicker than MTSO-initiated hand-off.

6.4.3 Japanese Digital System (JDC)

The Japanese digital system is remarkably similar to the American Digital Cellular (ADC) system based on IS 54. The only difference is the use of a 25 kHz RF channel bandwidth instead of the 30 kHz used in ADC. Once again, three voice channels are present in every RF channel. The modulation technique employed is $\pi/4$-DQPSK. The gross bit rate of 42 kbps is lower than the rate for ADC, resulting in a spectral efficiency of 1.68 bps/Hz. The difference of 2.8 kbps between 14 kbps/user and 11.2 kbps/user, for protected speech, results in an overhead rate of $2.8/14$ or 20%. The JDC specifications are given in Table 6.3.

6.4.4 GSM

Global System for Mobile (GSM) is a second-generation cellular system widely used throughout the world (Europe, Australia, and parts of Asia and the United States). Some specifications of GSM are given in Table 6.4. The spectral efficiency of GSM is less than the spectral efficiencies of systems based on linear modulation schemes such as the ones used in ADC and JDC.

TABLE 6.3 Characteristics of Japanese Digital Cellular

Parameter	*Specification*
Multiple access	TDMA/FDMA
Duplex	FDD
Channel bandwidth	25 kHz
Uplink (reverse channel 1) band	810–826 MHz
Downlink (forward channel 1) band	940–956 MHz
Uplink (reverse channel 2) band	1429–1441 MHz
Downlink (forward channel 2) band	1477–1489 MHz
Uplink (reverse channel 3) band	1453–1465 MHz
Downlink (forward channel 3) band	1501–1513 MHz
Modulation technique (voice)	$\pi/4$-DQPSK
Forward/reverse channel data rate	42 kbps
Spectral efficiency	1.68 bps/Hz
Number of users/channel	3 at 6.7 kbps/user
Equalizer	Unspecified

TABLE 6.4 Characteristics of GSM

Parameter	Specification
Multiple access	TDMA/FDMA
Duplex	FDD
Channel bandwidth	200 kHz
Uplink (reverse channel) band	890–915 MHz
Downlink (forward channel) band	935–960 MHz
Modulation technique (voice)	GMSK (BT = 0.3)
Forward/reverse channel data rate	270.83333 kbps
Spectral efficiency	1.35 bps/Hz
Number of users/channel	8

A unique feature of GSM is the slow frequency hopping provision. In this mode of operation, the radio carrier follows a frequency hopping pattern instead of occupying the same radio frequency. This can reduce the effects of distortion present at a given radio frequency. Frequency hopping can also reduce the effects of co-channel interference.

Hand-off Procedure The hand-off procedure followed in GSM is based on MAHO. The MU performs channel quality measurements, and hand-off is initiated in a fashion very similar to the approach discussed in connection with ADC.

6.5 NORTH AMERICAN DIGITAL CELLULAR SYSTEMS BASED ON CDMA

As discussed in Section 6.3, systems based on code division multiple access can achieve the goal of increased capacity. The CDMA system based on IS 95 became operational in 1996.

There are several differences between the wireless systems based on IS 54 and the ones based on CDMA (IS 95). In the second-generation wireless systems, the forward and reverse links use a similar mode of operation in terms of the modulation technique. CDMA-based systems use different forms of modulation as well as different modes of data encoding in the two directions (Whip 1994, Pado 1994). The other fundamental difference is the use of variable-bit-rate traffic in CDMA versus the constant data rate used in ADC, JDC, and GSM. Since CDMA uses a larger bandwidth, the question of frequency reuse does not normally arise. CDMA has a reuse factor of unity, as shown in Figure 6.22b, compared with the standard seven-cell reuse pattern shown in Figure 6.22a. The reverse link operates in the band 824–849 MHz, and the forward link operates in the band 869–894 MHz. The user data rate, which is variable, is spread at a rate of 1.2288 Mchips/s. For a maximum user data rate of 9.6 kbps, this provides a spreading of 128. We will now look at these and other details of the IS 95–based systems currently operating.

6.5.1 Forward CDMA Channel

The modulation format used in the forward channel is filtered QPSK. A pilot is also simultaneously transmitted so that the MU can use a coherent detection scheme for demodulation purposes. The pilot also provides a means of judging the signal strengths from different base stations, allowing decisions to be made on hand-off strategies.

6.5.2 Reverse CDMA Channel

Since the MU cannot afford to expend unnecessary power, the pilot signal is not transmitted in the reverse direction. The modulation format used in the reverse link is OQPSK.

6.5.3 Power Control in CDMA Systems

Since all the terminals (MUs) transmit over the same frequency band, power control is essential in the reverse channel. At the base station, the signal from the desired user is likely to be swamped by other users who are "nearby." This near/far problem is caused by the absence of complete decorrelation between the codes of different users. The problem can be alleviated through power control, which will ensure that all the terminals within a certain geographical region arrive at the base station with almost equal power. Note that the capacity of a CDMA-based system is limited by the interference. This interference can be reduced and capacity enhanced by using power control. Power control also plays a role in conserving transmitted signal power, thereby increasing the battery recharge cycle.

Power control is usually performed in two ways: using an "open-loop" method or using a "closed-loop" method. In the former approach, it is assumed that transmission characteristics are identical in the forward and reverse directions, and therefore the power loss should be the same in the two directions. The terminal keeps the total power, the sum of the transmitted and received power, at a constant level (−73 dBm). The constant monitoring of the received power allows control of the transmitted power from the terminal. In the closed-loop method, the base station monitors the power of the terminal and the commands to adjust the power in steps of ±1 dB at a very fast rate (~800 times/s).

Note also that IS 95 systems use voice-activated transmission. This means that a carrier signal is transmitted only when a voice signal is detected. This reduces overall interference by the ratio of this "silence time" to the whole speech period.

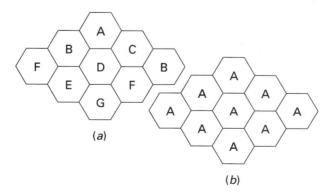

FIGURE 6.22 (*a*) Seven-cell reuse plan. (*b*) CDMA reuse plan.

6.5.4 Hand-off Procedure

The hand-off procedure in IS 95 systems is unique and is referred to as "soft hand-off" (Wong 1997, Garg 1999). As the mobile unit moves from cell to cell, it communicates simultaneously with the two base stations (Figure 6.23).

Essentially, there is no break in the transition from one base station to the next—hence the term *soft hand-off*. This not only makes the switch unnoticeable, but also reduces the chances of being dropped, distinguishing it from the MAHO technique employed in second-generation digital systems.

After initiation of the call, the terminal continues to monitor whether the signal from another cell is comparable to the signal from its own base station. If the signal is strong, the terminal informs the MTSO that the new cell site is strong and identifies the new site. Based on this, the MTSO starts the hand-off by linking to the new cell site while maintaining the connection to the existing base station. Thus for a short period, the terminal achieves a sense of diversity by having two base stations simultaneously communicating with it. This simultaneous connection with the two base stations also eliminates the "ping-pong" effect, which is a result of repeated requests to transfer back and forth between two base stations—a common occurrence in systems not having this unique "diversity."

6.5.5 Diversity in CDMA Systems

The high bandwidth of the signal (~1.25 MHz) provides a unique advantage in terms of time diversity. The chips are so narrow that multipath fading, which results in overlapping, unresolvable pulses in TDMA/FDMA-based systems, produces nonoverlapping multiple pulses, resulting in time diversity. A RAKE receiver can therefore be used to combine these multiple versions of the signal.

Unique Features of IS 95

Frequency reuse of unity

Automatic power control

Time diversity through a RAKE receiver

Soft hand-off

Voice-activated transmission

A detailed discussion of the capacity of CDMA-based systems is given in Section 6.6.1.

6.6 COMPARISON OF MULTIPLE-ACCESS SYSTEMS IN WIRELESS COMMUNICATIONS

One parameter used for comparison of the various multiple-access systems is the channel capacity (Lee 1991a, Rait 1991, Gilh 1991, Kcha 1993). Channel capacity, or radio capacity, is a quantitative measure of the ability of the access scheme to provide a maximum number of channels in a given bandwidth. As we saw in Chapter 4, we can certainly pack more channels into a given bandwidth if we ignore the effect of co-channel interference (CCI). Interference from cells operating at the same carrier frequency will degrade the performance of the communication system.

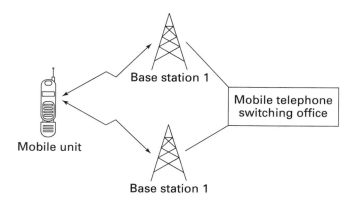

FIGURE 6.23 Hand-off procedure in CDMA systems.

If R is the radius of the cell and D is the distance from the mobile unit to the "interfering channel," or the base station responsible for the interfering channel, the co-channel interference reduction factor or frequency reuse factor is given by eq. (4.8),

$$q = \frac{D}{R} \qquad (6.18)$$

The signal-to-CCI ratio, C/I, is expressed as

$$\frac{C}{I} = \frac{R^{-\nu}}{6D^{-\nu}} = \left(\frac{1}{6}\right)\left(\frac{R}{D}\right)^{-\nu}, \qquad (6.19)$$

where it is assumed that all the interfering stations are separated from the desired channel by the same distance D and have the same loss exponent, ν. In any practical system, the actual C/I must be greater than some acceptable value of $(C/I)_{min}$. Using eqs. (4.7) and (4.8), the expression for the co-channel interference reduction factor, q, can be rewritten as

$$q = \left[6\left(\frac{C}{I}\right)_{min}\right]^{1/\nu}. \qquad (6.20)$$

The radio capacity of the system, m, is defined as

$$m = \frac{\text{total allocated spectrum } (B_t)}{\text{channel bandwith } (B_c) \times \text{number of cells } (N)} \text{ radio channels/cell.} \quad (6.21)$$

Making use of the relationship between q and N,

$$q = \sqrt{3N}, \qquad (6.22)$$

the expression for the radio capacity becomes

$$m = \frac{3B_t}{B_c\left[6\left(\frac{C}{I}\right)_{min}\right]^{2/\nu}}. \qquad (6.23)$$

Using a value of $\nu = 4$, the expression for the capacity becomes

$$m = \frac{3B_t}{B_c \sqrt{\frac{2}{3}\left(\frac{C}{I}\right)_{\min}}} \quad \text{radio channels/cell} \tag{6.24}$$

If we wish to achieve the same capacity for two different types of systems, we can write

$$B_{ci}\sqrt{\left(\frac{C}{I}\right)_{\min i}} = B_{cj}\sqrt{\left(\frac{C}{I}\right)_{\min j}}, \tag{6.25}$$

where i, j correspond to the two systems. In other words, if we have a narrowband system that operates with a lower bandwidth than another system, the signal-to-CCI ratio must be high. Conversely, if we have a constant number of users/radio channel, $(C/I)_{\min}$ increases by 4 when the bandwidth is halved.

We now compare the capacity of analog mobile systems and digital mobile systems. Even though the digital systems use less bandwidth, because of the increased value of $(C/I)_{\min}$, the capacities are the same. However, in practice, the capacities of digital systems are higher than those of analog systems because the use of speech and error-corrective coding permits operation with lower $(C/I)_{\min}$ values. Table 6.5 illustrates this aspect of capacity improvement in digital mobile systems. It has been assumed that each cell site has a 120° antenna. A GOS of 2% has been used. With the use of mobile-assisted hand-off (MAHO), the monitoring of the base stations becomes easier and the best station choice can easily be made. Indeed, MAHO allows the use of densely packed microcells, which leads to increased capacity.

6.6.1 Capacity of CDMA Systems

The capacity of TDMA and FDMA systems depends on the available bandwidth and the CCI ratio. The capacity in CDMA systems depends solely on the interference from the other users since the channel is shared simultaneously by all users (Jung

TABLE 6.5 Capacity Gains of Digital Systems over AMPS

Parameter	Cellular system			
	AMPS	**USDC**	**JDC**	**GSM**
Access method	FDMA	TDMA	TDMA	TDMA
Bandwidth (MHz), B_t	25	25	25	25
Carrier spacing/bandwidth (kHz), B_c	30	30	25	200
Users/carrier	1	3	3	8
Total number of voice channels	833	2500	3000	1000
Frequency reuse (cluster size)	7	7	7	4
Channels/site	119	357	429	250
Traffic* (Erl/km²)	11.9	41	50	27.7
Capacity gain (over AMPS)	1	3.45	4.20	2.33

* The cell radius (R) has been assumed to be 1 km. This gives rise to a (hexagonal) cell area of $2.6R^2 = 2.6$ km². A three-sector antenna provides, respectively, $119/3 = 40$, $357/3 = 120$, $429/3 = 143$, and $250/3 = 83$ channels, giving rise to a load of 30.99, 107.4, 130, and 71.5 Erl, respectively. The traffic value given is obtained by dividing the load by the area.

1993, Kim 1993, Pado 1994). Thus, if we can decrease the interference from other users, the capacity of a CDMA system is likely to increase.

Let us start by considering a single cell with N users who share the cell. The interference to any one of the users is a result of the $(N-1)$ other users sharing the band in the cell. If we have power control, all the terminals will be transmitting with the same power. If S is the signal power from the desired terminal, the signal-to-noise ratio due to interference can be expressed as

$$\text{SNR} = \frac{S}{(N-1)S} = \frac{1}{N-1}. \tag{6.26}$$

Instead of looking only at the signal-to-interference ratio, we can look at the signal-to-noise ratio in terms of the energy in the bit. The signal-to-noise ratio at the base station receiver, given by E/N_0, can be expressed as

$$\frac{E}{N_0} = \frac{S/R}{(N-1)^{S/B_t}} = \frac{B_t}{R}\frac{1}{(N-1)}, \tag{6.27}$$

where R is the information bit rate and B_t is the RF bandwidth. We have divided the signal power (S) in the numerator by the bandwidth (data rate R) of the message data, while the signal power in the denominator has been divided by the bandwidth (B_t) occupied by the interfering signal. The quantity (B_t/R) is the processing gain K of the CDMA processing, defined in connection with Figure 6.7.

Note that eq. (6.27) has assumed that the reception is influenced only by the interference from other users. However, in almost all systems, additional degradation results from the presence of thermal noise in the system. Including thermal noise of total power over the whole available bandwidth, η_0, the equation for the signal-to-noise ratio (6.27) can be expressed as

$$\frac{E}{N_0} = \frac{K}{(N-1) + \eta_0/S}. \tag{6.28a}$$

At this point, let us understand the meaning of eq. (6.28a). By including both the thermal noise and the interference noise, the left-hand side can be interpreted as the effective signal-to-noise ratio observed at the receiver. We can now explore ways of finding the maximum number of users that can be supported in the system if we expect to maintain an acceptable level of performance. This acceptable level will come from the acceptable value of the bit error rate, which in turn will dictate a minimum acceptable value of the signal-to-noise ratio. If the minimum signal-to-noise ratio required to maintain an acceptable error rate is $(E/N_0)_{\text{min}}$, the maximum number of users N_{max} that can be supported will be given by

$$\left(\frac{E}{N_0}\right)_{\text{min}} = K\left(N_{\text{max}} - 1 + \frac{\eta_0}{S}\right)^{-1}. \tag{6.28b}$$

However, for us to use eq. (6.28b), we need to convert η_0 to a more meaningful quantity.

If $N_0/2$ is the spectral density of the thermal noise, we can write

$$\eta_0 = B_t N_0. \tag{6.29}$$

We can rewrite eq. (6.28b) using eq. (6.29) as

$$\left(\frac{E}{N_0}\right)_{\text{min}} = K\left(N_{\text{max}} - 1 + \frac{B_t N_0}{S}\right)^{-1}. \tag{6.30}$$

We can now express eq. (6.30) in terms of the signal-to-noise ratio of the message signal for the single-user case. The signal-to-noise ratio (energy) can be expressed as S/N_0R. This is the signal-to-noise ratio if only a single signal is to be transmitted, and we will identify this as $(E/N_0)_S$. We can rewrite eq. (6.30) as

$$\left(\frac{E}{N_0}\right)_{min} = K\left(N_{max} - 1 + K\left(\frac{E}{N_0}\right)_S^{-1}\right)^{-1}. \tag{6.31}$$

Inverting to get the maximum number of users that can access the system with an acceptable level of performance, we get

$$N_{max} = 1 + K\left[\frac{1}{(E/N_0)_{min}} - \frac{1}{(E/N_0)_S}\right]. \tag{6.32}$$

In other words,

$$N_{max} = 1 + K\left[\frac{(E/N_0)_S - (E/N_0)_{min}}{(E/N_0)_{min}\,(E/N_0)_{min}}\right]. \tag{6.33}$$

If thermal noise is negligible, $(E/N_0)_S$ is very high, and the maximum number of users will be determined by the minimum acceptable performance level set by $(E/N_0)_{min}$. Otherwise, in the presence of thermal noise, the number of users will be reduced by a factor determined by the amount of noise present in the system.

EXAMPLE 6.4

In a DS-CDMA system, an input data stream is coming in at the rate of 2 kbits/s. If the chip rate is 200 kbits/s, what is the processing gain?

Answer   The processing gain is given by the ratio of the chip rate to the data rate:

$$K = 200/2 = 100, \text{ or } 20 \text{ dB}. \qquad\blacksquare$$

EXAMPLE 6.5

If there is negligible thermal noise in a system, what is the signal-to-noise ratio at the receiver? The number of users is 11. If the transmit power is increased, how will the signal-to-noise ratio be affected?

Answer   We have seen that the signal-to-noise ratio in the absence of receiver noise is given by eq. (6.27). The SNR ratio is

$$\frac{E}{N_0} = \frac{K}{N-1} = \frac{100}{10} = 10, \text{ or } 10 \text{ dB}.$$

When the transmit power is increased, the signal-to-noise ratio given in eq. (6.27) is *unaffected* (still 10 dB) because the ratio is dependent only on the processing gain and the number of users (both unaffected by the transmit power). $\blacksquare$

EXAMPLE 6.6

In the DS-CDMA system, the incoming data rate is 10 kbits/s. The chip rate is 10 Mchips/s. It has been determined that for acceptable performance, the signal-to-noise ratio (E/N_0) at the receiver must be greater than 10 dB. If only a single signal is transmitted, the signal-to-noise ratio will be 16 dB.

(a) What is the maximum number of users that can be supported under these conditions?

(b) Under ideal conditions, what is the maximum number of users that can be supported?

Answer

(a) We are given that the minimum value of SNR is 10 dB, or 10 in absolute units, i.e., $(E/N_0)_{min} = 10$. We are also given that signal-to-thermal noise ratio is 16 dB, or $10^{1.6} = 40$ in absolute units, i.e., $(E/N_0)_S = 40$. Note also that the processing gain $K = (10^7/10^4) = 1000$. Making use of eq. (6.33), the maximum number of users that can be supported is given by

$$N_{max} = 1 + 1000\left(\frac{40 - 10}{40 \times 10}\right) = 1 + 75 = 76.$$

(b) Under ideal conditions, $(E/N_0)_S$ is very high, and we make use of eq. (6.27), where the left-hand side now is the minimum acceptable signal-to-noise ratio. We see that

$$10 = \frac{1000}{N - 1}$$

or $N_{max} = 101$. We can clearly see the advantage of eliminating receiver noise. The maximum number of users goes up significantly when the system noise is eliminated. ■

EXAMPLE 6.7

In Example 6.6, if the transmit power is reduced by 2 dB, what is the maximum number of users that will be able to share the spectrum?

Answer In an ideal scenario (i.e., if we neglect thermal noise), there will be no change in the number of users (101). However, under realistic conditions, reduction in the signal-to-noise ratio will result when the transmit power is reduced because the thermal noise remains constant.

We are given $(E/N_0)_S = 14$ dB, or about 25 in absolute units. Making use of eq. (6.33), the maximum number of users when the power is reduced by 2 dB is given by

$$N_{max} = 1 + 1000\left(\frac{25 - 10}{250}\right) = 61 .$$

We see that we can support 15 fewer users when the transmit power is reduced by 2 dB. ■

We can certainly enhance the capacity if the signal-to-noise ratio is high. We see from eq. (6.33) that such an increase in capacity can be accomplished by reducing the number of interfering channels. Going from an omnidirectional antenna to a 120° sector antenna automatically brings down the number of interfering cells to one-third of the number in the onmidirectional antenna case, thereby increasing the capacity.

A second means of increasing the capacity is to incorporate a voice activity monitor and turn off the transmitter when speech is stopped. When we speak, we normally use pauses between words, and by turning off the tranmitter when no voice activity is detected, we can reduce the effective number of interfering users. It is known that speakers are active only 35–40% of the time. If we designate α as the voice activity factor, eq. (6.27) and the equations that follow can be rewritten. Specifically, eq. (6.28) becomes

$$\left(\frac{E}{N_0}\right) = \frac{K}{(N-1)\alpha + \eta_0/S}. \tag{6.34}$$

Typical values of α are around 3/8, and such values can bring down the effective number of interfering cells and increase the actual number of users.

It must be noted, however, that the capacity of a CDMA system is soft while the capacity of TDMA or FDMA systems is hard. In other words, CDMA will allow more users into the system at the cost of reduced quality, while TDMA/FDMA-based systems cannot allow any more users when no channels are available.

The capacity calculations presented here have ignored the fundamental differences between reverse-link and forward-link performance of CDMA. In the forward link (downlink), the MU has the availability of a pilot signal, making coherent demodulation possible. The reverse link, on the other hand, does not have the availability of a pilot tone to conserve the power of the MU. Thus, the capacity of the CDMA system is likely to be limited by the weaker of the two links, namely, the reverse link. We will derive approximate expressions for the capacity of the reverse-link CDMA, including multiple cells, the effects of which were not considered in arriving at eq. (6.33) or (6.34).

In CDMA systems, the reuse factor is unity and, hence, the same frequency is used in all the cells. This means that there will be interference occurring from users in other cells in addition to the users from the single cell under consideration. If we include an intercell interference factor f (Gilh 1991, Jung 1993) in eq. (6.34), we can rewrite the equation as

$$\frac{E}{N_0} = \frac{K}{(N-1)\alpha(1+f) + \eta_0/S}. \tag{6.35}$$

Equation (6.35) also includes the voice activity reduction factor α. The intercell interference factor f depends on a number of factors, such as the geometry of the serving cell and the neighboring cells, the path loss factor ν, the standard deviation of lognormal fading (σ_{dB}), and hand-off procedures used. Estimating the intercell factor f is also a bit difficult because of the lack of control on the part of the serving cell on the users in other cells. Still, a few assumptions can be made. The intercell interference factor f will be less if the path loss factor or exponent is high. It will also be less if the size of the cell is large. Small values of the standard deviation of fading also reduce the value of f. In addition, as stated above, it is influenced by the hand-off procedures employed. It may be in the range of 0.44 to 2. It is now possible to rewrite eq. (6.31) taking this additional interference into consideration:

$$\left(\frac{E}{N_0}\right)_{min} = K\left([N_{maxi} - 1]\alpha(1+f) + K\left(\frac{E}{N_0}\right)_S^{-1}\right), \tag{6.36}$$

where N_{maxi} is the maximum number of users in a cell, taking intercell interference into account. Solving for N_{maxi}, we get

$$N_{maxi} = 1 + K\left[\frac{(E/N_0)_S - (E/N_0)_{min}}{(E/N_0)_{min}(E/N_0)_S}\right]\frac{1}{\alpha(1+f)}. \tag{6.37}$$

If we define N_{id} to be the maximum number of users/cell in the absence of thermal noise ($(E/N_0)_S \gg (E/N_0)_{min}$) and intercell interference ($\alpha = 1, f = 0$), we obtain (from eq. (6.37)),

$$N_{id} = 1 + K\left(\frac{E}{N_0}\right)_{min}^{-1}. \tag{6.38}$$

Rewriting eq. (6.37) using eq. (6.38), we get

$$N_{maxi} \approx N_{id} \left[\frac{(E/N_0)_S - (E/N_0)_{min}}{(E/N_0)_S} \right] \frac{1}{\alpha(1+f)}, \tag{6.39}$$

where we have assumed that $N_{id} \approx k(E/N_0)_{min}^{-1}$. This is a valid assumption in most cases because $N_{id} \gg 1$.

If we define cell utilization N_u as the ratio of N_{maxi} to N_{id}, we can write

$$N_u = \left[\frac{(E/N_0)_S - (E/N_0)_{min}}{(E/N_0)_S} \right] \frac{1}{\alpha(1+f)}. \tag{6.40}$$

The limiting value of N_u in the absence of voice activity reduction ($\alpha = 1$) is realized when $(E/N_0)_S$ goes to ∞ and is given by

$$N_u\big|_{max} = \frac{1}{(1+f)}. \tag{6.41}$$

Thus, the ability of the CDMA system to provide access to a large number of users is limited by the interference introduced by the neighboring cells.

Equation (6.40) gives us a measure of the performance of the CDMA system in terms of the required signal-to-noise ratio $(E/N_0)_{min}$, the signal-to-thermal noise ratio $(E/N_0)_S$, the voice activity factor α, and the intercell interference factor f. Let us now explore this dependence of the capacity of the CDMA systems on these various parameters.

EXAMPLE 6.8

In a DS-CDMA system, the minimum signal-to-noise ratio $(E/N_0)_{min}$ for acceptable performance is 12 dB. For a single channel, the signal-to-noise ratio $(E/N_0)_S$ is 18 dB. If the incoming data rate is 10 kbits/s and the chip rate is 10 Mchips/s, calculate the maximum number of users supported. If an intercell interference parameter of 1.2 is observed, calculate the maximum number of users. Assume that no voice reduction factor is included in the calculation.

Answer
$$K = 1000$$

$$\left(\frac{E}{N_0}\right)_S = 18 \text{ dB , or } 63 \text{ in absolute units}$$

$$\left(\frac{E}{N_0}\right)_{min} = 12 \text{ dB , or } 15.85 \text{ in absolute units}$$

Using eq. (6.33),

$$N_{max} = 1 + 1000 \times \left[\frac{63 - 15.85}{63 \times 15.85} \right] = 148 \,.$$

Including the intercell interference factor, in terms of eq. (6.37),

$$N_{maxi} = 1 + 1000 \times \left[\frac{63 - 15.85}{63 \times 15.85} \right] \frac{1}{(1 + 1.2)} = 23 \,.$$

Intercell interference reduces the number of users by more than 50%. ∎

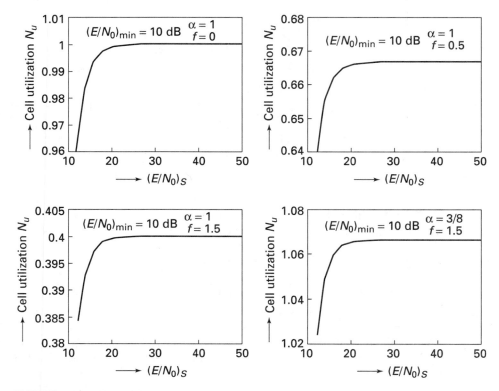

FIGURE 6.24 Plots of the cell utilization N_u for different parameters.

The cell utilization parameter N_u is plotted in Figure 6.24 for different conditions in the geographical area. It is clear that for intercell interference of 0.5, the utilization is only in the 60% range. It goes down to the 40% range when the interference factor goes up to 1.5. Once we incorporate the voice activity reduction factor, we see that the utilization exceeds 100%, pointing to the possibility of accommodating more users even in the presence of intercell interference.

A few statements are in order here. From Example 6.5, we see that an increase in thermal noise and an increase in intercell interference lead to a reduction in the number of users. This does not imply that there is a cap on the number of users. Indeed, when more users come into the system, nobody is turned away, and the performance level goes down. In other words, $(E/N_0)_{min}$ will no longer be the minimum acceptable value; it will be a floating value that keeps going down as the number of users entering the system goes up.

Let us explore this further. If acceptable performance is defined by $(E/N_0)_{min}$, the maximum number of users is given by N_{id}, given in eq. (6.38). We will also assume that there is no intercell interference. Rewriting eq. (6.32), we get

$$\frac{N-1}{K} + \left(\frac{E}{N_0}\right)_S^{-1} = \left(\frac{E}{N_0}\right)^{-1}. \tag{6.42}$$

We can now obtain an expression for the effective signal-to-noise ratio $(E/N_0)_{eff}$ as the number of users N increases:

$$\left(\frac{E}{N_0}\right)_{\text{eff}} = \frac{K}{N - 1 + K(E/N_0)_S^{-1}}. \tag{6.43}$$

We will look at this reduction in signal-to-noise ratio as more and more users come into the system (see Figure 6.25). If the number of users in the system is the maximum allowed (for the case in this figure, $N_{\text{max}} = 101$), the minimum acceptable signal-to-noise ratio is chosen to be 10 dB. As additional users enter the system, the actual or effective signal-to-noise ratio goes down.

The results for two values of the signal–to–thermal noise ratio $(E/N_0)_S$ are also shown. At a very low noise value (high value of the ratio = 50 dB), the SNR values are better than those at a high value of thermal noise (low value of the ratio = 18 dB). These results show that the introduction of additional users will reduce the performance level, though the level may still be good enough for use of the wireless system.

Figure 6.26 shows the improvement in signal-to-noise ratio at the receiver when the actual number of users is less than the maximum number based on acceptable performance. We see that when there are fewer users, it is possible to have a higher-than-required signal-to-noise ratio at the receiver. The effect of reducing thermal noise as seen here is similar to the one seen in Figure 6.25.

6.6.2 Comparison of Features and Standards

Table 6.6 compares some of the features of multiple-access schemes, and Table 6.7 compares the various standards on the basis of several parameters.

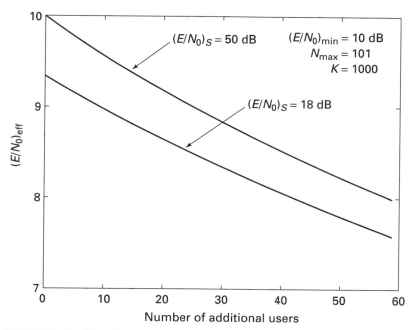

FIGURE 6.25 The effective signal-to-noise ratio is plotted as a function of the additional users beyond the maximum set by the acceptable SNR value at the receiver (10 dB).

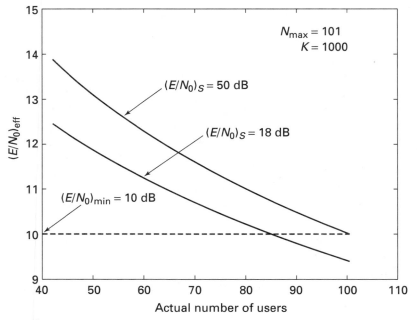

FIGURE 6.26 Enhancement in signal-to-noise ratio at the receiver as a function of the number of users in the cell.

6.7 CDMA2000

The multifold increase in the use of wireless services (both voice and data) has made it necessary to increase the capacity of the wireless communication systems. Combined with this need for increased capacity, there has been an interest in providing a greater variety of services, such as video-on-demand and other multimedia services (Miya 1997; Knis 1998a,b; Rao 1999; Garg 1999; Pras 1999). The goals of increased capacity, diversity of applications, and universal use of wireless systems are to be achieved through the IMT 2000 (International Mobile Telecommunications 2000) initiative. The major goals of IMT 2000 are the universal adoption of a single standard and global roaming capability, with deployability indoors, outdoors, in mobile or stationary environments,

TABLE 6.6 Features of FDMS, TDMA, and CDMA

Feature	FDMA	TDMA	CDMA
High carrier frequency stability	Required	Not necessary	Not necessary
Timing/synchronization	Not required	Required	Required
Power monitoring	Difficult	Easy	Easy
Near/far problem	No	No	Yes; power control required
Variable transmission rate	Difficult	Easy	Easy
Fading mitigation	Equalizer not needed	Equalizer may be required	RAKE receiver possible
Zone size (large/small)	Any size	Any size	Large size difficult because of power control

TABLE 6.7 Comparison of Standards

Parameter	IS 54	JDC	GSM	IS 95
Multiple access	TDMA/FDMA	TDMA/FDMA	TDMA/FDMA	CDMA
Duplex	FDD	FDD	FDD	FDD
Channel bandwidth	30 kHz	25 kHz	200 kHz	1.25 MHz
Uplink (reverse channel 1) band	824–849 MHz	810–826 MHz	890–915 MHz	824–849 MHz
Downlink (forward channel 1) band	869–894 MHz	940–956 MHz 1429–1441 MHz (U) 1477–1489 MHz (D) 1453–1465 MHz (U) 1501–1513 MHz (D)	935–960 MHz	869–894 MHz
Modulation technique (voice)	$\pi/4$-DQPSK	$\pi/4$-DQPSK	GMSK	QPSK/OQPSK
	Roll-off factor = 0.3	Roll-off factor = 0.5	$BT = 0.3$	
Forward/reverse channel data rate	48.6 kbps	42 kbps	270.833 kbps	
Forward/reverse channel symbol rate	24.3 ksymbols/s	21 ksymbols/s	270.833 kbps	1.288 kchips/s
Users/channel	3 at 7.95 kbps/user	3 at 6.7 kbps/user	8 at 13 kbps/user	Variable
Number of voice channels	2500	3000	1000	
Spectral efficiency	1.62 bps/Hz	1.68 bps/Hz	1.35 bps/Hz	
Equalizer	Adaptive	Adaptive	Adaptive	

in megacells (i.e., global cells) (>35 km radius), in macrocells (1–35 km radius), in microcells (up to 1 km radius), and in picocells (a few meters radius). The planned scope of IMT 2000 is sketched in Figure 6.27 (Shaf 1998, Shum 1998, Knis 1998b).

The third-generation (3G) systems based on IMT 2000/UMTS (Universal Mobile Telecommunication Systems) are being put in place in Europe, Asia, and other parts of the world (Sasa 1998, Samu 1998, Wee 1998, Rao 1999). The North American system designated as cdma2000 can support all the IMT 2000 requirements as well as any future enhancements. The cdma2000 radio transmission technology (RTT) is a wideband spread-spectrum-based interface using the code division multiple access (CDMA) approach. It also provides backward compatibility with IS 95 CDMA systems currently in operation. While a new spectral band around 1900 MHz is being used for these systems in most of the world, the PCS band of 1900 MHz, which is currently used for the second-generation digital systems, is being used in North America for cdma2000.

Some of the unique features of cdma2000 are as follows:

Compatibility with IS 95

Possibility of multiple bandwidths: 1.25 MHz, 3.75 MHz, 7.5 MHz, 11.25 MHz, and 15 MHz in multiples (1, 3, 6, 9, and 12) of 1.25 MHz

Chip rates of $N \times 1.2228$ Mchips, with $N = 1, 3, 6, 9, 12$

Multicarrier transmission in the forward (down) link, making frequency diversity possible

Pilot-based coherent detection in both forward and reverse links

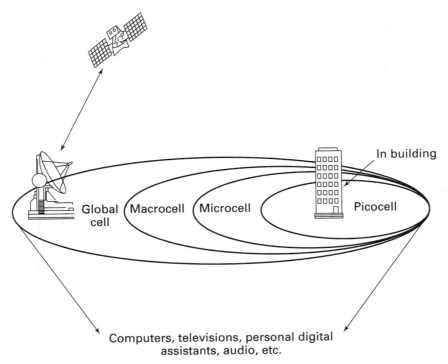

Computers, televisions, personal digital
assistants, audio, etc.

FIGURE 6.27 Vision and scope of IMT 2000.

We will now review some of these characteristics and compare them with those of W-CDMA, the prime candidate for IMT 2000 implementation in Europe.

Before we can compare the features of cdma2000 and W-CDMA, we need to look at the similarities between the existing CDMA-based systems operating in North America, namely, the features associated with IS 95 and cdma2000. Keeping in mind the need to provide a bridge between IS 95 and cdma2000, modifications to the original IS 95 (known as IS 95A) were made and designated as IS 95B. This modified IS 95B supports user data rates much higher than the ones offered by IS 95A (14.4 kbps). These higher data rates are 9.6–76.8 kbps in rate set 1 and 14.4–115.2 kbps in rate set 2. The updates have been provided without changing the physical layer associated with IS 95A. These changes have made compatibility with cdma2000 possible. Table 6.8 provides the details of this comparison.

Let us briefly look at two unique features of cdma2000 (Rao 1999; Pras 1999; Garg 1999, 2000). One of these features is the ability to use direct spreading or multicarrier transmission with the availability of bandwidths of 5 MHz or more. The concept for forward link transmission is shown in Figure 6.28 for two spreading options. In the multicarrier approach (a form of orthogonal frequency division multiplexing), three or more carriers are used, each with a spreading rate of 1.2288 Mchips/s. Figure 6.28 shows such an approach with three carriers. It also shows the second approach, in which the whole bandwidth is used for direct spreading at a rate of $N \times 1.2288$ Mchips/s. The diagram shows spreading at a rate of 3.6864 Mchips/s, three times the spreading of one of the channels in the multicarrier approach. Use of such multicarriers is equivalent to frequency diversity. The direct spreading over the whole bandwidth also provides the equivalent of frequency diversity.

TABLE 6.8 Comparison of IS 95 and cdma2000

Feature	IS 95 (A & B)	cdma2000
RF channel width (MHz)	1.25	1.25/5/10/15/30
Chip rate (Mchips/s)	1.2288	1.2288/3.6864/7.7328/11.0592/14.7456
Single-user data rate (kbps)	9.6–115.2	9.6–2400
Modulation	BPSK with quadrature spreading	QPSK with quadrature spreading
Pilot-based coherent detection	Forward link: yes Reverse link: no	Forward link: yes Reverse link: yes
Channel coding	Convolutional code	Convolutional code
Fast forward power control	No	Yes
Fast reverse power control	Yes	Yes
Forward link transmit diversity	No	Yes for BW of 5 MHz or more
Use of turbo codes to lower operating SNR levels	No	Yes

The second feature is the inherent diversity associated with the use of multiple carriers (see Appendix B, Section B.1). These multiple carriers can be transmitted using what is known as a multi-antenna. Two multi-antenna configurations are shown in Figure 6.29.

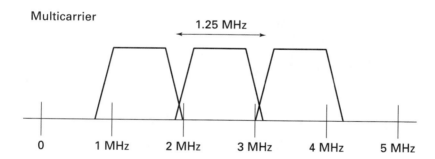

Multicarrier

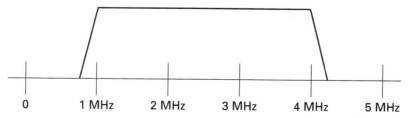

Direct spread

FIGURE 6.28 Two approaches to forward link transmission.

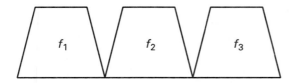

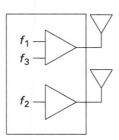

Two-antenna
configuration

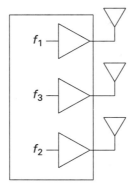

Three-antenna
configuration

FIGURE 6.29 Transmit diversity configuration for the forward link ($N = 3$).

In one of the configurations, two nonconsecutive frequencies, f_1 and f_3, are transmitted on one antenna while f_2 is transmitted on a second antenna. In the second configuration, three separate antennae are used to transmit the three frequencies f_1, f_2, and f_3. Use of diversity allows cdma2000 to mitigate fading effects and therefore improves the performance, leading to increased forward link capacity.

We can now compare the two schemes cdma2000 and W-CDMA (Fuka 1996; Knis 1998a,b). The two share a number of unique features that do not exist in the current IS 95 and GSM-based systems. Some of the important common features are the following:

- Coherent forward and reverse links
- Fast forward link power control and fast reverse link power control
- Turbo codes for higher rates
- Variable spreading factor to achieve higher data rates
- QPSK modulation on both forward and reverse links

There are also a few key differences between the two systems. They are listed in Table 6.9.

TABLE 6.9 Comparison of cdma2000 and W-CDMA

Feature	cdma2000	W-CDMA
Chip rate	3.6864 Mchips/s	4.096 Mchips/s
Network support	IS 41	GSM-MAP (mobile application part)
Base station synchronization	Synchronous	Asynchronous
Multicarrier spreading	Yes	No

6.8 OTHER NEW DEVELOPMENTS: BLUETOOTH NETWORKS

With the increasing use of computers, printers, fax machines, cell phones, and PDAs (personal digital assistants), it becomes necessary to exchange data among these units without wires or cables, as shown in Figure 6.30. Networks based on infrared (IR) systems are one option to accomplish this goal. However, the use of such systems is limited due to their high cost and line-of-sight and directional characteristics. Because of the necessity for links that operate smoothly and efficiently, these networks must possess the following unique features:

- The network must operate globally. This means that the frequency band must be available globally. It must also be license free and open to all radio systems.
- The network connections can be made on an ad hoc basis. Anyone should be able to bring in any unit and achieve connectivity.
- The network must support voice and data.
- The radio transceiver must be small and operate at a low power level. Such operation will keep the interference low. Low interference is a must because license-free frequency bands are likely to be used anywhere, anytime. The typical range will be from 0.1 to 10 m.

Bluetooth was developed by Ericsson to meet these goals (Gilb 2000, Haar 2000, Schn 2000, Zurb 2000). Bluetooth-enabled products will automatically seek each other and configure themselves into networks. We will briefly look at some of the characteristic features of the Bluetooth wireless network systems.

The frequency range that meets the requirement is the ISM (industrial-scientific-medical) band in the 2.4 GHz range. Since the ISM band is open to anyone, systems operating at this band must cope with several sources of interference from such units as baby monitors, garage door openers, cordless phones, and microwave ovens. To overcome the interference, the frequency-hopping spread-spectrum format is used. The modulation format used is Gaussian-shaped frequency shift keying (GFSK).

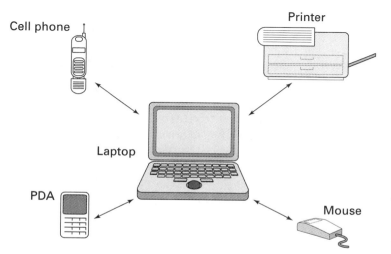

FIGURE 6.30
The Bluetooth concept. Wireless connectivity of all the diverse units (mouse, printer, cell phone, PDA, etc.) is the key.

TABLE 6.10 Characteristic Features of Bluetooth Wireless Networks

Carrier frequency (MHz)	2400–2483.5 (ISM band) · France: 2446–2483.5 · Japan: 2471–2497 · Spain: 2445–2475
Modulation	GFSK at a line rate of 1 Mbps · Modulation index 0.32 · Peak deviation 175 kHz · $BT = 0.5$
Frequency hopping	1600 hops/s · Four special hopping sequences reserved for connection setup · Periodicity of the sequence: 23 hours, 18 minutes
Transmit power (mW)	· Class 1: 1–100 · Class 2: 0.25–2.5 · Class 3: 1
Operating range (m)	· 0.1 to 10 · 100 with Class 1

GFSK is a form of digital frequency modulation using a Gaussian pulse with a modulation index that can be varied (if the index is equal to 0.5, GFSK becomes GMSK). In Bluetooth systems, a modulation index of 0.32 and a BT product of 0.5 are used. Other specifications of Bluetooth are given in Table 6.10 (Schn 2000). It is expected that Bluetooth-based products will be in the market shortly.

6.9 SUMMARY

The three multiple-access schemes, FDMA, TDMA, and CDMA, were discussed in this chapter. The two duplexing approaches, FDD and TDD, were briefly reviewed. Code division multiple access and the modulation schemes associated with CDMA, both direct-sequence and frequency hopping, were presented. The capacities of FDMA, TDMA, and CDMA were evaluated. The various modulation schemes and standards used in North America and Europe were compared. The G3 systems, cdma2000/IMT 2000, were briefly discussed. An overview of Bluetooth systems was also provided.

- FDMA is a simple scheme. Each channel is allocated a frequency band. Different frequency bands are separated by a guardband. In the FDMA/FDD format, uplinks and downlinks use nonoverlapping frequency bands.
- FDMA is not compatible with variable-rate transmission. Power level monitoring also is not easy. In addition, the transmitters and receivers must have a high Q value to ensure channel selectivity.
- TDMA systems are compatible with variable-rate transmission. However, synchronization issues are critical. TDMA schemes are also susceptible to fading.
- In CDMA each user is assigned a unique PN code. Each code consists of K chips, each with a duration of T_c, and $KT_c = T$, the bit duration. Thus, CDMA uses a much larger bandwidth than TDMA or FDMA. All users share the same bandwidth all the time.

- PN sequences are almost orthogonal to each other. This allows CDMA systems to have a "soft capacity limit" or no limit at all. However, as more and more users come into the system, the performance of the system goes down because of an increased level of interference.
- Since the chips are of extremely short duration, when multipath conditions exist the various delayed signals are resolvable, creating "multipath diversity." A RAKE receiver may be used to take advantage of the existence of multipath components.
- In DS-CDMA systems, BPSK may be used as the fundamental modulation scheme.
- In FH-CDMA, BFSK or MFSK is used.
- Fast frequency hopping occurs if more than one frequency hop is present during each transmitted symbol.
- Slow frequency hopping occurs if more than one symbol is transmitted between frequency hops.
- The capacity of CDMA is soft. It is possible for more users to come into the network. This may lead to a reduction in performance levels.
- The capacity of CDMA is enhanced by the incorporation of voice activity detection schemes.
- The capacity of CDMA goes down when intercell interference is present.
- The future of wireless communications, cdma2000, is already here.

PROBLEMS

1. Design a FDMA scheme that uses SSB-SC. If the voice message is assumed to have a 4 kHz bandwidth, draw the spectrum of the FDMA signal if each channel is allocated 5 kHz, including the guardband. The carrier frequency range is from 100 to 500 kHz. How many signals can be multiplexed?

2. Give a block diagram of a receiver to recover the signals generated in Problem 1. Show all the equations necessary to recover the message signal.

3. Consider data being generated at a rate of 1 kbps. This data set is to be transmitted using DS-CDMA with a processing gain of 8. Plot the data stream, a typical chip sequence, and the DS-CDMA transmitted sequence.

4. Plot the spectra of the bit and the chip. Note the number of zero crossings in the spectra.

5. Using results of Problem 4, generate a BPSK-DSSS signal.

6. In a DS-CDMA cell, there are 64 equal power channels that share a common frequency band. The signal is transmitted using a BPSK format. The data rate is 4800 bps. A coherent receiver is used for recovering the data. Assuming the

receiver noise to be negligible, calculate the chip rate to maintain a bit error rate of 10^{-3}.

7. In a DS-CDMA system, the processing gain offered is 1000. The modulation format used is BPSK. If a bit error rate of 10^{-6} is required, how many users can share the system? Assume that the receiver noise is negligible.

8. A DS-CDMA system is expected to have a processing gain of 20 dB. The expected data rate is 9600 bps. What should be the chip rate? If BPSK modulation will be employed, what is the bandwidth required for transmission using a null-to-null criterion?

9. A shift register with 10 taps is used for generation of the PN sequence. If the chip duration is 0.1 μs, calculate

(a) The length of the PN sequence
(b) The bit duration
(c) The processing gain

10. A RAKE receiver will be used to take advantage of the multipath effects in the channel. If the minimum delay difference is 200 m, what is the minimum chip rate necessary to successfully operate the RAKE receiver?

11. Obtain an expression for the autocorrelation function of a PN sequence of length $K = 7$. Assume that the chip duration is T_c.

12. Use the results of Problem 11 to obtain the expression for the power spectral density.

13. Consider the following binary PN sequence: $1 -1 -1 1 -1 1 1$. Assuming any duration, obtain the autocorrelation function of this sequence. Note that you may need a few of these sequences to get a good plot of the autocorrelation function.

14. From the results of Problem 13 verify that between the periods of the autocorrelation function, the value of the function is equal to $-1/K$, where K is the number of chips in the sequence. Verify the time value at which the autocorrelation function crosses the time axis.

15. Obtain the power spectral density of the PN sequence.

16. Obtain the spectrum of an unmodulated PN sequence, i.e., the spectrum of the PN sequence in Problem 13 multiplied by $\cos(2\pi f_0 t)$.

17. Repeat Problem 13 by interchanging $+1$ and -1.

18. Repeat Problem 13 replacing -1 with 0. Comment on the autocorrelation function.

19. A frequency-hopping spread-spectrum system has the following characteristics: number of bits/MFSK symbol $= 3$ and number of MSFK symbols/hop $= 5$. Is this a slow hopping or fast hopping system? What is the processing gain?

20. Use MATLAB to plot curves similar to Figure 6.24 for $f = 1.2$ and 2.0. Ignore the voice activity reduction.

21. Consider data occupying a bandwidth of 8 kHz and a spread bandwidth of 512 kHz. Use MATLAB to generate a plot similar to Figure 6.25.

22. In the DS-CDMA system, the minimum signal-to-noise ratio $(E/N_0)_{min}$ for acceptable performance is 11 dB. For a single channel, the signal-to-noise ratio $(E/N_0)_S$ is 15 dB. If the incoming data rate is 1 kbit/s and the chip rate is 1 Mchip/s, calculate the maximum number of users supported. If an intercell interference parameter of 0.8 is observed, calculate the maximum number of users. Assume that no voice reduction factor is included in the calculation.

TOPICS IN SIGNALS AND SYSTEMS

A.1 FOURIER TRANSFORMS

There are generally two types of signals encountered in communication systems, deterministic and random (Dave 1958, Gagl 1988, Couc 1997, Hayk 2001). A signal is identified as deterministic if there are no uncertainties about its value at any time. A random signal is one that does have uncertainties; its values are random. Initially, we will describe some of the properties of deterministic signals, and then we will introduce the concept of probability and random variables. We can write the expression for a deterministic signal in an explicit form, such as $x(t) = 5\sin(3t)$. We may classify the deterministic signals further as *periodic* or *aperiodic*. If $x(t)$ is periodic, there exists a constant T such that

$$x(t) = x(t + T), \quad -\infty \leq t \leq \infty. \tag{A.1.1}$$

The smallest value of T that satisfies this condition is identified as the *period* of the signal. A signal for which no value of T exists that can satisfy the condition given in eq. (A.1.1) is an *aperiodic* or nonperiodic signal.

Periodic signals are sometimes referred to as power signals. A signal is referred to as a power signal if, and only if, it has finite but nonzero power, P, given by

$$P = \lim_{T \to \infty} \frac{1}{T} \int_{-T/2}^{T/2} x^2(t)\, dt. \tag{A.1.2}$$

Nonperiodic signals are sometimes referred to as energy signals. An energy signal has a nonzero finite energy, E,

$$E = \int_{-\infty}^{\infty} x^2(t)\, dt. \tag{A.1.3}$$

Note that energy and power classifications are mutually exclusive. An energy signal has zero average power and finite energy, and a power signal has finite power and infinite energy. Let us now look at deterministic signals in detail. These signals can be understood using the concepts of Fourier series and Fourier transforms.

An energy signal, $s(t)$, and its Fourier transform, $S(f)$, are related through the Fourier transform pair,

$$S(f) = \int_{-\infty}^{\infty} s(t)\exp(-j2\pi ft)\, dt \tag{A.1.4}$$

249

and

$$S(t) = \int_{-\infty}^{\infty} S(f) \exp(j2\pi ft) \, dt. \tag{A.1.5}$$

We can write these two equations as

$$\mathcal{T}^{-1}S(f) = s(t) \leftrightarrow S(f) = \mathcal{T}\{s(t)\}, \tag{A.1.6}$$

where $\mathcal{T}$ represents the Fourier transform and $\mathcal{T}^{-1}$ represents the inverse Fourier transform. Even though the Fourier transform is defined for energy signals, it is possible to obtain the Fourier transform for a periodic signal as well.

If a periodic signal $g(t)$ is given by

$$g(t) = \sum_{n=-\infty}^{\infty} a_n \exp jn2\pi f_0 t, \tag{A.1.7}$$

its Fourier transform $G(f)$ can be expressed as

$$G(f) = \sum_{n=-\infty}^{\infty} a_n \delta(f - nf_0). \tag{A.1.8}$$

Thus the Fourier transform of a periodic function will consist of impulses at the harmonics of the fundamental of strength equal to a_n.

The energy density, E, of a signal $s(t)$ is given by

$$E = \int_{-\infty}^{\infty} |s(t)|^2 \, dt = \int_{-\infty}^{\infty} |S(f)|^2 \, df. \tag{A.1.9}$$

The quantity $|S(f)|^2$ is known as the energy density spectrum.

Some of the standard functions are defined in Table A.1.

Some of the properties of Fourier transforms are given in Table A.2, and some of the commonly used Fourier transform pairs are given in Table A.3.

TABLE A.1 List of Standard Functions

Rectangular function: $\text{rect}\left(\dfrac{t}{T}\right) = \Pi\left(\dfrac{t}{T}\right) = \begin{cases} 1 & \left(-\dfrac{T}{2} \leq t \leq \dfrac{T}{2}\right) \\ 0 & |t| > \dfrac{T}{2} \end{cases}$

Triangular function: $\text{tria}\left(\dfrac{t}{T}\right) = \Lambda\left(\dfrac{t}{T}\right) = \begin{cases} 1 & -T \leq t \leq T \\ 0 & |t| > T \end{cases}$

Unit step function: $U(t) = \begin{cases} 1 & t \geq 0 \\ 0 & t < 0 \end{cases}$

Signum function: $\text{sgn}(t) = \begin{cases} 1 & t > 0 \\ -1 & t < 0 \end{cases}$

Sampling function: $\text{Sa}(t) = \text{sinc}(t) = \dfrac{\sin(\pi t)}{\pi t}$

Delta function: $\delta(t) = 0$ if $t \neq 0$ and $\displaystyle\int_{-\infty}^{\infty} \delta(t) dt = 1$ or $\displaystyle\int_{-\infty}^{\infty} g(t)\delta(t - t_0) \, dt = g(t_0)$

TABLE A.2 Properties of Fourier Transforms

Operation	$g(t)$	$G(f)$
Scaling	$g(at)$	$\frac{1}{\lvert a \rvert}G\left(\frac{f}{a}\right)$
Time shifting	$g(t - t_0)$	$G(f)\exp(-j2\pi ft_0)$
Frequency shifting	$g(t)\exp(j2\pi f_0 t)$	$G(f - f_0)$
Time differentiation	$\dfrac{d^n g(t)}{dt^n}$	$(j2\pi f)^n G(f)$
Frequency differentiation	$(-j2\pi t)^n g(t)$	$\dfrac{d^n G(f)}{df^n}$
Time integration	$\displaystyle\int_{-\infty}^{t} g(\tau)\,d\tau$	$\dfrac{1}{j2\pi f}G(f) + \dfrac{1}{2}G(0)\delta(f)$
Time convolution	$g(t) \circledast h(t)$	$G(f)H(f)$
Frequency convolution	$g(t)h(t)$	$G(f) \circledast H(f)$

Parseval's theorem

$$\int_{-\infty}^{\infty} g^*(t)h(t)\,dt = \int_{-\infty}^{\infty} G^*(f)H(f)df \quad \text{or}$$

$$\int_{-\infty}^{\infty} \lvert g(t) \rvert^2\,dt = \int_{-\infty}^{\infty} \lvert G(f) \rvert^2\,df$$

A.2 LINEAR NETWORKS AND IMPULSE RESPONSE

For the case of an ideal network, the output of the network must have undergone no distortion. The output may have undergone some scaling, but must have the same shape as the input signal. If $x(t)$ is the input and $y(t)$ is the output,

$$y(t) = \int_{-\infty}^{\infty} x(\tau)h(t;\tau)\,dt. \tag{A.2.1}$$

For time invariance, the impulse response $h(t;\tau)$ must not depend on the time of occurrence, and hence

$$h(t;\tau) = h(t - \tau). \tag{A.2.2}$$

Using eq. (A.2.2), we get the output, $y(t)$, to be

$$y(t) = \int_{-\infty}^{\infty} x(\tau)h(t - \tau)\,dt. \tag{A.2.3}$$

For an ideal network,

$$y(t) = Kx(t - t_0), \tag{A.2.4}$$

where K is a scaling factor and t_0 is a constant delay. Taking the Fourier transform,

$$Y(f) = KX(f)\exp(-j2\pi ft_0). \tag{A.2.5}$$

This means that we require a constant amplitude response and a phase shift that is linear with the frequency to realize an *ideal distortionless transmission* through the network or filter. In other words, constant magnitude at all frequencies at the output alone will not guarantee an undistorted signal; all frequency components at the output must also arrive with identical time delays. If either of these conditions is not met, additional equalization will be required to remove the distortion.

Ideal Filter   It is not possible to design a filter having ideal characteristics over the entire frequency band spanning DC to infinity. However, it is possible to have ideal low-pass, high-pass, and bandpass filters.

TABLE A.3   Fourier Transform Pairs

Time function	Fourier transform
$\delta(t)$	1
1	$\delta(f)$
$\cos(2\pi f_0 t)$	$\frac{1}{2}[\delta(f+f_0) + \delta(f-f_0)]$
$\sin(2\pi f_0 t)$	$\frac{1}{2j}[\delta(f+f_0) - \delta(f-f_0)]$
$\delta(t-t_0)$	$\exp(-j2\pi f t_0)$
$\exp(j2\pi f_0 t)$	$\delta(f-f_0)$
$\text{rect}\left(\frac{t}{T}\right)$	$T\,\text{sinc}(fT)$
$\text{triang}\left(\frac{t}{T}\right)$	$T\,\text{sinc}^2(fT)$
$U(t)$	$\frac{1}{2}\delta(f) + \frac{1}{j2\pi f}$
$\exp(-a|t|), \quad a > 0$	$\dfrac{2a}{a^2 + (2\pi f)^2}$
$\exp\left[-\pi\left(\frac{t}{T}\right)^2\right]$	$T\exp[-\pi(fT)^2]$
$\exp(-at)U(t), \quad a > 0$	$\dfrac{1}{a + j2\pi f}$
$t\exp(-at)U(t), \quad a > 0$	$\left(\dfrac{1}{a + j2\pi f}\right)^2$
$g(t)\cos(2\pi f_0 t)$	$\frac{1}{2}G(f+f_0) + G(f-f_0)$
$\displaystyle\sum_{n=-\infty}^{\infty} \delta(t-nT)$	$\dfrac{1}{T}\displaystyle\sum_{n=-\infty}^{\infty} \delta\left(f - \frac{n}{T}\right)$
$B\,\text{sinc}(Bt)$	$\Pi\left(\dfrac{f}{B}\right)$

Ideal Low-Pass Filter   The transfer function, $H(f)$, of the filter is given by

$$H(f) = |H(f)| \exp[-j\theta(f)], \qquad (A.2.6)$$

with

$$|H(f)| = \begin{cases} 1 & |f| \le B \\ 0 & |f| > B \end{cases} \qquad (A.2.7)$$

and

$$\theta(f) = \exp(-j2\pi f t_0). \qquad (A.2.8)$$

Taking the Fourier transforms, the impulse response of the low-pass filter becomes

$$h(t) = 2B \operatorname{sinc}[2B(t - t_0)]. \qquad (A.2.9)$$

The impulse response and transfer function of the ideal low-pass filter are shown in Figure A.2.1.

The transfer functions of an ideal bandpass filter of bandwidth B_{bp} and an ideal high-pass filter with cutoff B_h are shown in Figure A.2.2.

A.3   ORTHOGONAL FUNCTIONS

It is often necessary to represent a signal using a number of basis functions having some clearly defined numerical values (Taub 1986; Proa 1994, 1995; Glov 1998; Hayk 2001). The choice of these basis functions is dictated by mathematical convenience and the specific application involved. The signal is usually represented as

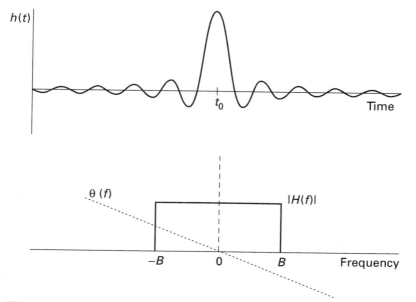

FIGURE A.2.1   The impulse response, $h(t)$, and transfer function, $H(f)$, of an ideal low-pass filter.

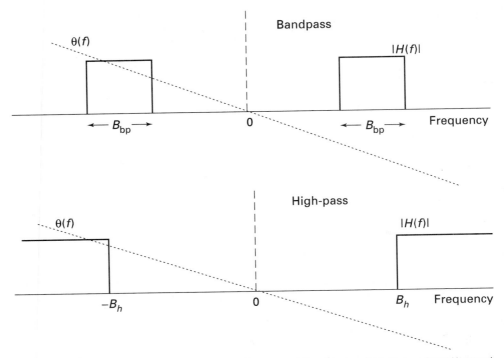

FIGURE A.2.2 Transfer functions of an ideal bandpass filter (*top*) and high-pass filter (*bottom*).

a linear combination of these basis functions. For example, consider a signal $s(t)$ that can be expressed as follows:

$$s(t) = \sum_{n=-N}^{N} a_n \phi_n(t), \tag{A.3.1}$$

where $\phi_n(t)$ form the basis function set and a_n form the weighting factors or coefficients. The number N may be infinity or may take only positive values. The goal is to find the best basis functions and corresponding weighting factors to suit the application. We would like to ensure that we can determine any one of the weighting factors without any knowledge about the rest of them. To realize this goal, the basis function must be orthogonal over the time interval.

Any two functions $\phi_n(t)$ and $\phi_m(t)$ are said to be orthogonal with respect to each other in an interval $\{a < t < b\}$ if

$$\int_a^b \phi_n(t) \phi_m^*(t) \, dt = 0 \quad \text{if} \quad n \neq m. \tag{A.3.2}$$

If functions in the set $\{\phi_n(t)\}$ are orthogonal,

$$\int_a^b \phi_n(t) \phi_m^*(t) \, dt = \begin{cases} 0 & n \neq m \\ K_n & n = m \end{cases} = K_n \delta_{nm}, \tag{A.3.3}$$

where δ_{mn} is the Kronecker delta function. If the constant K_n is unity, we have a set of orthonormal functions.

Going back to eq. (A.3.1), if we multiply both sides by $\phi_m^*(t)$ and integrate, we get

$$\int \phi_m^*(t)s(t)\,dt = \int \sum_{n=-N}^{N} a_n \phi_n(t)\phi_m^*(t)\,dt = a_n K_n \qquad (A.3.4)$$

by virtue of the orthogonal property of the functions.

Examples of orthogonal functions are sines, cosines, and complex exponentials. The usefulness of the orthogonal representation can be found in the computation of the power or energy of signal. The energy E_s in a signal $s(t)$ is given by

$$E_s = \int_a^b s^2(t)\,dt = \int_a^b \sum_{n=-N}^{N} a_n \phi_n(t) \sum_{m=-N}^{N} a_m \phi_m(t)\,dt = \sum_{n=-N}^{N} a_n^2 K_n. \qquad (A.3.5)$$

Each term in the summation is the energy associated with the basis functions, and the total energy is obtained by summation of the energy associated with different components. It is possible to see the potential use of an orthogonal series representation of signals.

Orthogonal functions that have found extensive use in communication systems are the Walsh functions. A typical set of these functions is shown in Figure A.3.1. These are orthonormal, and this property can be easily observed.

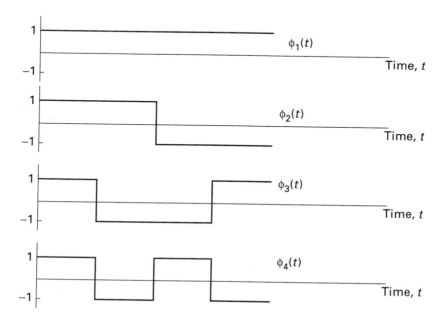

FIGURE A.3.1 Orthogonal functions.

A.4 PROBABILITY, RANDOM VARIABLES, AND RANDOM PROCESSES

A.4.1 Probability

Consider an experiment where the outcomes are random. Examples of such experiments include the roll of a die, a coin toss, and measurement of the voltage drop across a resistor. These outcomes may be discrete, as is the case for a coin toss (heads or tails) or the roll of a die (face showing 1, 2, 3, 4, 5, or 6), or continuous, as is the case with measurement of the voltage across a resistor (voltage lying between 1 and 1.2 V or 1.0 and 1.001 V, etc.). All the possible outcomes of any one of these experiments form a sample space, with the outcomes identified as elements of the sample space. We can refer to outcomes of the experiment as events belonging to the sample space. For example, in the roll of a die, the observation of a 5 is an event. In the experiment where voltage developed across a resistor is measured, an event may be the observation of a voltage between 3 V and 4 V. The reasons for the choice of a range of values in the case of a voltage will become obvious when we discuss random variables.

For each of the events in the sample space, we may be able to provide in a quantitative fashion the likelihood of occurrence of that particular event. Consider, for example, that we roll a die a number of times, and we note how many times a 5 was observed. If this experiment, i.e., the roll of a die, was repeated N times and a 5 was observed m times, we can safely say that the frequency f_5 of a 5 occurring is given by

$$f_5 = \frac{m}{N}. \tag{A.4.1}$$

If we now assume that the number of times the experiment is repeated is increased to infinity, we say that the probability p of a 5 showing up during the roll of a die can be expressed as

$$p = \frac{m}{N}, \quad N \longrightarrow \infty. \tag{A.4.2}$$

If we identify an event by A, the probability of event A occurring is written as $P(A)$. Another interesting observation can be made if we examine the experiment of a coin toss. Let A define the event of observing heads, and B the event of observing tails. In this case the sample space S contains only two events, A and B. If out of the N tosses we observe heads h times, the probabilities of the events A and B can be written as

$$P(A) = \frac{h}{N}; \quad P(B) = \frac{N-h}{N} = 1 - \frac{h}{N}, \tag{A.4.3}$$

making

$$P(A) + P(B) = 1. \tag{A.4.4}$$

The outcomes that are not part of the event can be identified as belonging to the complementary event, A^c. In the case of a coin toss, $B = A^c$. This is illustrated in Figure A.4.1.

Based on the preceding discussion, we can now state the axioms of probability (Dave 1958, Papo 1991):

1. The probability of any event is always a nonnegative number, i.e., $P(A) \geq 0$.

2. The probability of all events in the sample space $P(S) = 1$.

3. If the sample space contains a number of events $(A_1, A_2, A_3, \ldots)$ that are mutually exclusive—i.e., the occurrence of one event precludes the occurrence of

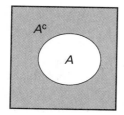

FIGURE A.4.1 The event A and the complementary event A^c. The box contains the sample space consisting of all possible outcomes.

the others (for example, the roll of a die)—the probability of observing the outcomes of (for example) A_1, A_2, and A_3 will be

$$P(A_1 + A_2 + A_3) = P(A_1) + P(A_2) + P(A_3). \tag{A.4.5}$$

Joint Probability Consider a case of two experiments going on at the same time. For example, suppose we are tossing a coin and rolling a die simultaneously. Let event A be identified as the observation of {heads} for the coin and B be identified as the observation of {4} for the die. We can now create a new event C that is the joint observation of A and B,

$$\{C\} = \{A, B\} = \{\text{heads}, 4\} = \{4, \text{heads}\}.$$

The probability of the event $\{C\}$ is referred to as the *joint probability* of A and B. We can now extend the concept to a number of experiments or outcomes. If we have a number of events A_1, A_2, A_3, ..., A_N, we can express the joint probability of all these events as $P(A_1, A_2, A_3, ..., A_N)$.

Two events, A and B, are said to be *statistically independent* if the joint probability $P(A,B)$ can be expressed as the product of the two marginal probabilities $P(A)$ and $P(B)$,

$$P(A, B) = P(A)P(B). \tag{A.4.6}$$

For example, if we assume that we have a fair coin, i.e., $P\{\text{heads}\} = P\{\text{tails}\} = 0.5$, and we have a fair or unbiased die, i.e., $P\{4\} = P\{3\} = (1/6)$ and so on, in a joint experiment the probability of observing heads and a 4 is $0.5(1/6) = (1/12)$.

In general, if the events A_1, A_2, A_3, ..., A_N are statistically independent, their joint probability can be expressed as the product of the marginal probabilities:

$$P(A_1, A_2, A_3, ..., A_N) = P(A_1)P(A_2)P(A_3)\cdots P(A_N). \tag{A.4.7}$$

Two events, A and B, are said to be *mutually exclusive* if their joint probability is zero, i.e.,

$$P(A, B) = 0: A \text{ and } B \text{ are mutually exclusive.} \tag{A.4.8}$$

Note that $P(A,B) = P(B,A) = P(AB) = P(BA)$.

The total probability, $P(A + B)$, is the probability that either A or B or both occur and is given by

$$P(A + B) = P(A) + P(B) - P(AB). \tag{A.4.9}$$

If A and B are mutually exclusive,

$$P(A + B) = P(A) + P(B). \tag{A.4.10}$$

If A and B are independent,

$$P(A + B) = P(A) + P(B) - P(A)P(B). \tag{A.4.11}$$

In some experiments, the likelihood of one event (A) depends on another event (B). This requires that we find the *conditional probability* of one event, given that the other event has taken place. For example, consider that we have two manufacturers (I and II) supplying processors for computers. Supplier I manufactures processors with a very high degree of tolerance, and only 2% of the processors fail. Supplier II manufactures processors with a tolerance such that 3% of the processors fail. The question now is, if we pick a computer at random, what is the likelihood that it breaks down because of a faulty processor?

If the event F is picking a faulty processor, we are given that P(F given that we have a processor from I) = 0.02 and P(F given that we have a processor from II) = 0.03. These are referred to as the conditional probabilities $P(F \mid I)$ and $P(F \mid II)$, respectively.

In general, for any two events A and B, we can express the conditional probability $P(A \mid B)$ as

$$P(A|B) = \frac{P(AB)}{P(B)}. \tag{A.4.12}$$

We can rewrite this as

$$P(AB) = P(A|B)P(B). \tag{A.4.13}$$

Reversing the roles of A and B, we can also write

$$P(B|A) = \frac{P(AB)}{P(A)}. \tag{A.4.14}$$

Combining eqs. (A.4.12) and (A.4.14), we can write

$$P(A|B)P(B) = P(B|A)P(A) = P(AB). \tag{A.4.15}$$

It is clearly seen that, if the events A and B are statistically independent,

$$P(A|B) = P(A); \qquad P(B|A) = P(B). \tag{A.4.16}$$

Consider now the case of a sample space S consisting of five events $A_1, A_2, \ldots, A_5$, as shown in Figure A.4.2. An arbitrary event B (shown by the ellipse) is also shown. The event B can be expressed as

$$\{B\} = \{BA_1\} + \{BA_2\} + \{BA_3\} + \{BA_4\} + \{BA_5\}. \tag{A.4.17}$$

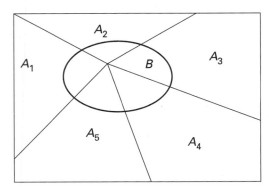

FIGURE A.4.2 An arbitrary event B and five partitioned events $A_1, \ldots, A_5$.

The probability of event B can now be expressed as

$$P(B) = P(BA_1) + P(BA_2) + P(BA_3) + P(BA_4) + P(BA_5) = \sum_{k=1}^{5} P(BA_k). \quad \text{(A.4.18)}$$

Using the relationship between conditional probability and joint probability, we can rewrite eq. (A.4.18) as

$$P(B) = \sum_{k=1}^{5} P(B|A_k)P(A_k). \quad \text{(A.4.19)}$$

This equation can be generalized as

$$P(B) = \sum_{k=1}^{M} P(B|A_k)P(A_k), \quad \text{(A.4.20)}$$

where M is the number of partitions of the event A. Equation (A.4.20) is known as the *total probability theorem*. Note that $P(B \mid A_k)$ is the conditional probability and $P(A_k \mid B)$ is known as the *a posteriori* probability. The probability of the event A_k, $P(A_k)$, is known as the *a priori* probability. Using eq. (A.4.15), we can obtain an expression for the *a posteriori* probability as

$$P(A_k|B) = \frac{P(A_kB)}{P(B)} = \frac{P(B|A_k)P(A_k)}{P(B)}. \quad \text{(A.4.21)}$$

This relationship is known as *Bayes' theorem*.

A.4.2 Random Variables

Outcomes of random experiments need to be represented numerically so that we can have some idea about averages and variances of outcomes. Random variables are a powerful tool to take us from a set of outcomes, numerical or otherwise, to a set of numbers. A random variable is defined as a real function (Dave 1958, Papo 1991) that maps the outcomes of an experiment onto a set of numbers, as shown in Figure A.4.3. We use capital letters to represent random variables. For example, for the coin toss experiment, we can create a random variable X such that $\{X = 1\}$ represents heads and $\{X = 0\}$ represents tails. For the case of a die roll, $\{X = 1\}$, $\{X = 2\}$, … will constitute the events associated with the random variable X.

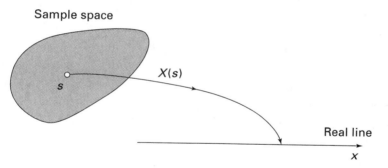

Sample space

$X(s)$

Real line

x

FIGURE A.4.3
Concept of a random variable. To each outcome s, the function $X(s)$ assigns a number. X is the random variable, and x is the value it takes.

We can now identify $\{X = a\}$, $\{X < a\}$, etc. as events. Since these are events, they have probabilities associated with them. The probability that the random variable X takes a value less than or equal to x is referred to as the *cumulative distribution function* (CDF) of the random variable and is represented by $F_X(x)$, i.e.,

$$F_X(x) = P\{X \le x\}. \tag{A.4.22}$$

The cumulative distribution function has the following properties:

$$F_X(-\infty) = P\{X \le -\infty\} = 0$$

$$F_X(\infty) = P\{X \le \infty\} = 1$$

$$0 \le F_X(x) \le 1 \tag{A.4.23}$$

$$P\{x_1 \le X \le x_2\} = F_X(x_2) - F_X(x_1)$$

The rate of change of the cumulative distribution function is referred to as the *probability density function* (pdf) $f_X(x)$,

$$f_X(x) = \frac{d}{dx}[F_X(x)]. \tag{A.4.24}$$

The pdf has the following properties:

$$f_X(x) \ge 0$$

$$F_X(x) = \int_{-\infty}^{x} f_X(\xi)\, d\xi \tag{A.4.25}$$

$$P\{x_1 \le X \le x_2\} = \int_{-\infty}^{x_2} f_X(\xi)\, d\xi - \int_{-\infty}^{x_1} f_X(\xi)\, d\xi = \int_{x_1}^{x_2} f_X(\xi)\, d\xi$$

If the outcomes are discrete, we have a discrete random variable, and if the outcomes are continuous, we have a continuous random variable. For example, the random variable formed by the roll of a die is a discrete random variable, while the random variable formed by observing the voltage developed across a resistor or the strain induced by an applied force is a continuous random variable. The expressions for the CDF and pdf of a discrete random variable with N outcomes can be expressed as

$$F_X(x) = \sum_{i=1}^{N} P(X = x_i) U(x - x_i) \tag{A.4.26}$$

and

$$f_X(x) = \sum_{i=1}^{N} P(X = x_i)\delta(x - x_i), \tag{A.4.27}$$

where U is the unit step function, given by

$$U(x) = \begin{cases} 1 & x \ge 0 \\ 0 & x < 0 \end{cases}. \tag{A.4.28}$$

Multiple Random Variables For multiple random variables, say, X, Y, Z, we can write the expression for the joint CDF as

$$F_{X,Y,Z}(x, y, z) = P(X \leq x, \ Y \leq y, \ Z \leq z). \qquad (A.4.29)$$

The joint pdf is given by

$$f_{X,Y,Z}(x, y, z) = \frac{\partial^3}{\partial x \ \partial y \ \partial z}[F_{X,Y,Z}(x, y, z)]. \qquad (A.4.30)$$

In general, we may omit the subscript for the CDF and pdf with the understanding of their implied existence.

If X, Y, and Z are independent random variables, their joint pdf is the product of marginal probability density functions, i.e.,

$$f(x, y, z) = f(x)f(y)f(z). \qquad (A.4.31)$$

Their joint cumulative distribution function is the product of the three marginal cumulative distribution functions,

$$F(x, y, z) = F(x)F(y)F(z). \qquad (A.4.32)$$

Statistical Averages and Moments of Random Variables If $g(X)$ is a function of a random variable, the expected value of this function is given by

$$E[g(X)] = \int_{-\infty}^{\infty} g(x)f(x) \ dx. \qquad (A.4.33)$$

If the random variable is discrete, the expected value is given by

$$E[g(X)] = \sum_{k=1}^{M} g(x_k)P(X = x_k). \qquad (A.4.34)$$

In general, if we have a function of several random variables, $X_1, X_2, \ldots, X_N$, the expected value of $g(X_1, X_2, \ldots, X_N)$ is given by

$$E[g(X_1, X_2, \ldots, X_N)] = \int_{-\infty}^{\infty} \cdots \int_{-\infty}^{\infty} g(x_1, \ x_2, \ \ldots, \ x_N)f(x_1, \ x_2, \ \ldots, \ x_N)(dx_1 \ dx_2 \ \ldots \ dx_N). \qquad (A.4.35)$$

The statistical average or mean of a random variable, denoted by μ, is given by

$$\mu = E[X] = \int_{-\infty}^{\infty} xf(x) \ dx = \sum_{k=1}^{M} x_k P(X = x_k). \qquad (A.4.36)$$

The mean (also known as expected value) is a weighted average of the possible outcomes, the weights being the probabilities of the respective outcomes. The concept of the average is analogous to the concept of "center of gravity," with probabilities acting as masses and the possible outcome values (x) acting as locations.

In addition to the expected value of a random variable, we are interested in knowing the spread of the values (around the mean) that we are likely to observe when an experiment is conducted. Since the spread could be on either side of the mean, we normally look at the squared value, which is calculated as the weighted average of

$[X - \mu]^2$. This is known as the variance, denoted by σ_x^2 and expressed as

$$\sigma_x^2 = \int_{-\infty}^{\infty} (x - \mu)^2 f(x) \, dx = \sum_{k=1}^{M} (x_k - \mu)^2 P(X = x_k). \qquad \text{(A.4.37)}$$

The deviation from the mean is the standard deviation, given by σ. If the standard deviation is high, we expect significant fluctuations in the observed outcomes, while small values of σ signify a very tight set of observed outcome values.

For the case of two random variables, the joint moment, $E[XY]$, given by

$$E[XY] = \int_{-\infty}^{\infty} \int_{-\infty}^{\infty} xy f(x, y) \, dx \, dy, \qquad \text{(A.4.38)}$$

is known as the correlation of the two random variables X and Y, represented by R_{XY}.

The covariance, C_{XY}, of the two random variables is given by

$$C_{XY} = E[x - E[X]]E[y - E[Y]]$$

$$= \int_{-\infty}^{\infty} \int_{-\infty}^{\infty} (x - E[X])(y - E[Y]) f(x, y) \, dx \, dy. \qquad \text{(A.4.39)}$$

The covariance can be expressed as

$$C_{XY} = R_{XY} - E[X]E[Y]. \qquad \text{(A.4.40)}$$

Central Limit Theorem If we have a number of independent random variables, the density function of the sum of these random variables will approach a normal distribution under certain conditions. If $X_1, X_2, X_3, \ldots, X_N$ are a number of independent identically distributed (iid) random variables with finite variances, then

$$Z = X_1 + X_2 + X_3 + \cdots + X_N \qquad \text{(A.4.41)}$$

will have a pdf given by

$$f(z) = \frac{1}{\sqrt{2 \pi \sigma_z^2}} \exp\left[-\frac{(x - \mu_z)^2}{2 \sigma_z^2} \right]. \qquad \text{(A.4.42)}$$

Note that if the random variables are not identical and none of them has a variance that is overwhelmingly high, the pdf of the sum will approach a normal distribution (normality). If the random variables are iid, conditions for normality are satisfied when $N = 6$. If the random variables are not identical, N will have to be much higher to achieve the normality condition.

Some Common Probability Density Functions (Papo 1991, Eva 2000)

Uniform Distribution The uniform distribution arises when we are studying engineering problems dealing with cases where the outcomes are equally likely. For example, we are asked to pick a real number in the range of 0 to 5. Note that this is a continuous random variable, and hence the probability that the number is equal to a specific value is zero. However, we can consider the case where this number lies in the range of 2 to 2.5. Now consider another set of numbers that lie in the range of 3.7

to 4.2. For a uniformly distributed random variable, these probabilities are equal; they are determined by the range of values and not by the values themselves. Another example is the case of a pointer turning on a wheel. The probability that it will be come to rest at a range of angles between 45° and 55° is the same as the probability that it will come to rest in the range of 205° to 215°. Note that the whole range is 360°. In the first case, we say that the random variable is uniformly distributed in the range 0 to 5, and for the latter it is uniform in the range 0° to 360° or 0 to 2π rad. A uniform pdf is used in communication theory to describe the statistics of the phase of the signal. In general, if a random variable is uniform in the range a to b, the density function $f(x)$ is

$$f(x) = \begin{cases} \dfrac{1}{b-a}, & a \leq X \leq b \\ 0 & \text{otherwise} \end{cases}.$$

The CDF is given by

$$F(x) = \begin{cases} 0 & x \leq 0 \\ \dfrac{x-a}{b-a} & a \leq x \leq b. \\ 1 & x \geq b \end{cases}$$

The mean and variance are given by

$$E[X] = \frac{a+b}{2}, \qquad \text{Var}[X] = \frac{(b-a)^2}{12}.$$

Normal or Gaussian Distribution The Gaussian distribution is used to model a number of naturally occurring phenomena. These include the noise observed in electrical circuits, characteristics of a biological population such as the weights and heights of students, and the statistical fluctuations observed during the manufacturing of products (the diameter of a precision nut, for example). It is also used to model the output generated from the contributions from a number of experiments (Central Limit Theorem). In communication theory, the noise observed is modeled using the Gaussian or normal distribution. The normal density function is given by

$$f(x) = \frac{1}{\sqrt{2\pi\sigma_z^2}} \exp\left[-\frac{(x-\mu)^2}{2\sigma_x^2}\right], \qquad -\infty \leq X \leq \infty.$$

The CDF is given by

$$F_X(x) = \int_{-\infty}^{x} \frac{1}{\sqrt{2\pi\sigma_x^2}} \exp-\left[\frac{(y-\mu)^2}{2\sigma_x^2}\right] dy = \frac{1}{2} + \frac{1}{2}\text{erf}\left(\frac{x-\mu}{\sqrt{2\sigma_x^2}}\right),$$

where erf is the error function (see Appendix B).

The mean and variance are given by

$$E[X] = \mu, \qquad \text{Var}[X] = \sigma_x^2 .$$

Rayleigh Distribution   The Rayleigh density function arises when we are interested in finding the statistics of

$$X = \sqrt{U^2 + W^2} ,$$

where U and W are iid Gaussian random variables of zero mean. Thus, the Rayleigh distribution is used to describe the fading observed in wireless communications. The Rayleigh density function is given by

$$f(x) = \frac{x}{a^2} \exp\left(-\frac{x^2}{2a^2}\right) U(x) .$$

The CDF is given by

$$F(x) = \begin{cases} 0 & x < 0 \\ 1 - \exp\left(-\frac{x^2}{2a^2}\right) & x \geq 0 \end{cases} .$$

The mean and variance are given by

$$E[X] = a\sqrt{\frac{\pi}{2}}, \quad \text{Var}[X] = \left(2 - \frac{\pi}{2}\right) a^2 .$$

Exponential Distribution   The exponential density function and the Rayleigh density function are related. While the Rayleigh density function models the envelope, the exponential density function models the power, i.e., $U^2 + W^2$, where U and W are once again zero-mean independent Gaussian random variables of equal variance. In other words, an exponential random variable is the square of a Rayleigh random variable. The exponential density function is given by

$$f(x) = \frac{1}{2a^2} \exp\left(-\frac{x}{2a^2}\right) U(x) .$$

The CDF is given by

$$F(x) = \begin{cases} 0 & x < 0 \\ 1 - \exp\left(-\frac{x}{2a^2}\right) & x \geq 0 \end{cases} .$$

The mean and variance are given by

$$E[X] = 2a^2, \quad \text{Var}[X] = \left(2a^2\right)^2 .$$

Rician Distribution While the Rayleigh density function arises from two Gaussian random variables having zero means and identical variances, the Rician density function arises when the one of the two random variables has a nonzero mean. The Rician random variable is given by $\sqrt{U_1^2 + W^2}$, where U_1 and W are independent Gaussian random variables with equal variances, but with W having a zero mean and U_1 having a nonzero mean. Whenever there is a direct path between the transmitter and receiver in wireless communications, the envelope of the signal is Rician distributed. The Rician density function is given by

$$f(x) = \frac{x}{\sigma^2}\exp\left(-\frac{x^2 + A_0^2}{2\sigma^2}\right)I_0\left(\frac{xA_0}{\sigma^2}\right),$$

where A_0 is the mean of U_1, I_0 is the modified Bessel function of zeroth order, and σ^2 is the variance of W (or U_1).

The CDF is given by

$$F(x) = 1 - Q_1\left[\frac{A_0}{\sigma}, \frac{x}{\sigma}\right],$$

where $Q_1[u, v]$ is the Marcum's Q function (Grad 1979, Proa 1995), given by

$$Q_1[u, v] = \exp\left(-\frac{u^2 + v^2}{2}\right)\sum_{k=0}^{\infty}\left(\frac{u}{v}\right)^k I_k(uv).$$

I_k is the modified Bessel function of kth order.

The mean and variance of the Rician distribution can be expressed in terms of the parameter k_0, given by

$$k_0^2 = \frac{A_0^2}{2\sigma^2}.$$

$$E[X] = \sigma\sqrt{\frac{\pi}{2}}e^{-k_0^2/2}\left[\left(1 + k_0^2\right)I_0\left(\frac{k_0^2}{2}\right) + k_0^2 I_1\left(\frac{k_0^2}{2}\right)\right]$$

$$\mathrm{Var}[X] = \sigma^2\left(2 + 2k_0^2\right) - [E(x)]^2$$

The Rician distribution becomes Rayleigh when $A_0 \longrightarrow 0$ and becomes Gaussian when A_0 is very large.

Lognormal Density Function The lognormal density function arises in situations where the value being observed is a result of the product of a number of quantities. An example is the sound measured at a location, where the sound has been reflected from a number of surfaces or objects. A similar situation arises in wireless communications systems where the signal reaching the receiver undergoes multiple reflections, leading to long-term fading or "shadowing." Thus, the lognormal distribution is used to model shadowing in wireless communication systems. The probability density function of the lognormal random variable is given by

$$f_X(x) = \frac{1}{\sqrt{2\pi\sigma^2 x^2}} \exp - \frac{1}{2\sigma^2}(\ln x - \mu)^2 U(x).$$

The CDF is given by

$$F_X(x) = \begin{cases} \int_0^x \frac{1}{\sqrt{2\pi\sigma^2 y^2}} \exp - \frac{1}{2\sigma^2}(\ln y - \mu)^2 \, dy & x \geq 0 \\ 0 & x < 0 \end{cases}.$$

The mean and variance are given by

$$E[X] = \exp\left(\mu + \tfrac{1}{2}\sigma^2\right)$$

$$\mathrm{Var}[X] = \left[\exp\left(\sigma^2\right) - 1\right]\exp\left(\sigma^2 + 2\mu\right).$$

The lognormal and normal density functions are related, as we observed in Chapter 4. When power expressed in mW is lognormal, power expressed in dBm is Gaussian.

Chi-Square Distribution The chi-square–distributed random variable arises when we take the sum of the squares of a number of iid zero-mean Gaussian random variables. If X is defined as

$$X = \sum_{k=1}^{n} X_k^2,$$

where the X_k are iid zero-mean random variables, each with a variance of σ^2, the chi-square density function is given by

$$f(x) = \frac{1}{\left(\sqrt{2\sigma^2}\right)^n \Gamma(n/2)} x^{(n/2-1)} \exp\left(-\frac{x}{2\sigma^2}\right) U(x),$$

where Γ is the gamma function, given by

$$\Gamma(z) = \int_0^\infty w^{z-1} e^{-z} \, dz.$$

When n Gaussian random variables are used to generate the chi-square random variable, the chi-square distribution has n degrees of freedom.

The CDF is given by

$$F(x) = 1 - \exp\left(-\frac{x}{2\sigma^2}\right) \sum_{k=0}^{\infty} \frac{1}{k!}\left(\frac{x}{2\sigma^2}\right)^k.$$

The mean and variance are given by

$$E[X] = n\sigma^2$$

$$\text{Var}[X] = 2n\sigma^4.$$

Poisson Distribution   The Poisson distribution is often used to model the number of accidental deaths in a given period for insurance purposes, the number of earthquakes taking place over a certain time period, the number of traffic accidents during peak hours, etc. In communications, the Poisson distribution is used to model telephone traffic. The number of calls or connections made in a given period is Poisson distributed. The density function of the Poisson-distributed random variable is given by

$$f(x) = \sum_{k=0}^{\infty} \frac{a^k \exp(-a)}{k!} \delta(x - k).$$

The CDF is given by

$$F(x) = \sum_{k=0}^{\infty} \frac{a^k \exp(-a)}{k!} U(x - k).$$

The mean and variance are given by

$$E[X] = a, \qquad \text{Var}[X] = a.$$

Gamma Distribution   The gamma density function is used to model the lifetimes of electronic components, computer parts, etc. The density function of the gamma-distributed random variable is given by

$$f(x) = \frac{x^{c-1}}{b^c \Gamma(c)} \exp\left(-\frac{x}{b}\right) U(x).$$

The CDF is given in terms of the incomplete gamma function $I(u, v)$ as

$$F_X(x) = I\left(\frac{x}{b\sqrt{c}}, c - 1\right).$$

The incomplete gamma function (Grad 1979) is given by

$$I(u, v) = \frac{1}{\Gamma(v + 1)} \int_0^{u\sqrt{v+1}} x^v e^{-x} \, dx.$$

The mean and variance are given by

$$E[X] = bc, \qquad \text{Var}[X] = b^2 c.$$

For $c = 1$, the gamma density function becomes the exponential density function. The gamma density function is also related to the Nakagami density function, and some researchers have used the gamma density function to model fading in wireless systems. The Nakagami random variable Y can be expressed as $Y = \sqrt{X}$, where X is a gamma-distributed random variable.

Comments on the Generation of Random Numbers A number of the problems in Chapters 2–6 require the generation of random numbers. Most of the random numbers can be generated using the Statistics Toolbox in MATLAB. It is also possible to generate these random numbers starting from a set of uniformly distributed random numbers. We will briefly review the process of generating different types of random numbers. A uniformly distributed random number u_i is the starting element of most approaches for the generation. Consider a random variable with a distribution function $F_X(x)$. It can be shown (Papo 1991, Hahn 1994, Evan 2000) that if the function $F_X(x)$ can be inverted, we can write

$$x_i = F_X^{-1}(u_i).$$

Thus, for each uniformly distributed random number u_i in the range $(0, 1)$, we can get a random number x_i. The sequence x_i will have a distribution corresponding to $F_X(x)$.

We will look at some examples. The following is an exponential random number set:

$$F_X(x) = 1 - \exp\left(-\frac{x}{2a^2}\right) = u$$

Rewriting the distribution function, we get

$$x_i = -2a^2 \ln(1 - u_i).$$

Noting that $(1 - u_i)$ is also uniform in the range $(0, 1)$, an exponentially distributed random number sequence is obtained as

$$x_i = -2a^2 \ln(u_i), \quad i = 1, 2, \ldots, N.$$

Following is a Rayleigh random number set:

$$F_X(x) = 1 - \exp\left(-\frac{x^2}{2a^2}\right) = u.$$

Rewriting, we get the Rayleigh-distributed sequence

$$x_i = a\sqrt{-2\ln(u_i)}, \quad i = 1, 2, \ldots, N.$$

It can be easily seen that we get an exponentially distributed sequence by taking the square of the Rayleigh-distributed sequence.

Gaussian random numbers can be generated either directly from MATLAB or by making use of the relationship (Papo 1991) between the Rayleigh distribution, the uniform distribution, and the Gaussian distribution. If r_i forms a Rayleigh-distributed sequence of parameter $2a^2$ and ϕ_i forms a uniformly distributed sequence in the range $(0, 2\pi)$, the random number sequences z_i and w_i given by

$$z_i = r_i \cos(\phi_i), \qquad w_i = r_i \sin(\phi_i)$$

will be Gaussian and independent of each other. Making use of the relationship between Rayleigh-distributed random variables and exponentially distributed random variables, z_i and w_i can be expressed as

$$z_i = a\sqrt{-2\ln(u_i)}\cos[\pi(2u_i - 1)]$$

$$w_i = a\sqrt{-2\ln(u_i)}\sin[\pi(2u_i - 1)].$$

If $a = 1$, we get two independent sequences of Gaussian-distributed random numbers, s_{xi} and s_{yi}, with zero mean and standard deviation of unity:

$$s_{xi} = \sqrt{-2\ln(u_i)}\cos[\pi(2u_i - 1)]$$

$$s_{yi} = \sqrt{-2\ln(u_i)}\sin[\pi(2u_i - 1)].$$

Reversing, it is possible to show that $\sqrt{s_{xi}^2 + s_{yi}^2}$ will be a Rayleigh-distributed random number sequence.

A.4.3 Random Processes

A random process $X(\xi, t)$ may be viewed as a function of two variables, time and random outcomes ξ. Consider a simple experiment measuring the voltage across a resistor, R, which is random. For a specific value of the resistance, for example, R_1, we now have a function of time, $X_1(t)$, as shown in Figure A.4.4. For another value of resistance, say, R_2, we have another time function, $X_2(t)$. If we repeat this procedure for all possible outcome values (as far as the resistance R is concerned), we will have a set

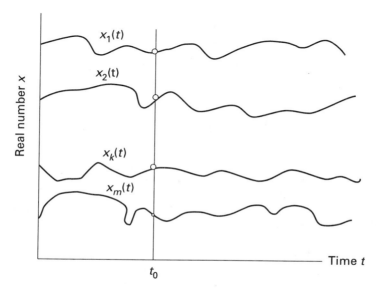

FIGURE A.4.4
Concept of a random process.

of sample functions corresponding to each of the outcomes. All of these sample functions together form an "ensemble." For a specific time $t = t_0$ we have a random variable $X(t_0)$; for a specific outcome, for example, R_m, we have a mere time function, $X_m(t)$. When we consider all possible times and outcomes, we have a random process, $X(t)$. For a given R_k and fixed time t_1, $X_k(t_1)$ is a mere number.

For a specific time, for example, $t = t$, the random process becomes a random variable, and we can associate a cumulative distribution function,

$$F(x, t) = \text{prob}\{X(t) \le x\}. \tag{A.4.43}$$

This is known as the first-order CDF of the random process. The corresponding first-order probability density function of the random process is given by

$$f(x, t) = \frac{\partial F(x, t)}{\partial x}. \tag{A.4.44}$$

The second-order CDF is given by

$$F(x_1, x_2; t_1, t_2) = \text{prob}\{X(t_1) \le x_1, X(t_2) \le x_2\}, \tag{A.4.45}$$

where t_1 and t_2 are two time instants.

The corresponding density function is given by

$$f(x_1, x_2; t_1, t_2) = \frac{\partial^2 F(x_1, x_2; t_1, t_2)}{\partial x_1 \partial x_2}. \tag{A.4.46}$$

We can similarly define the nth-order CDF and pdf.

The mean of the random process, $\eta(t)$, is given by

$$\eta(t) = E\{X(t)\} = \int_{-\infty}^{\infty} xf(x, t) \, dx. \tag{A.4.47}$$

The mean is also known as the *ensemble average*.

The autocorrelation, $R(t_1, t_2)$, of the random process is given by

$$R(t_1, t_2) = E\{X(t_1)X(t_2)\}$$

$$= \int_{-\infty}^{\infty} \int_{-\infty}^{\infty} x_1 x_2 f(x_1, x_2; t_1, t_2) \, dx_1 \, dx_2. \tag{A.4.48}$$

The value of the autocorrelation function (also known as the *ensemble autocorrelation*) for $t_1 = t_2$ gives the average power of $X(t)$ as

$$E\{X^2(t)\} = R(t, t). \tag{A.4.49}$$

The autocovariance of the random process, $C(t_1, t_2)$ is given by

$$C(t_1, t_2) = R(t_1, t_2) - \eta(t_1)\eta(t_2). \tag{A.4.50}$$

The variance of $X(t)$ is given by $C(t, t)$.

A stochastic process $X(t)$ is called a *strict-sense stationary* process if the statistical properties of $X(t)$ are unaffected by a shift of origin (time). A strict-sense stationary process has a pdf of nth order such that

$$f(x_1, x_2, \ldots, x_m; t_1, t_2, \ldots, t_n) = f(x_1, x_2, \ldots, x_m; t_1 + \tau, t_2 + \tau, \ldots, t_n + \tau) \tag{A.4.51}$$

for any τ.

First-order stationarity implies that the first-order pdf, $f(x; t)$, is independent of time, t,

$$f(x; t) = f(x; t + \tau) = f(x) \qquad (A.4.52)$$

for any τ.

Second-order stationarity implies that the second-order pdf, $f(x_1, x_2; t_1 + c, t_2 + c)$, is independent of c,

$$f(x_1, x_2; t_1, t_2) = f(x_1, x_2; \tau), \qquad (A.4.53)$$

where $\tau = t_1 - t_2$. This means that the second-order density function depends only on the time difference and not on the time instants.

A process is called *wide-sense stationary* (WSS) if the mean of the random process is independent of time and its autocorrelation depends only on the time difference. This means that

$$E\{X(t)\} = \eta(t) = \eta \qquad (A.4.54)$$

and

$$E\{X(t)X(t + \tau)\} = R(t, t + \tau) = R_{xx}(\tau). \qquad (A.4.55)$$

Note that strict-sense stationarity implies wide-sense stationarity, but the converse is not true.

We may need to estimate the mean and variance of a stochastic process, but the pdf of the process may not be readily available. It may be possible to determine the mean, variance, and other statistical attributes of the random process under certain conditions by examining the time domain characteristics instead of dealing with the ensemble. These conditions are contained in the concept of *ergodicity*.

A random process $X(t)$ is called *ergodic* if the ensemble average equals the time average. The time average of a random process is given by

$$\overline{X(t)} = \frac{1}{2T} \int_{-T}^{T} x(t) \, dt. \qquad (A.4.56)$$

We say that the random process $X(t)$ is *ergodic in the mean* if

$$\overline{X(t)} = \eta \quad \text{as} \quad T \longrightarrow \infty, \qquad (A.4.57)$$

where η is the ensemble average given in the equation.

We say that the random process $X(t)$ is *ergodic in the autocorrelation* if the temporal autocorrelation given by

$$\overline{X(t)X(t + \tau)} = \frac{1}{2T} \int_{-T}^{T} x(t)x(t + \tau) \, dt, \qquad (A.4.58)$$

is equal to the ensemble autocorrelation when $T \longrightarrow \infty$, i.e.,

$$\overline{X(t)X(t + \tau)} = R_{xx}(\tau), \quad T \longrightarrow \infty. \qquad (A.4.59)$$

The assumption of *ergodicity* allows us to get estimates of the mean and variance of the random process using observations made in time.

The power spectrum or spectral density, $S_{xx}(f)$ or $G_{xx}(f)$, of a random process $X(t)$ is the Fourier transform of the autocorrelation, $R_{xx}(\tau)$, given by

$$S_{xx}(f) = \int_{-\infty}^{\infty} R_{xx}(\tau)\exp(-j2\pi f\tau)\ dt\ . \tag{A.4.60}$$

Properties of Power Spectral Density

$S_{xx}(f)$ is always real-valued and positive.

$S_{xx}(f)$ is even if $X(t)$ is real-valued.

$S_{xx}(f)$ and $R_{xx}(\tau)$ form a Fourier transform pair.

A typical random binary pattern, $X(t)$, is shown in Figure A.4.5 along with its autocorrelation. Its power spectrum is shown in Figure A.4.6. The shaded area gives the total average power given by $\int_{-\infty}^{\infty} S_{xx}(f)df$.

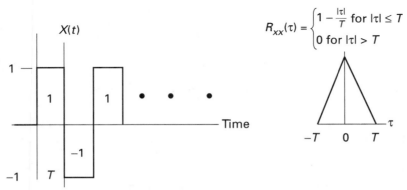

FIGURE A.4.5 A random bit pattern and its autocorrelation.

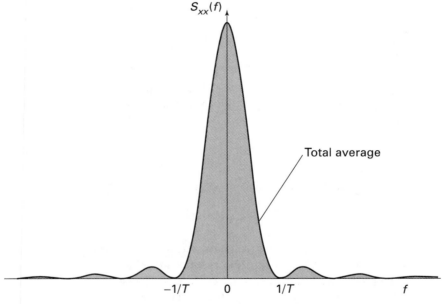

FIGURE A.4.6 The power spectral density of the signal in Figure A.4.5.

A.5 NOISE IN COMMUNICATION SYSTEMS

The noise in communication systems is generally referred to as "white Gaussian noise." The noise is caused by the thermal motion of the electrons in all the components present in the electronic circuits (Dave 1958, Papo 1991, Ziem 1995). The thermal noise has been modeled as a zero-mean Gaussian random process. The Gaussian noise process, $n(t)$, is characterized by a Gaussian density function for n, the noise at any arbitrary time, t:

$$f(n) = \frac{1}{\sqrt{2\pi\sigma^2}} \exp\left(-\frac{n^2}{2\sigma^2}\right). \tag{A.5.1}$$

Another characteristic of the thermal noise is that all the frequency components have equal powers. In other words, its spectral density, $S_{nn}(f)$, is white and is given by

$$S_{xx}(f) = \frac{N_0}{2}, \quad -\infty \leq f \leq \infty. \tag{A.5.2}$$

The corresponding autocorrelation function, $R_{nn}(\tau)$, is a delta function and is expressed as

$$R_{nn}(\tau) = \frac{N_0}{2}\delta(\tau). \tag{A.5.3}$$

The power spectrum of the white noise and the corresponding autocorrelation function are shown in Figure A.5.1. Note that the average power of white noise is infinite; however, the noise is typically band-limited to B Hz, and hence we can treat the energy density as finite and given by N_0B, as shown in Figure A.5.1c.

A.6 POWER SPECTRA OF DIGITAL SIGNALS

The power spectra of digital signals may be evaluated starting from the fundamental definition of the PSD of a stochastic process (Dave 1958, Proa 1995), given in eqs. (A.4.59) and (A.4.60). Consider the input data stream $s(t)$ given by

$$s(t) = \sum_{k=-\infty}^{\infty} a_k g(t - kT), \tag{A.6.1}$$

where a_k represents the symbols taking appropriate values and T is the symbol duration. The linearly modulated signal $v(t)$ can be expressed as

$$v(t) = \text{Re}\left\{s(t)e^{j2\pi f_0 t}\right\}, \tag{A.6.2}$$

where f_0 is the carrier frequency. Note that while $s(t)$ is a low-pass signal, $v(t)$ is a bandpass signal. The representation given in eq. (A.6.2) can apply to ASK, PSK, and QAM. If $R_{vv}(\tau)$ is the autocorrelation function of $g(t)$,

$$R_{vv}(\tau) = \text{Re}\left\{R_{ss}(\tau)e^{j2\pi f_0 t}\right\}, \tag{A.6.3}$$

where $R_{ss}(\tau)$ is the autocorrelation of $s(t)$ given by

$$R_{ss}(\tau) = E\{s(t)*s(t+\tau)\} = \sum_{j=-\infty}^{\infty}\sum_{k=-\infty}^{\infty} E(a^*_j a_k)g^*(t-jT)g(t+\tau-kT), \tag{A.6.4}$$

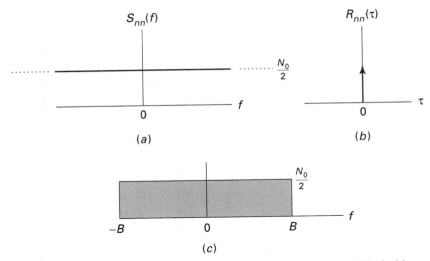

FIGURE A.5.1 The power spectral density (a) and autocorrelation (b) of white noise. The power spectral density of band-limited white noise is shown in (c).

where $E()$ is the ensemble average. We assume that the sequence of symbols is wide-sense stationary with a mean of η_a and discrete autocorrelation of $R_{aa}(k)$, given by

$$R_{aa}(k) = E(a^*_j a_{j+k}). \tag{A.6.5}$$

Using eq. (A.6.5), eq. (A.6.4) can be rewritten as

$$R_{ss}(\tau) = \sum_{j=-\infty}^{\infty} \sum_{k=-\infty}^{\infty} R_{aa}(k-j)g^*(t-jT)g(t+\tau-kT). \tag{A.6.6}$$

Defining $(k - j)$ as m, eq. (A.6.6) can be expressed as

$$R_{ss}(\tau) = \sum_{m=-\infty}^{\infty} R_{aa}(m) \sum_{k=-\infty}^{\infty} g^*(t-kT)g(t+\tau-kT-mT). \tag{A.6.7}$$

Note that the autocorrelation function of $s(t)$ is periodic with period T. The mean of $s(t)$ is also periodic, thus making $s(t)$ a cyclostationary process. The power spectrum of such a signal can be obtained by eliminating the dependence of the autocorrelation function on t (and not τ). This is done by averaging the autocorrelation function, $R_{ss}(\tau)$, over period T. The average autocorrelation function is given by

$$\overline{R}_{ss}(\tau) = \frac{1}{T} \int_{-T/2}^{T/2} R_{ss}(\tau)\, dt$$

$$= \sum_{m=-\infty}^{\infty} R_{aa}(m) \sum_{k=-\infty}^{\infty} \frac{1}{T} \int_{-T/2}^{T/2} g^*(t-kT)g(t+\tau-kT-mT)\, dt \tag{A.6.8}$$

$$= \sum_{m=-\infty}^{\infty} R_{aa}(m) \sum_{k=-\infty}^{\infty} \frac{1}{T} \int_{-T/2-kT}^{T/2-kT} g^*(t)g(t+\tau-mT)\, dt.$$

If we define $R_{gg}(t)$ as the autocorrelation of the pulse shape given by

$$R_{gg}(\tau) = \int_{-\infty}^{\infty} g^*(t)g(t+\tau)\,dt\,, \tag{A.6.9}$$

eq. (A.6.8) becomes

$$\bar{R}_{ss}(\tau) = \frac{1}{T}\sum_{m=-\infty}^{\infty} R_{aa}(m)R_{gg}(\tau - mT)\,. \tag{A.6.10}$$

Taking the Fourier transform of eq. (A.6.10), the power spectral density $S_{ss}(f)$ of $s(t)$ is given by

$$S_{ss}(f) = \frac{1}{T}|G(f)|^2 S_{aa}(f)\,, \tag{A.6.11}$$

where $S_{aa}(f)$ is the power spectral density of the symbol sequence and $G(f)$ is the Fourier transform of the pulse shape. Note that the symbol sequence is periodic, and hence we can use the concepts of exponential Fourier series to express

$$S_{aa}(f) = \sum_{m=-\infty}^{\infty} R_{aa}(m)e^{+j2\pi fmT} \tag{A.6.12}$$

and

$$R_{aa}(m) = T\int_{-1/2T}^{1/2T} S_{aa}(f)e^{+j2\pi fmT}\,df\,. \tag{A.6.13}$$

If the symbols are real and uncorrelated with a variance of σ^2, the autocorrelation function $R_{aa}(m)$ can be expressed as

$$R_{aa}(m) = \begin{cases} \eta^2 + \sigma^2 & m = 0 \\ \eta^2 & m \neq 0 \end{cases}. \tag{A.6.14}$$

Equation (A.6.12) can now be expressed using eq. (A.6.14) as

$$S_{aa}(f) = \sigma^2 + \eta^2 \sum_{m=-\infty}^{\infty} e^{-j2\pi fmT}$$

$$= \sigma^2 + \eta^2\left(\frac{1}{T}\right)\sum_{m=-\infty}^{\infty} \delta\left(f - \frac{m}{T}\right). \tag{A.6.15}$$

The spectral density $S_{ss}(f)$ now becomes

$$S_{ss}(f) = \sigma^2\left(\frac{1}{T}\right)|G(f)|^2 + \eta^2\left(\frac{1}{T}\right)^2 \sum_{m=-\infty}^{\infty}\left|G\left(\frac{m}{T}\right)\right|^2\delta\left(f - \frac{m}{T}\right). \tag{A.6.16}$$

The first term in eq. (A.6.16) is continuous and depends on the spectral characteristics of the pulse shape $g(t)$. The second term is composed of discrete frequency components spaced $1/T$ Hz apart. In most digital communication systems, such as PSK, there are equal numbers of symbols of either type, so that the mean value η is zero and the power spectral density $S_{ss}(f)$ retains only the continuous part,

$$S_{ss}(f) = \sigma^2\left(\frac{1}{T}\right)|G(f)|^2. \tag{A.6.17}$$

TOPICS IN COMMUNICATIONS THEORY

B.1 ORTHOGONAL FREQUENCY DIVISION MULTIPLEXING

Wireless and mobile communication systems are adversely affected by frequency-selective fading, as discussed in Chapters 2 and 5. We examined several ways of mitigating the effects of flat and frequency-selective fading through diversity and equalizing (Chapter 5). It is possible to reduce the effects of frequency-selective fading if we reduce the data rate by splitting the data stream into several parallel blocks and then transmit these blocks. Such transmission of information at a lower baud rate will mitigate the effects of fading and intersymbol interference (ISI). The transmission of these blocks of data can be undertaken using a multicarrier modulation (MCM) scheme (Cimi 1985; Le Fl 1989; Bing 1990; Hara 1993, 1997; Salt 1998) as shown in Figure B.1.1. This technique is also known as orthogonal frequency division multiplexing (OFDM).

The incoming data at a rate of $R = N/T$ symbols/s is split into N separate blocks. This reduces the data rate in each group to $1/T$, in effect increasing the symbol duration in these blocks. This increase in the symbol duration is primarily responsible for the reduction in distortion. Each of these blocks of data is modulated by a separate carrier signal, f_n, $n = 1, 2, \ldots, N$, with

$$f_n = f_0 + \frac{n}{T}. \tag{B.1.1}$$

The frequency separation between the carrier frequencies is $1/T$, making the carrier frequencies (subcarriers) orthogonal to each other. If the QAM format is used (Cimi 1985, Wils 1996, Slim 1998), the transmitted signal can be expressed as

$$D(t) = \sum_{n=1}^{N} \{a_n \cos(2\pi f_n t) + b_n \sin(2\pi f_n t)\}, \tag{B.1.2}$$

where a_n and b_n are the in-phase and quadrature components of the data symbol d_n, given by

$$d_n = a_n \pm jb_n. \tag{B.1.3}$$

The conceptual block diagram shown in Figure B.1.1 may be difficult to implement because of the complexity of the equipment (such as filters and modulators) required. However, with the advances in DFT (discrete Fourier transform) techniques,

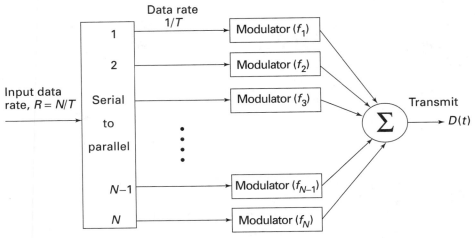

FIGURE B.1.1 Concept of OFDM. The input serial data stream is split into N parallel blocks. Each block modulates a separate carrier.

both the transmitter and receiver can be implemented using DFT processors (Hiro 1981). To see how the concept of DFT can be invoked, let us rewrite eq. (B.1.2) as

$$D(t) = \text{Re}\left[\sum_{n=1}^{N} d_n e^{\pm j2\pi f_n t}\right]. \tag{B.1.4}$$

If we now substitute, $t = m\Delta t$, where $\Delta t = T/N$, we see that eq. (B.1.4) represents a sampled signal. The sampled sequence $D(m)$ is the real part of the discrete Fourier transform of the sequence d_n. We have treated d_n as being in the frequency domain and $D(m)$ as being in the time domain. In other words, the transmitter consisting of a number of filters and modulators can be replaced by an IDFT (inverse discrete Fourier transform) or IFFT (inverse fast Fourier transform) element. The transmitted data symbols, $D(m)$, can be represented as

$$D(m) = \text{IFFT}(d_n) = \sum_{n=1}^{N} d_n e^{j(2\pi/N)nm}. \tag{B.1.5}$$

Reception is accomplished through the use of a DFT (or FFT) element or a processor. Block diagrams for the transmission and reception of OFDM signals (Chen 1999) are shown in Figure B.1.2.

Multicarrier approaches using OFDM have also been proposed and implemented for CDMA systems based on their ability to mitigate the effects of fading. Generally, three different approaches have been used to combine CDMA and OFDM techniques (Chou 1993, Kosh 1999, Chen 1999): multicarrier CDMA (MC-CDMA), multicarrier direct-sequence CDMA (MC-DS-CDMA), and multitone CDMA (MT-CDMA). The original symbol stream is broken down into smaller groups, and each of these subgroups is multiplied by the spreading code. Each of these spread groups then modulates a separate subcarrier signal of frequency f_n, similar to the case shown in Figure B.1.1.

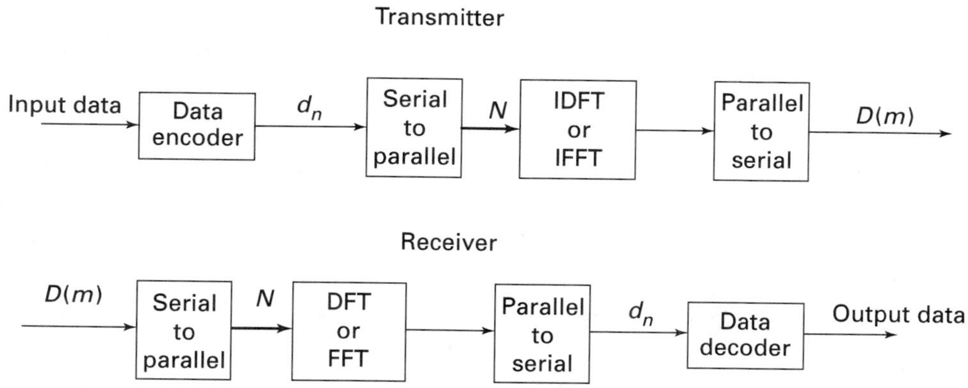

Transmitter
Input data
Data encoder
d_n
Serial to parallel
N
IDFT or IFFT
Parallel to serial
D(m)
Receiver
D(m)
Serial to parallel
N
DFT or FFT
Parallel to serial
d_n
Data decoder
Output data

FIGURE B.1.2 The DFT implementation of OFDM. IDFT is used in transmission, and DFT is used in reception.

B.2 TELETRAFFIC CONSIDERATIONS

Consider the case of a telephone network with a limited number of trunks (or lines, circuits, etc.) available, as shown in Figure B.2.1. Assume this number C to be much larger than the estimated number of users who may wish to use the network at any given time (Papo 1991, Stee 1999, Rapp 1996b). This means that depending on whether or not an empty line is available, a few calls will go through in the order in which the attempts are made. Some of the callers will wait for a line to be connected, and others will not even have a place in the "queue." These call attempts will be blocked. Note also that every user will occupy the line for a certain period of time. The arrival rate (call attempts per second) and the service rate (line usage per second)

Network
Call attempts
Waiting
Connection
Connection
Connection
Connection
Connection
Call completed
Blocked calls

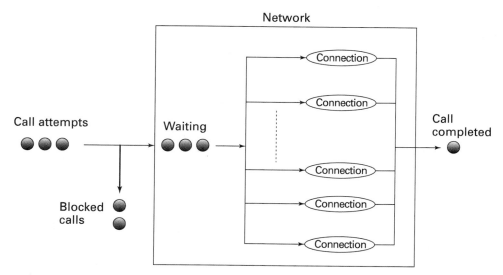

FIGURE B.2.1 Network operation showing call attempts, call waiting, and blocked calls.

are random. If we denote λ as the mean arrival rate and μ as the service rate, we can write expressions for the average holding time H and the average call attempts U as

$$U = \frac{1}{\lambda} \qquad (B.2.1)$$

and

$$H = \frac{1}{\mu}. \qquad (B.2.2)$$

Consider the scenario where call attempts are being made and calls are being terminated. In a steady state, the number of transitions going in will be equal to those going out of the state. The state of the system (also known as a Markov chain) is described by the number of channels that are held, as shown in Figure B.2.2. Let P_r denote the probability that r channels per line or circuit are held. Over a small interval of time Δt, the rate of flow into the state must be equal to the rate of flow out of the system. Consider the case where no channels are held:

$$\lambda \Delta t P_0 = \mu \Delta t P_1. \qquad (B.2.3)$$

The flow considerations lead to similar service rate equations for held channels numbering $1, 2, \ldots, r$:

$$(\lambda + \mu)\Delta t P_1 = \lambda \Delta t P_0 + 2\mu \Delta t P_2$$

$$(\lambda + 2\mu)\Delta t P_2 = \lambda \Delta t P_1 + 3\mu \Delta t P_3 \qquad (B.2.4)$$

$$\vdots$$

$$(\lambda + r\mu)\Delta t P_r = \lambda \Delta t P_{r-1} + (r+1)\mu \Delta t P_{r+1}.$$

These equations can be simplified to get

$$\lambda P_{r-1} = r\mu P_r. \qquad (B.2.5)$$

Noting that the total probability must satisfy

$$\sum_{r=1}^{C} P_r = 1, \qquad (B.2.6)$$

the probability that r circuits are occupied, P_r, is given by

$$P_r = \frac{1}{\displaystyle\sum_{n=1}^{C} \frac{(\lambda/\mu)^n}{n!}} \left[\frac{(\lambda/\mu)^r}{r!} \right]. \qquad (B.2.7)$$

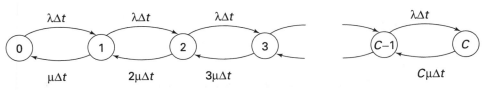

FIGURE B.2.2 State of the system (Markov chain) showing the transitions.

The quantity (λ/μ) is the offered traffic, A. The blocking probability (i.e., the likelihood that all circuits are occupied) is given by

$$P_C = \frac{1}{\sum\limits_{n=1}^{C} \dfrac{(\lambda/\mu)^n}{n!}} \left[\frac{(\lambda/\mu)^C}{C!} \right] . \tag{B.2.8}$$

This is known as the Erlang B, the blocking probability, $p(B)$. The carried traffic, A_c, is given by

$$A_c = A[1 - P(B)] \text{ Erl} . \tag{B.2.9}$$

This equation for the Erlang B has been obtained under the condition that no waiting is allowed in the system. If no circuits or channels are available, the user is blocked out. This condition is known as *blocked calls cleared* (BCC).

Let us now consider the case where we still have only C channels available, but no attempt is blocked. The users are in a queue waiting to be served, and the length of the queue is infinite. This means that the users either are connected or wait in line to be connected. Since the number of users or attempts can now exceed C, the equilibrium flow conditions can now be expressed as

$$\lambda P_{r-1} = r\mu P_r, \qquad r < C$$

$$\vdots \tag{B.2.10}$$

$$\lambda P_{r-1} = C\mu P_r, \qquad r \geq C$$

This case is shown in Figure B.2.3.

Rewriting, we can express the probability that r channels are busy, P_r, as

$$P_r = \frac{\lambda}{r\mu} P_{r-1} = \left(\frac{\lambda}{\mu}\right)^r \frac{1}{r!} P_0 \qquad r \leq C$$

$$\tag{B.2.11}$$

$$P_r = \frac{\lambda}{C\mu} P_{r-1} = \frac{1}{C^{k-C}} \left(\frac{\lambda}{\mu}\right)^r \frac{1}{C!} P_0 \qquad r \geq C.$$

Since the number of users waiting to be connected is infinite, eq. (B.2.6) can be rewritten as

$$\sum_{r=0}^{\infty} P_r = 1 . \tag{B.2.12}$$

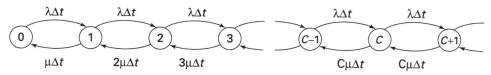

FIGURE B.2.3 State of the system showing the transitions when waiting is allowed.

Equation (B.2.12) can be expanded using eq. (B.2.11) to give

$$P_0\left[1 + \left(\frac{\lambda}{\mu}\right) + \frac{1}{2!}\left(\frac{\lambda}{\mu}\right)^2 + \cdots + \frac{1}{C!}\left(\frac{\lambda}{\mu}\right)^C + \frac{1}{C(C+1)!}\left(\frac{\lambda}{\mu}\right)^{C+1} + \cdots\right] = 1 \quad \text{(B.2.13)}$$

or

$$P_0 = \frac{1}{\left[1 + \displaystyle\sum_{r=1}^{C-1} \frac{1}{r!}\left(\frac{\lambda}{\mu}\right)^r + \displaystyle\sum_{r=C}^{\infty} \frac{1}{C!}\left(\frac{\lambda}{\mu}\right)^r \frac{1}{C^{k-C}}\right]}. \quad \text{(B.2.14)}$$

The probability that an arriving call has to wait in a queue arises when all channels are busy:

$$P\{C \text{ channels are busy}\} = P\{r \geq C\}$$

$$= \sum_{r=C}^{\infty} P_r = P_0 \sum_{r=C}^{\infty} \frac{1}{C!}\left(\frac{\lambda}{\mu}\right)^r \frac{1}{C^{k-C}} \quad \text{(B.2.15)}$$

$$= P_0 \frac{1}{C!}\left(\frac{\lambda}{\mu}\right)^C \frac{1}{1 - \lambda/\mu C} = P_0 \frac{1}{C!} A^C \frac{1}{1 - A/C}$$

This is known as the Erlang C probability.

There is another, more general case of traffic analysis where the length of the queue is finite. That case is beyond the scope of this book.

B.3 CODING

Speech and channel coding are essential in wireless systems. Speech coding techniques produce digital versions of the analog information that have to be transmitted across the wireless channels (Atal 1967, 1970; Span 1994), and channel coding is necessary to mitigate fading and noise. The coding algorithms used will vary depending on the standards being used for wireless systems. We will briefly look at the speech coding algorithms and then present an overview of channel coding algorithms.

B.3.1 Speech Coding

The simplest form of signal or waveform encoding is pulse code modulation (PCM). A block diagram of the PCM system is shown in Figure B.3.1. The analog signal (voice) is sampled, quantized, and encoded.

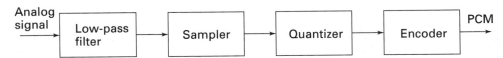

FIGURE B.3.1 Block diagram of a simple PCM system.

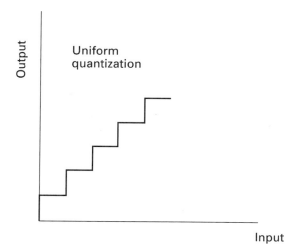

FIGURE B.3.2 Response of a uniform quantizer.

The analog signal is filtered through a low-pass filter to limit the highest frequency components. Based on the highest frequency present in the signal, it is sampled at a rate corresponding to twice the maximum frequency present in the system. This sampled version of the signal is quantized, providing a signal that is discrete both in time and amplitude. Normally a uniform quantizer is used, with a typical response shown in Figure B.3.2.

However, such a uniform quantizer may be less suited to digitizing the human voice. To understand this, let us look at a typical statistical distribution of human speech as shown in Figure B.3.3. It is clear that large amplitude values are rare in the human voice. This means that use of a uniform quantizer will be less efficient since many of the levels are unlikely to be used. Furthermore, in a system that uses uniform quantization, noise resulting from quantization is the same for all signal amplitudes. Noise depends only on the step size of the quantization, which leads to poor signal-to-quantization noise ratios for low-amplitude signals. This problem can be overcome through the use of nonuniform quantization using smaller step sizes (finer quantization) for low amplitudes and larger step sizes (coarser quantization) for high amplitudes. Such nonuniform quantization can be achieved using a compression algorithm. There are two compression algorithms commonly used, the μ law and the A law. Each produces an output similar to the one shown in Figure B.3.4. The low amplitudes are scaled up while high amplitudes are scaled down. If we use a uniform quantizer after the compression, we achieve a nonuniform quantization, which reduces some of the problems. A reverse compression operation is employed at the receiver to bring the signal back to its original form.

Because human speech lies in the range of 300 to 3300 Hz, a sampling rate of 8 kHz is used. The minimum number of quantization levels to achieve acceptable performance is 256 (2^8), resulting in a coding or data rate of 64 kbps. Such a high data rate would require prohibitively high transmissions bandwidths. In wireless communications, where bandwidth is at a premium, PCM is seldom used.

We need to look for other techniques to achieve speech coding at rates lower than 64 kbps without sacrificing the quality. It is possible to reduce the coding rates for speech to 32 kbps by using adaptive differential pulse code modulation (ADPCM). In this

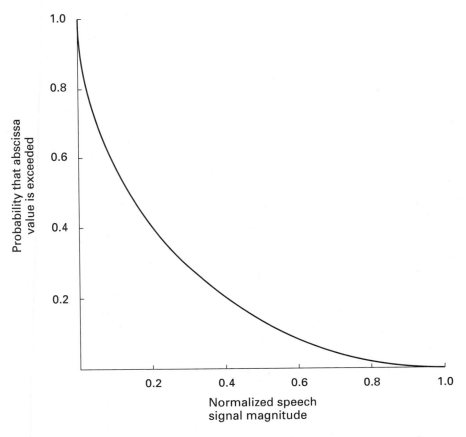

FIGURE B.3.3 Statistical distribution of human speech.

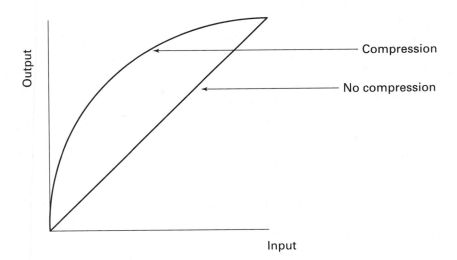

FIGURE B.3.4 The input-output relationship of a system with and without compression.

approach a multilevel quantizer and an adaptive prediction filter are used. The step size of the quantizer is varied in an adaptive fashion, depending on the speaker and time. Instead of encoding the difference between adjacent samples, as done in differential pulse code modulation (DPCM), a predictor is used to get the current sample. The difference between the predicted sample and the actual sample is then encoded for transmission.

Mobile communication systems generally use some form of "vocoders" to accomplish the task of speech coding. Vocoders operate by analyzing the voice signals, and then generating and transmitting parameters based on the analysis. At the receiver, the voice is synthesized from these parameters. Vocoders operate both in the time and frequency domains.

We will look briefly at some of the time domain vocoders used in mobile communications.

Linear Predictive Coder (LPC) and Multipulse Excited LPC

In a linear predictive coder the attempt is made to synthesize certain features of the voice message (time waveform). The assumption here is that the speech is produced by a source-filter combination (Schr 1966, 1985). The vocal tract is modeled as an all-pole filter (i.e., a filter whose transfer function has poles only). The input to this filter is either a single pulse or white noise, depending on whether the speech is voice or nonvoice. The coefficients of the filter are obtained using linear prediction techniques (Atal 1967, 1970; Schr 1966, 1985). A block diagram of the encoder is shown in Figure B.3.5.

The encoder shown in Figure B.3.5 consists of three main parts: a synthesis filter, an excitation generator, and an error weighting/error minimization filter. The synthesis filter is an all-pole predictor that models the short-term spectral envelope of the speech waveform. The parameters of the filter are determined from short speech samples (10–30 ms duration) outside the optimization loop. The excitation generator produces the excitation pulse that is applied to the synthesis filter. The error minimization optimizes the weighted error between the original speech and the synthesized speech using a mean-square-error criterion. The filter parameters and excitation parameters that minimize the error constitute the transmitted signal.

The decoder is shown in Figure B.3.6. Using the filter parameters and excitation parameters, the speech is synthesized at the receiver.

Single-pulse excitation does not produce a good-quality audio signal. We can also use a number of pulses per pitch period, instead of only a single pulse. Through the use of eight pulses per pitch period and by adjusting the individual pulse positions and amplitudes, it is possible to reduce the minimum mean squared error. This results in better-quality speech reproduction.

Code Excited Linear Predictor (CELP)

Instead of using multiple pulses, it is possible to use a predetermined set of codes from a codebook. The codebook consists of a set of stochastic excitation signals, ensembles of zero-mean Gaussian noise. At the transmitter, the codebook containing stochastic signals is searched to find the best perceptual match to the voice signal used as an excitation to the linear predictive coder (LPC). The block diagram of this arrangement is shown in Figure B.3.7. The index of the codebook with the best match is transmitted. The receiver, having a matching codebook, uses this index to pick the correct excitation signal to synthesize the voice at the receiver. Often the gain factor used to minimize the perceptual error is transmitted together with the codebook index.

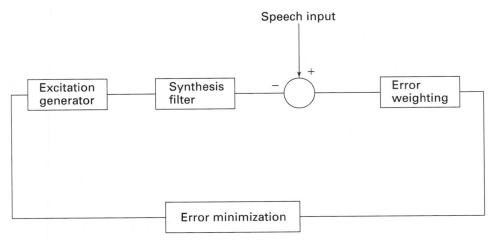

FIGURE B.3.5 Block diagram of an encoder.

FIGURE B.3.6 Block diagram of a decoder.

CELP systems are computationally very complex because of the exhaustive search necessary to identify the appropriate excitation signal (code). Simultaneously with the search, computations for the weights to minimize the perceptual error have to be made, making CELP systems computationally very intensive. However, these coders can provide good-quality speech even at a rate of 4.8 kbps, which typically uses less than one bit per sample.

The complexity and computations of the CELP can be reduced using a vector sum excited linear predictive coders (VSELP). Multiple codebooks are used and organized so as to reduce the exhaustive search and computations.

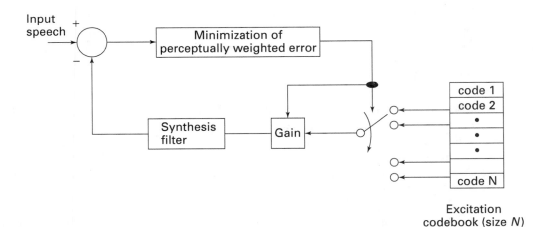

FIGURE B.3.7 Block diagram of a CELP system.

B.3.2 Channel Coding

The presence of additive noise and fading in mobile communication systems increases the error rates to unacceptably high levels (Berl 1987). These systems also suffer from errors occurring in bursts due to deep fades (from Doppler). While error rates of 10^{-2} to 10^{-3} may be tolerable in voice communications, such error rates are unacceptable when the communication involves sensitive information in health, defense-related, or other highly critical systems. Some form of mitigation is necessary to overcome the high error rates observed in wireless systems. Channel coding techniques (i.e., methods to arrange or rearrange the data) provide coding gain in terms of a lower signal-to-noise ratio required to maintain an acceptable bit error rate (Taub 1986, Proa 1995, Hayk 2001). Channel coding can be easily understood by revisiting the concepts of diversity techniques used to mitigate the effects of fading.

We have seen that diversity techniques lead to lower error rates in fading systems. Instead of using frequency, space, or other forms of diversity discussed in Chapter 5, it is possible to imagine that we can transmit more information than what is needed as a minimum. In other words, for every three bits of data we wish to transmit, if we add another two bits of redundant data, we can achieve conditions similar to diversity. Such techniques constitute what are known as error detection and error control coding schemes (Berl 1987). In general, channel coding schemes encompass a broad range of signal transformations directed toward improving the performance of digital communication systems, either through a reduction in error rates or a reduction in the signal-to-noise ratio required to achieve a minimum acceptable bit error rate. Note that these signal transformation schemes will come at the cost of increased bandwidth for transmission.

Before we look at the different methods of error coding, we will try to understand the rationale for error reduction by adding redundant data. The equation for channel capacity from Chapter 3 will be reproduced here for this discussion. The channel capacity C (bits/s) is given by

$$C = B \log_2\left(1 + \frac{S}{N}\right). \tag{B.3.1}$$

Consider the addition of a few bits of redundant data to the raw data stream. The increase in data rate caused by these redundant bits will increase the bandwidth (B) and reduce the bandwidth efficiency. However, if sufficient signal-to-noise ratio (S/N) is available, this increase in bandwidth will reduce the error rate. This was discussed in Chapter 3, where it was argued that wider bandwidths (at the expense of reduced bandwidth efficiency) improve the performance of digital communication systems. The addition of redundant bits and the consequent increase in bandwidth can therefore lead to lower error rates.

The introduction of redundancies forms the basis of channel coding techniques. Redundancies may be used to detect errors in transmission and correct a few such errors. Forms of channel coding can be grouped into three categories:

1. Error detection coding.
2. Forward error correction (FEC) codes. These are codes that can detect and correct errors. These do not require a return path.
3. Automatic repeat request (ARQ) schemes. These require a return path.

Error Detection Coding The simplest form of error detection code is a parity check code (Skla 1988). This concept is illustrated in Table B.3.1. A stream of raw data bits is indicated by a_1, a_2, a_3. The bit c at the end of the stream is the check bit (parity check bit). The parity is checked by verifying that the modulo-2 addition of the coded signal bits is zero. The parity check bit c is given by

$$c = \sum_{i=1}^{n} a_i \, (\text{mod } 2). \tag{B.3.2}$$

If we denote the received bits by b_i, we can determine the existence of an error by checking whether $\sum_{i=1}^{n+1} b_i$ is equal to zero or not. Note that this check includes the parity check bit (summation extends to $n + 1$). We can also apply the parity checking bits to vertical arrays of data bits as shown in Table B.3.2.

This simple means of detecting errors can be extended to include error correction, as described in the next section. Before we look at error control coding techniques, we will briefly discuss two different structured ways of arranging the bits: block codes and convolutional codes (Proa 1995, Hayk 2001).

In the case of block coding, k bits are encoded into n bits $(n > k)$. This means that there are $(n - k)$ redundant bits added to each block of raw data consisting of k bits. Such a code is referred to as an (n, k) code. The *rate* of the code is defined as k/n, and the *redundancy* of the code is given by $(n - k)/k$. The code rate can be understood as the "information bearing" fraction of the encoded bits. We can now go back to Shannon's coding theorem and see that a half-rate code ($k/n = 1/2$) will require double the bandwidth of an uncoded system. If a three-fourths rate code is used, redundancy is 1/3 and the bandwidth expansion is only 4/3.

Convolutional codes are different from block codes. In convolutional encoders, information bits are accepted in a continuous fashion, and the encoder generates a continuous stream of bits at a rate higher than the input raw data rate. While block codes are represented by two integers (n, k), the convolutional coders are represented by three integers, n, k, and K. The encoder is fed k bits at a time, and the output is n bits. K is known as the *constraint length*. It reflects the fact that the convolutional encoder has memory, and K is the number of input bits on which the current output depends. The numbers n and k are small (1 to 6), and K is varied to control the redundancy.

TABLE B.3.1 Parity Check Code (Horizontal)

a_1	a_2	a_3	c
1	1	1	1
1	1	0	0
1	0	0	1
0	0	1	1
0	1	0	1
0	0	0	0

TABLE B.3.2 Parity
Check Code (Vertical)

a_{11}	a_{12}	a_{13}
a_{21}	a_{22}	a_{23}
a_{31}	a_{32}	a_{33}
c_1	c_2	c_3

Forward Error Correction Codes The ability to correct the errors in a block coding scheme is determined primarily by the *code distance*. The code distance d is defined as the number of elements by which two code words C_i and C_j differ,

$$d(C_i, C_j) = \sum_{l=1}^{M} C_{i,l} \oplus C_{j,l} \ (\text{mod } q). \tag{B.3.3}$$

The quantity q is the number of possible values of C_i and C_j (for a binary code, $q = 2$). This distance is called the *Hamming distance* (d_H) if the code is binary. The minimum Hamming distance is the smallest of the Hamming distances between two valid code words. The number of nonzero elements in the code word is identified as the weight of the code. A number of different code families exist that can provide various levels of error correction capability. One class of codes that uses wireless systems are the Reed-Solomon (RS) codes. These are nonbinary cyclic codes (i.e., any cyclic shift of a code word in a code is also a code word), and they possess a well-defined mathematical structure that makes it easy to decode them. A RS (n, k) code is used to encode m-bit symbols into blocks consisting of $n = 2^m - 1$ symbols. The number of parity symbols that need to be added to correct t errors is $2t$. The properties of these RS codes can be summarized as

Block length	$n = 2^m - 1$ symbols
Message size	k symbols
Parity check size	$n - k = 2t$ symbols
Minimum distance	$2t + 1$ symbols

Various decoding techniques are available for use with Reed-Solomon codes. Note that redundancy is used to accomplish error correction. Other codes available include Hamming codes, Hadamard or Walsh codes, and Golay codes.

We will briefly look at convolutional encoders. A typical convolutional encoder is shown in Figure B.3.8. One bit of data $(k = 1)$ enters the encoder, and two bits of data $(n = 2)$ are generated at the output. The constraint length K is 7 (six shift registers), indicating the number of input bits that the current output is dependent on (Berl 1987).

When convolutional encoding is used rather than block encoding, decoding is a bit more complicated. The decoder attempts to estimate the encoded information using a rule that minimizes the possible number of errors in the system. One such decoder is based on the Viterbi algorithm (Vite 1971, Forn 1973), which performs a maximum likelihood decoding of the convolutional codes. IS 95 systems use convolutional encoders with a constraint length of 9. While the rate of the uplink (k/n) is 1/3, the rate of the downlink is 1/2.

ARQ Schemes The forward error correction (FEC) methods provide no return path to the transmitter. ARQ schemes use a return path. While FEC methods detect errors and correct them using some algorithms, ARQ systems detect errors and then send a

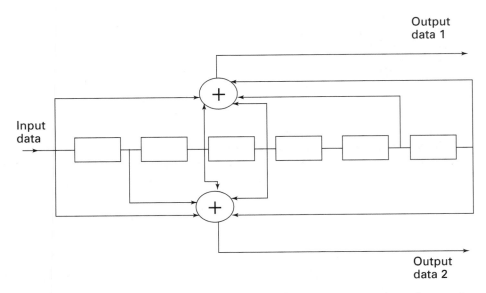

FIGURE B.3.8 A convolutional encoder. One bit of data ($k = 1$) enters the register, and two bits ($n = 2$) of data are obtained. The constraint length K is 7.

request to the transmitter to resend parts of the message or the whole message, making it essential to have a return path. The need to resend the message also means that there may be latency built into the system, and this makes ARQ schemes less suitable for wireless and mobile systems where information transmission, such as a conversation, is taking place on a real-time basis. On the other hand, ARQ schemes are useful when data are being transferred between computers and little or no harm is done by some delay in reception when retransmission is necessary.

There are also hybrid schemes that use both FEC and ARQ systems.

Interleaving   Errors in transmission occur randomly and in bursts. As discussed earlier, error bursts occur when the signal undergoes deep fades. The ability of coding schemes to correct a long string of errors is limited. This limitation can be minimized by interleaving the data. Interleaving can be described as a form of scrambling. Consider the following scenario. We are transmitting a long stream of data in the order in which they are received, as shown in Figure B.3.9a. At the receiver, we will see a stream of data with random errors as well as burst errors ($a_{15} - a_{19}$).

When the received data stream is unscrambled, errors are more or less distributed and it may be easier to correct them. This is seen in Figure B.3.9b. The interleaver is usually used between the encoder and modulator at the transmitter, as illustrated in Figure B.3.10.

At the receiver the deinterleaver is inserted between the demodulator and FEC decoder. Interleaving is done first by storing the data in a two-dimensional table (rows and columns). Data are written in the horizontal direction (rows) and read and transmitted in the vertical direction. At the receiver, data are written and read in the opposite manner. This process takes burst errors and turns them into random bit errors as shown in Figure B.3.9b.

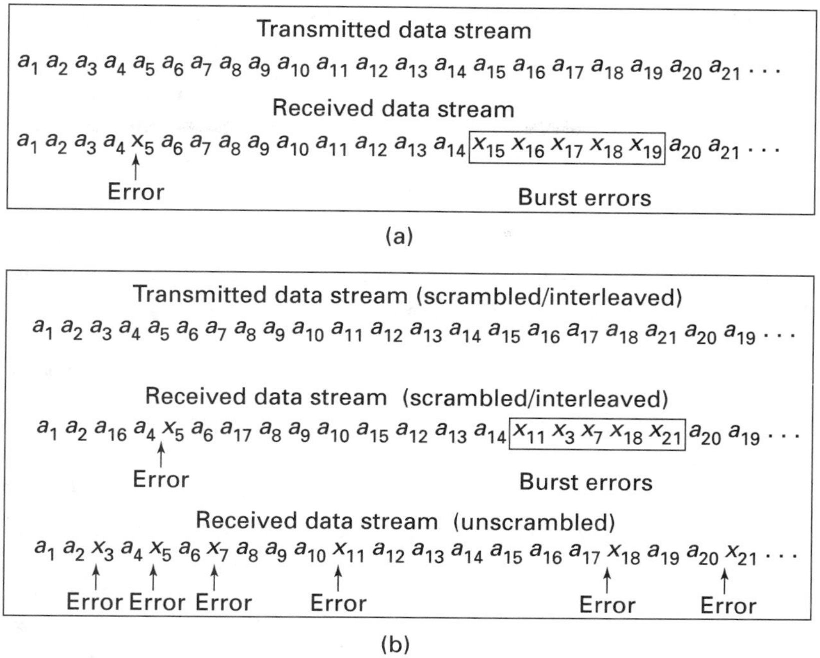

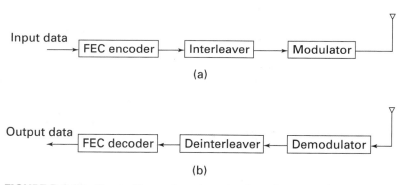

Transmitted data stream

$a_1\ a_2\ a_3\ a_4\ a_5\ a_6\ a_7\ a_8\ a_9\ a_{10}\ a_{11}\ a_{12}\ a_{13}\ a_{14}\ a_{15}\ a_{16}\ a_{17}\ a_{18}\ a_{19}\ a_{20}\ a_{21}\ \cdots$

Received data stream

$a_1\ a_2\ a_3\ a_4\ x_5\ a_6\ a_7\ a_8\ a_9\ a_{10}\ a_{11}\ a_{12}\ a_{13}\ a_{14}\ \boxed{x_{15}\ x_{16}\ x_{17}\ x_{18}\ x_{19}}\ a_{20}\ a_{21}\ \cdots$

↑ Error   Burst errors

(a)

Transmitted data stream (scrambled/interleaved)

$a_1\ a_2\ a_3\ a_4\ a_5\ a_6\ a_7\ a_8\ a_9\ a_{10}\ a_{11}\ a_{12}\ a_{13}\ a_{14}\ a_{15}\ a_{16}\ a_{17}\ a_{18}\ a_{21}\ a_{20}\ a_{19}\ \cdots$

Received data stream (scrambled/interleaved)

$a_1\ a_2\ a_{16}\ a_4\ x_5\ a_6\ a_{17}\ a_8\ a_9\ a_{10}\ a_{15}\ a_{12}\ a_{13}\ a_{14}\ \boxed{x_{11}\ x_3\ x_7\ x_{18}\ x_{21}}\ a_{20}\ a_{19}\ \cdots$

↑ Error   Burst errors

Received data stream (unscrambled)

$a_1\ a_2\ x_3\ a_4\ x_5\ a_6\ x_7\ a_8\ a_9\ a_{10}\ x_{11}\ a_{12}\ a_{13}\ a_{14}\ a_{15}\ a_{16}\ a_{17}\ x_{18}\ a_{19}\ a_{20}\ x_{21}\ \cdots$

↑ Error ↑ Error ↑ Error   ↑ Error   ↑ Error   ↑ Error

(b)

FIGURE B.3.9 The principle of interleaving is shown through the example of a scrambler. (*a*) No scrambling/interleaving. (*b*) Scrambling/interleaving.

Input data → FEC encoder → Interleaver → Modulator → (antenna)

(a)

Output data ← FEC decoder ← Deinterleaver ← Demodulator ← (antenna)

(b)

FIGURE B.3.10 Transmitter and receiver structures incorporating interleaver and FEC schemes. (a) Transmitter. (b) Receiver.

B.4 SYNCHRONIZATION

Digital modulation schemes that need a coherent receiver to recover the information carried by the modulated signals require that the receiver and the transmitter be synchronized. In other words, there must be strict one-to-one correspondence between the carrier frequency, the carrier phase, and the timing of the symbols at the receiver and the corresponding quantities at the transmitter. The operation by which this perfect matching is achieved is called *synchronization* (Proa 1995, Ande 1999).

Carrier frequency and phase synchronization are undertaken at the same time through the process of carrier synchronization. Synchronization of the start/end times of the symbols is undertaken through the process of clock recovery or symbol synchronization.

B.4.1 Carrier Synchronization or Carrier Recovery

In digital modulation systems employing a coherent demodulator at the receiver, there is a need to know precisely the frequency and phase of the carrier wave. For example, consider the reception of a BPSK signal. The signal at the input to the receiver is given by

$$r(t) = m(t) \cos(2\pi f_0 t + \theta) + n(t), \qquad \text{(B.4.1)}$$

where $m(t) = \pm 1$ and $n(t)$ is the white Gaussian noise. The spectrum of $r(t)$ does not contain a delta function at the carrier frequency, and therefore, the precise value of the carrier frequency f_0 or the carrier phase θ is not known. To extract the carrier information (frequency and phase), the incoming signal is squared to produce

$$r^2(t) = \frac{1}{2} + \frac{1}{2} \cos(2\pi f_0 t + 2\theta) + n^2(t) + 2n(t)m(t) \cos(2\pi f_0 t + \theta). \quad \text{(B.4.2)}$$

The second term is the carrier at twice the angular frequency and phase, which can be passed through a phase-locked loop (PLL) circuit, shown in Figure B.4.1.

Note that the approach shown in Figure B.4.1 is not without problems. Since the phase term has been doubled, any uncertainty in phase (phase noise) and phase jitter are also doubled. Even though this doubling is finally taken care of in the frequency divider, the phase noise doubling will affect the signal-to-noise ratio at the PLL, affecting its performance. The two noise terms after the squaring produce still more noise, reducing the overall signal-to-noise ratio.

The carrier circuit for an M-ary PSK will be slightly different. Instead of a squaring device, an Mth-power device is used. The frequency divider (by 2) is replaced by a higher-order divider (by M).

B.4.2 Symbol Synchronization

To achieve perfect demodulation, the digital receiver must be synchronized with the incoming digital transitions. In other words, we need to know with some precision the sampling time instants. This is the process of symbol synchronization or timing recovery.

One of the simplest ways to achieve symbol synchronization is to transmit, in some multiplexed version, a clock signal alongside the information-bearing data signal. At the

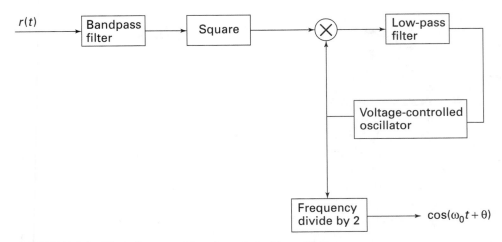

FIGURE B.4.1 Block diagram of the phase-locked loop (PLL).

receiver the clock signal may be recovered using appropriate demultiplexing and filtering. Even though this procedure is simple and straightforward, transmission of a clock signal leads to waste of transmitted power because part of the power is used to carry the clock signal.

There are several methods for obtaining timing information (Proa 1994). We will briefly look at one technique that takes advantage of the symmetry property of the signal at the output of a matched filter. Consider a rectangular pulse at the input of a matched filter, shown in Figure B.4.2a. The matched filter output corresponding to this input pulse will be of a triangular shape, as shown in Figure B.4.2b. This triangular pulse ensures that at two points, separated by ΔT on either side of the exact sampling instant T, the magnitudes of the sampled values are equal. This means that the point midway between two sampling instants that have equal values will constitute the proper sampling instant. This principle forms the basis of the synchronizer known as the *early-late symbol synchronizer*. Figure B.4.3 shows a block diagram of the early-late symbol synchronizer.

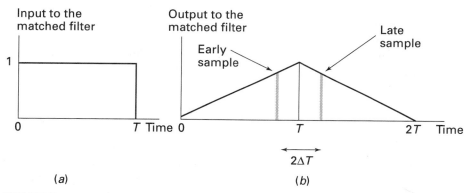

FIGURE B.4.2 Concept of a synchronizer that uses a matched filter. (*a*) Rectangular pulse input to the matched filter. (*b*) Output of the matched filter with ideal sampling at T, an early sample at $T - \Delta T$, and a late sample at $T + \Delta T$.

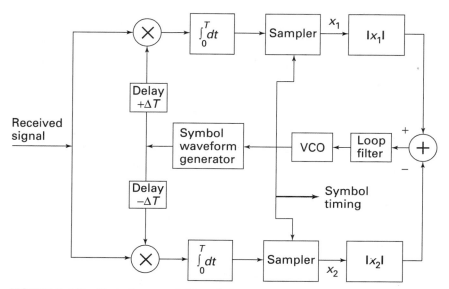

FIGURE B.4.3 Block diagram of the early-late synchronizer.

Making use of their similarity to matched filters (Chapter 3), correlators are used, as shown in Figure B.4.3. The correlators integrate over an interval of T; one correlator integrates ΔT seconds early, relative to the estimated sampling instant, while the other integrates ΔT seconds after the estimated sampling instant. The difference between the two correlator output magnitudes is filtered to remove any noise and is applied to a voltage-controlled oscillator (VCO). Depending on the sign of the difference between the magnitudes, the clock is advanced or delayed. The VCO output signal is also used to provide the timing information to the correlators.

B.5 CHANNEL CAPACITY IN FADING CHANNELS

We looked at the channel capacity for a Gaussian channel in Chapter 3. However, those results may not be valid when transmission of the signal is taking place in a fading channel. The results for channel capacity need to be modified to account for the random nature of the signal received because of Rayleigh fading (Lee 1990, Gunt 1996). Channel capacity in a Rayleigh channel can be calculated in a manner similar to the way in which the average error probabilities in fading are calculated.

For a Gaussian channel, the capacity C (bits/s) is given as

$$C = B \log_2\left(1 + \frac{S}{N}\right) \text{bps}, \tag{B.5.1}$$

where B is the bandwidth. By virtue of fading, the signal-to-noise ratio becomes an exponentially distributed random variable, as discussed in Chapter 5. If γ represents the signal-to-noise ratio in a fading channel, we can rewrite eq. (B.5.1) as

$$\frac{C}{B} = \log_2(1 + \gamma) \text{ bps/Hz}, \tag{B.5.2}$$

which is the expression for spectral efficiency. The average value of spectral efficiency is given by

$$\left[\frac{C}{B}\right]_{av} = \int_0^\infty \log_2(1 + \gamma) f(\gamma) \, d\gamma$$

$$= \int_0^\infty \log_2(1 + \gamma) \frac{1}{\gamma_0} \exp\left(-\frac{\gamma}{\gamma_0}\right) d\gamma, \tag{B.5.3}$$

where γ_0 is the average signal-to-noise ratio. The pdf of $\gamma, f(\gamma)$, was given in Chapter 5.

It is also possible to see the effects of diversity by replacing $f(\gamma)$ with the pdf of the signal-to-noise ratio of the diversity combiner. The spectral efficiency after diversity combining, $[C/B]_{avd}$, is given by

$$\left[\frac{C}{B}\right]_{avd} = \int_0^\infty \log_2(1 + \gamma) f(\gamma) \, d\gamma$$

$$= \int_0^\infty \log_2(1 + \gamma) \frac{M^M \gamma^{M-1}}{(M-1)! \gamma_0^M} \exp\left(-\frac{M\gamma}{\gamma_0}\right) d\gamma. \tag{B.5.4}$$

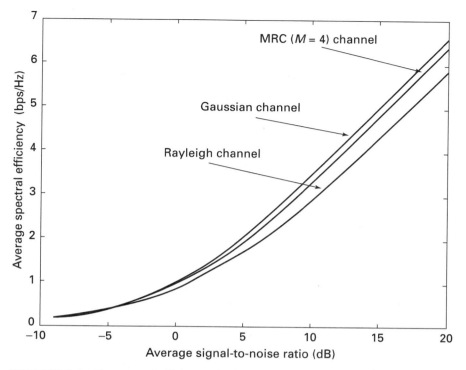

FIGURE B.5.1 The spectral efficiency of a Gaussian channel compared with the efficiencies of a Rayleigh channel with and without diversity.

In eq. (B.5.4) we have used the pdf of the envelope after diversity combining (Chapter 5). The spectral efficiencies are plotted in Figure B.5.1 for the case of a Gaussian channel, a Rayleigh channel, and an M-branch diversity channel. The effect of fading on the spectral efficiency is apparent in that the values for the latter two channels are less than those for the Gaussian channel. The use of the diversity techniques brings the spectral efficiency closer to that of an ideal (i.e., Gaussian) channel.

B.6 INTERSYMBOL INTERFERENCE AND EYE PATTERNS

When pulses are transmitted through physical channels, the pulses may be distorted. The distortion arises from the fact that the channels do not have infinite bandwidth, and hence all frequency components present in the pulse will not experience the same gain. The distortion leads to pulse spreading, with energy or information from neighboring pulses occurring in the pulse being observed.

Figure B.6.1 shows the transmitted pulse and a distorted pulse. The spreading on either side of the pulse leads to intersymbol interference (ISI). One of the ways in which the channel characteristics can be studied is through the use of "eye patterns" (Skla 1988, Hayk 2001, Ande 1999), which allow us to study not only the distortion caused by the channel, but also the effect of noise on the pulses being received. An eye pattern is nothing but a synchronized superposition of all possible realizations of

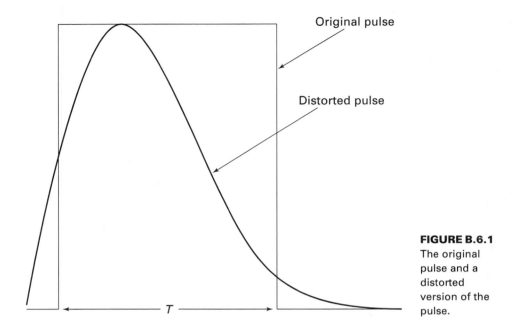

FIGURE B.6.1
The original pulse and a distorted version of the pulse.

the signal by the various bit patterns. A typical eye pattern is shown in Figure B.6.2. It is so called because it resembles the human eye. A number of characteristics of the eye pattern are also shown.

The width of the eye opening provides information on the time interval over which the received signal can be sampled without error from intersymbol interference. A set of computer-generated eye patterns is shown in Figure B.6.3. The figure shows the eye patterns for two values of BT, where B is the 3 dB bandwidth of a low-pass filter and T is the bit duration. The low value of BT shows narrowing of the eye, making it difficult to sample. Once noise is added, the situation gets worse. At the higher value of BT ($=0.5$), it becomes easier to sample the signal, since the eye is almost fully open.

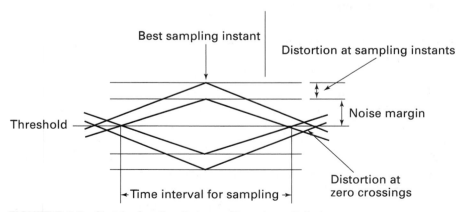

FIGURE B.6.2 Sketch of an "eye" along with various attributes.

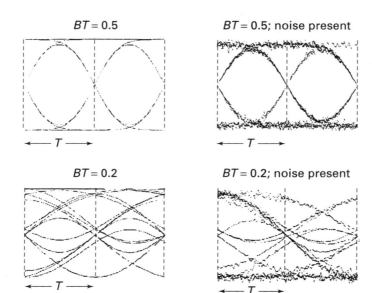

BT = 0.5

$\longleftarrow T \longrightarrow$

BT = 0.5; noise present

$\longleftarrow T \longrightarrow$

BT = 0.2

$\longleftarrow T \longrightarrow$

BT = 0.2; noise present

$\longleftarrow T \longrightarrow$

FIGURE B.6.3
A set of computer-generated eye patterns.

B.7 ERROR FUNCTION

The error function occurs in engineering and science applications when we are dealing with the Gaussian or normal probability density functions. As we have seen in Chapter 3, error probabilities are often expressed in terms of the error function (Skla 1988).

The error function of u, represented by erf(u), is given by the integral

$$\text{erf}(u) = \frac{2}{\sqrt{\pi}} \int_0^u \exp(-x^2) \, dx . \tag{B.7.1}$$

Another, related function is the complementary error function, erfc(u), given by

$$\text{erfc}(u) = \frac{2}{\sqrt{\pi}} \int_u^\infty \exp(-x^2) \, dx . \tag{B.7.2}$$

The error function and complementary error function are related through

$$\text{erfc}(u) + \text{erf}(u) = 1 . \tag{B.7.3}$$

A Gaussian curve showing the region corresponding to the complementary error function is given in Figure B.7.1.

Some properties of the error function are as follows:

$$\text{erf}(-u) = -\text{erf}(u) \tag{B.7.4}$$

As u approaches infinity, erf(u) approaches unity, i.e.,

$$\text{erf}(\infty) = \frac{2}{\sqrt{\pi}} \int_0^\infty \exp(-x^2) \, dx = 1 . \tag{B.7.5}$$

Another, related function that is commonly used in probability theory is the Q function, given by

$$Q(v) = \frac{1}{2\sqrt{\pi}} \int_v^\infty \exp(-x^2) \, dx . \tag{B.7.6}$$

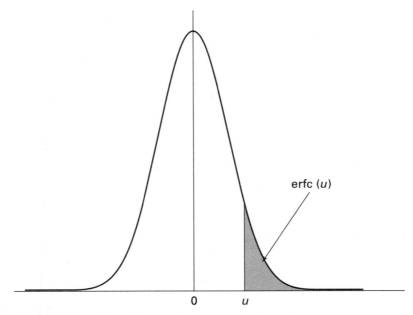

FIGURE B.7.1   Plot of the complementary error function.

Comparing eqs. (B.7.5) and (B.7.6), we can write

$$\text{erfc}(u) = 2Q(\sqrt{2}u) \tag{B.7.7}$$

or

$$Q(v) = \frac{1}{2}\,\text{erfc}\!\left(\frac{v}{\sqrt{2}}\right). \tag{B.7.8}$$

TOPICS IN ATTENUATION AND FADING

C.1 COHERENCE BANDWIDTH AND COHERENCE TIME

We have seen that a mobile channel exhibits both time dispersion and frequency dispersion. While time dispersion is caused by frequency-selective fading, frequency dispersion is caused by time-selective fading. The former results from multipath effects, while the latter is a direct result of Doppler spreading. The equations for the channel bandwidth (expressed in terms of the coherent bandwidth of the channel) and the coherent time can be derived easily from fundamental concepts of correlation.

A channel considered to be frequency selective has many frequency components that take different times to arrive at the receiver and undergo different attenuation levels. Let us examine the case where we are transmitting two very closely spaced tones through a frequency-selective channel (Kenn 1969). Since the frequencies are very close, they will undergo very similar degrees of attenuation. If we increase the frequency separation, the frequencies take different amounts of time to arrive at the receiver and therefore undergo different degrees of attenuation.

If we transmit a broadband signal instead of two distinct frequencies, the frequencies at the extrema of the spectrum will undergo different degrees of attenuation. We can argue that the frequency band over which the attenuation is constant provides us with a frequency region where all the frequency components behave identically. In other words, we can say that coherence exists over this band, and we will identify this frequency band as the *coherence bandwidth* of the channel (Kenn 1969, Stee 1999).

Consider now the case of a mobile channel, i.e., transmission and reception of a signal while the MU is in motion. The motion of the vehicle produces Doppler effect, and this leads to change in the frequency characteristics of the channel, which leads to either compression of the pulse (positive Doppler shift) or elongation of the pulse (negative Doppler shift). In other words, the pulse shape is likely to change, and we can describe this phenomenon in terms of frequency dispersion. This leads to a stretching (or narrowing) of the bandwidth of the received signal compared with the bandwidth of the transmitted signal. If we are transmitting data at a very high rate (large-bandwidth signal) through this time-varying channel, the change in the bandwidth caused by the Doppler shift may be negligible, and no distortion will result. If the data rate is very low, the change in bandwidth brought on by the Doppler shift may be significant, affecting the pulse. Thus, as the duration of the pulse increases (lower signal bandwidth), the channel changes as the pulse is traversing, distorting the signal. Analogous to coherence bandwidth, *coherence time* can be defined as the time over which the channel can be assumed to be constant (Jake 1974, Stei 1987).

However, wireless communication channels are random, and these two terms, coherence bandwidth and coherence time, are best described using the statistical characteristics of the channel. We are interested in finding how similar the envelopes are at two different frequencies (coherence bandwidth) and at two different time instants (coherence time). In other words, we must know how correlated these two envelopes are, either in the frequency domain or in the time domain. The correlation coefficient, which measures this degree of similarity can be used to calculate both the coherence bandwidth and coherence time: the frequency band over which the channel has similar (or constant) properties, and the time duration over which the channel has similar (or constant) properties, respectively. The envelope correlation coefficient, $\rho(\Delta f, \Delta t)$, is given by

$$\rho(\Delta f, \Delta t) = \frac{\langle A_1 A_2 \rangle - \langle A_1 \rangle \langle A_2 \rangle}{\sqrt{\left[\langle A_1^2 \rangle - \langle A_1 \rangle^2\right]\left[\langle A_2^2 \rangle - \langle A_2 \rangle^2\right]}}. \tag{C.1.1}$$

A_1 is the envelope at frequency f_1 and time instant t_1, while A_2 is the envelope at frequency f_2 and time instant t_2. In eq. (C.1.1), $\langle\rangle$ represents the statistical average over the delay time, or the time taken by the signal to reach the receiver from the transmitter.

To derive the correlation coefficient, we need to model the probability density function of the delay τ of the signal. This has been approximately modeled using an exponential distribution (Jake 1974, Stee 1999), i.e.,

$$p(\tau) = \frac{1}{\sigma} \exp\left(-\frac{\tau}{\sigma}\right) U(\tau). \tag{C.1.2}$$

The quantity σ is the delay spread of the channel, given by

$$\sigma = \int_0^\infty (\tau - \langle \tau \rangle)^2 p(\tau)\, d\tau, \tag{C.1.3}$$

where

$$\langle \tau \rangle = \int_0^\infty \tau p(\tau)\, d\tau. \tag{C.1.4}$$

Defining $\Delta f = |f_1 - f_2|$ and $\Delta t = |t_1 - t_2|$, the correlation coefficient is derived as

$$\rho(\Delta f, \Delta t) = \frac{J_0^2(2\pi f_d \Delta t)}{1 + (2\pi\Delta f)^2 \sigma^2}, \tag{C.1.5}$$

where $J_0()$ is the zeroth-order Bessel function. To observe the correlation between two signals separated by Δf, we set $\Delta t = 0$ in eq. (C.1.5). The frequency correlation coefficient, $\rho(\Delta f, 0)$, is

$$\rho(\Delta f, \Delta t)_{\text{at}\Delta t = 0} = \rho(\Delta f, 0) = \frac{1}{1 + (2\pi\Delta f)^2 \sigma^2}. \tag{C.1.6}$$

The frequency correlation coefficient is plotted in Figure C.1.1a. As the frequency separation Δf increases, the correlation coefficient decreases. The coherence

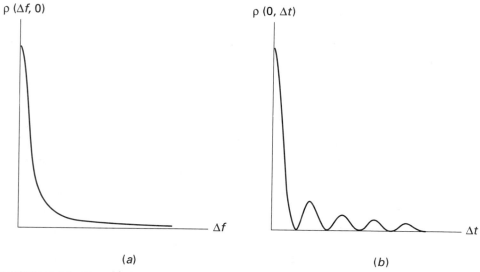

$\rho\ (\Delta f,\ 0)$

$\rho\ (0,\ \Delta t)$

Δf

Δt

(a)

(b)

FIGURE C.1.1 The correlation functions. (a) Frequency correlation. (b) Time correlation.

bandwidth, B_c, is taken to correspond to a correlation coefficient of 0.5, leading to a value of

$$B_c = \frac{1}{2\pi\sigma}. \qquad (C.1.7)$$

The choice of correlation value of 0.5 is arbitrary, and some researchers have used a value of 0.9. In either case, the coherence bandwidth provides us with a quantitative measure of the dispersive nature of the channel.

The time correlation can be obtained by setting $\Delta f = 0$ in eq. (C.1.5) and is plotted in Figure C.1.1b. The correlation coefficient shows a decaying oscillatory behavior as a function of time difference. The coherence time is taken to correspond to a correlation coefficient of 0.5, leading to

$$T_c = \Delta t_{\rho = 0.5} \approx \frac{9}{16\pi f_d}. \qquad (C.1.8)$$

C.1.1 Dispersion versus Distortion

One of the consequences of a multipath propagation is that the channel becomes dispersive. In other words, the channel can become frequency selective. Some confusion exists in literature between distortion and dispersion (Kenn 1969, Stee 1999). To understand the differences and similarities between the two phenomena, let us consider the case where we have a signal $x(t)$, shown in Figure C.1.2a, that is confined such that

$$x(t) = \begin{cases} \text{finite} & |t| \le T/2 \\ 0 & |t| > T/2 \end{cases}. \qquad (C.1.9)$$

Let us also assume that this signal is band-limited to $B/2$, so that

$$X(f) = \begin{cases} \text{finite} & |f| \le B/2 \\ 0 & |f| > B/2 \end{cases}. \qquad (C.1.10)$$

Consider now a simple case where the range of delays is limited to $\pm T_0/2$. If the bandwidth is such that B^{-1} is much larger than T_0 and the duration of the signal T is also much larger than T_0, we have no change in the signal characteristics—no distortion and no dispersion. This is because the temporal changes in the signal are very slow $(B^{-1} > T_0)$ within the range of delay values.

- No distortion, no dispersion: $BT_0 \ll 1$, $T_0/T \ll 1$

If B^{-1} is smaller than T_0, changes in the signal may occur, with some of the multipath components arriving with different phases, and distortion can occur in the received signal from the interference of the multipath components. If the received signal is spread over $T + T_0$, there will be very little or no spreading of the waveform, $y(t)$, as shown in Figure C.1.2b, if $T_0 \ll T$. Thus, we can say that the signal is distorted but not dispersed in time (i.e., no spreading in time).

- Distortion, no dispersion: $BT_0 \gg 1$, $T_0/T \ll 1$

Consider now the case where $B^{-1} \ll T_0$ and T_0 is larger than T. This means that the received waveform, $z(t)$, is now stretched, as shown in Figure C.1.2d, and will occupy a duration of $T + T_0$, resulting in a distorted and dispersed waveform.

- Distortion and dispersion: $BT_0 \gg 1$, $T_0/T \gg 1$

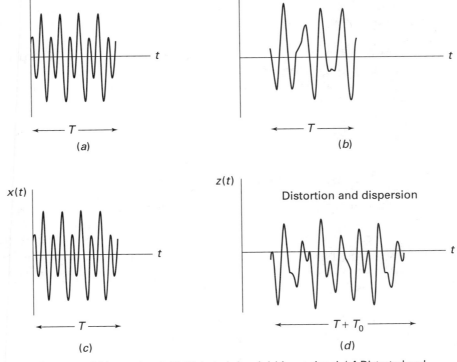

FIGURE C.1.2   (a) Input signal. (b) Distorted signal. (c) Input signal. (d) Distorted and dispersed signal.

Thus, dispersion is always accompanied by distortion.

It is possible to extend this argument to the case of frequency dispersion occurring when the MU is in transit. Let the Doppler spread be f_d (maximum Doppler frequency). If the bandwidth of the signal B is much larger than f_d, nothing happens to the signal; no frequency distortion and no frequency dispersion occur. If the Doppler spread f_d is much larger than the bandwidth of the signal B, the channel spreads the spectrum of the received signal and hence becomes dispersed in frequency. Consider a third case where the Doppler spread is much smaller than the bandwidth of the signal, but f_dT is larger than unity. In other words, there may be distortion of the spectrum, but little or no spreading of the spectrum.

- No frequency distortion and no frequency dispersion: $f_dT \ll 1, f_d/B \ll 1$
- Frequency distortion, no dispersion: $f_dT \gg 1, f_d/B \ll 1$
- Frequency distortion and frequency dispersion: $f_dT \gg 1, f_d/B \gg 1$

C.2 PATH LOSS MODELS FOR MICROCELLS AND PCS SYSTEMS

C.2.1 Path Loss Calculations in PCS systems

The Hata model for path loss predictions discussed in Chapter 2 is applicable to distances greater than 1 km. This lower range limit makes the model unsuited for loss calculations in microcells and picocells, where the distances between the transmitters and receivers vary from a few meters to a kilometer. Capacity problems in the major metropolitan areas are overcome through cell splitting (Chapter 4) and, consequently, the use of microcells and picocells is essential. A number of different models have been proposed to predict path loss over very short distances. We will look at two of these models (Walf 1988, Harl 1989, Ikeg 1991, Har 1999).

One of the models is based on a study conducted in San Francisco and Manhattan. The model proposed by Har, Xia, and Bertoni (Har 1999) can be used to calculate path loss for distances in the range of a few meters to 3 km. The orientation of the streets, heights of the buildings, and the heights of the MU and BS antennae are all taken into account to predict the path loss. Three different types of environments are considered to encompass all the possible areas of coverage: low rise, high rise, and low rise plus high rise. The path loss formulas are also available for the various street orientations: N-LOS, LOS, lateral, and combined staircase and transverse (ST) routes (Figure C.2.1a). The model takes into account the relative heights of the buildings and antennae in terms of Δh_b and Δh_m, given by

$$\Delta h_b = h_b - h_{\text{bdav}}$$

and

$$\Delta h_m = h_{\text{bdr}} - h_{\text{mu}},$$

where h_{bdav} is the average height of the building and h_{bdr} is the height of the last building relative to the MU. Another parameter of the model is d_h, the distance (in meters) between the MU and the last rooftop. The distances and heights are shown in Figure C.2.1b.

The path loss for the low-rise environment for N-LOS routes is given by

$$L_p \text{ (dB)} = \left[139.01 + 42.59 \log_{10}\left(\frac{f}{1000}\right) \right]$$

$$- \left[14.97 + 4.99 \log_{10}\left(\frac{f}{1000}\right) \right] \text{sgn}(\Delta h_b)\log_{10}(1 + |\Delta h_b|)$$

$$+ \left[40.67 - 4.57 \text{sgn}(\Delta h_b)\log_{10}(1 + |\Delta h_b|) \right] \log_{10}(d)$$

$$+ 20 \log_{10}\left(\frac{\Delta h_m}{7.8}\right) + 10 \log_{10}\left(\frac{20}{d_h}\right). \tag{C.2.1}$$

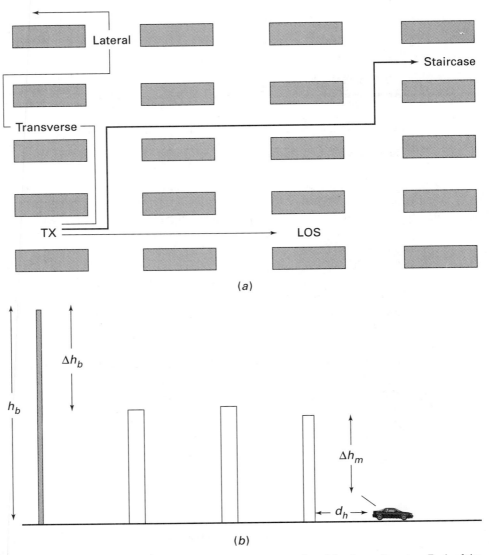

FIGURE C.2.1 (a) The concepts of "lateral," "transverse," and "staircase" routes. Each of the blocks represents a building. (b) The geometry of the various distances used in the model.

In eq. (C.2.1), the distance d is in kilometers and all the other distances/heights and the wavelength (λ) are in meters. The frequency f is in megahertz. The model is valid for distances of $0.05 \leq d \leq 3$ km and frequencies of $900 \leq f \leq 1900$ MHz.

For a high-rise environment ($h_{mu} = 1.6$ m), the path loss for the lateral routes and combined ST are respectively given by

$$L_p(dB)\big|_{lat} = 135.41 + 12.49 \log_{10}\left(\frac{f}{1000}\right) - 4.99 \log_{10}(h_b)$$
$$+ [46.84 + 2.34 \log_{10}(h_b)] \log_{10}(d) \qquad\qquad (C.2.2)$$

and

$$L_p(dB)\big|_{ST} = 143.21 + 29.74 \log_{10}\left(\frac{f}{1000}\right) - 0.99 \log_{10}(h_b)$$
$$+ [47.23 + 3.72 \log_{10}(h_b)] \log_{10}(d). \qquad\qquad (C.2.3)$$

For the case of low rise plus high rise, the path loss for the LOS route is given by

$$L_p(dB) = \begin{cases} 81.14 + 39.40 \log_{10}\left(\frac{f}{1000}\right) - 0.09 \log_{10}(h_b) \\[4pt] \qquad + [15.80 - 5.73 \log_{10}(h_b)] \log_{10}(d) \qquad\qquad d \leq d_k \\[8pt] 48.38 - 32.1 \log_{10}(d_k) + 45.7 \log_{10}\left(\frac{f}{1000}\right) \\[4pt] \qquad + [25.34 - 13.90 \log_{10}(d_k)] \log_{10}(h_b) \\[4pt] \qquad + [32.10 + 13.9 \log_{10}(h_b)] \log_{10}(d) \qquad\qquad d > d_k \end{cases} \qquad (C.2.4)$$

where

$$d_k = \frac{4h_b h_{mu}}{1000\lambda} \text{ (km)}. \qquad\qquad (C.2.5)$$

This model permits the computation of path losses in third-generation (3G) wireless systems, systems such as IMT 2000, which operate in the frequency band around 1900 MHz. At distances for which the Hata model is applicable, the Har, Xia, and Bertoni model gives very good results. Note, however, that the model is applicable even for very short distances.

Walfisch and Bertoni (Walf 1988) have also proposed a model to calculate the path loss in microcells. This model, in combination with the model proposed by Ike-gami (Ikeg 1991), results in an expression for path loss that is the sum of three components: a loss arising from free-space propagation, a loss arising from roof-to-street diffraction and scatter (L_{rts}), and a multiscreen loss (L_{ms}). The loss calculations take into account the average separation of buildings, width of the streets, etc. The expression for path loss can be written as (Walf 1988, Ikeg 1991, Garg 1997, Saun 1999)

$$L_p(dB) = \overbrace{32.4 + 20 \log_{10}(d) + 20 \log_{10}(f)}^{\text{free-space loss}} + L_{rts} + L_{ms}. \qquad (C.2.6)$$

The roof-to-street diffraction and scatter loss L_{rts} has been expressed as

$$L_{rts}(dB) = -16.9 - 10 \log_{10}(W) + 10 \log_{10}(f) + 20 \log_{10}(\Delta h_m) + L(\phi), \quad (C.2.7)$$

where W is the width of the street in meters. The loss term $L(\phi)$ is determined by the street orientation angle ϕ. For most applications, approximate results are obtained by

$$L(\phi) = L(90°) = 4 - 0.114 \times 45 = -1.13. \quad (C.2.8)$$

The multiscreen loss L_{ms} is given by

$$L_{ms} = L_b + k_a + k_d \log_{10}(d) + k_f \log_{10}(f) - 9 \log_{10}(b), \quad (C.2.9)$$

where

$$b = \text{separation of the buildings (m)}$$

$$L_b = \begin{cases} -18 \log_{10}(1 + \Delta h_b) & \Delta h_b > 0 \\ 0 & \Delta h_b \le 0 \end{cases} \quad (C.2.10)$$

$$k_a = \begin{cases} 54 & \Delta h_b \ge 0 \\ 54 - 0.8\Delta h_b & \Delta h_b < 0 \text{ and } d > 0.5 \text{ km} \\ 54 - 0.8d\Delta h_b & \Delta h_b < 0 \text{ and } d \le 0.5 \text{ km} \end{cases} \quad (C.2.11)$$

$$k_d = \begin{cases} 18 & \Delta h_b > 0 \\ 18 - 15\dfrac{\Delta h_b}{h_{bd}} & \Delta h_b \le 0 \end{cases} \quad (C.2.12)$$

$$k_f = -4 + 0.7\left(\frac{f}{925} - 1\right) \quad \text{for medium-sized city and suburban areas with medium tree density}$$

$$k_f = -4 + 1.5\left(\frac{f}{925} - 1\right) \quad \text{for metropolitan areas.}$$

The model is applicable to the frequency range of 800–2000 MHz, MU antenna heights in the range of 1–3 m, BS antenna heights in the range of 4–50 m, and distances in the range of 0.02–5 km. This makes the model useful for path loss calculations in microcells and macrocells, over a broad frequency range of applicability in wireless systems, including PCS and third-generation systems. Note that this model, sometimes referred to as the Walfisch-Ikegami model, requires more extensive calculations than the Har, Xia, and Bertoni model. This model has been adopted by the European Cooperation in the field of Scientific and Technical Research (COST 231) for the 3G systems.

In Figure C.2.2 the path loss predictions from the Hata model are compared with those from the Walfisch-Ikegami model. The parameters used were the following:

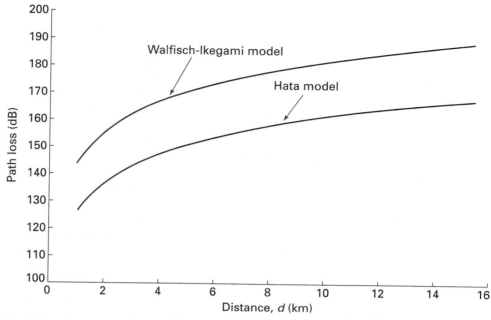

FIGURE C.2.2   Path loss predictions using the Hata model and the Walfisch-Ikegami model.

Carrier frequency, $f = 900$ MHz

Height of MU antenna, $h_{mu} = 1.5$ m

Height of BS antenna, $h_b = 30$ m

Average height of building, $h_{bd} = 30$ m

Width of street, $W = 15$ m

Separation of buildings, $b = 30$ m

Street orientation angle, $\phi = 90°$

The path loss calculations plotted in Figure C.2.2 show that the Hata model predicts lower values of loss. The Walfisch-Ikegami model is a bit more accurate than the Hata model because it includes the heights of the buildings, separation of the buildings, and street orientation angle.

C.3 MACRODIVERSITY SYSTEMS

The concepts of macrodiversity and microdiversity have a number of similarities (Klei 1996, Wang 1999). In either case, the goal is to create multiple independent versions of the signal and combine them using some algorithm. Whereas microdiversity mitigates short-term fading (Rayleigh, Rician, or Nakagami), macrodiversity mitigates the effects of long-term fading or shadowing (lognormal). This essential difference between the two leads to changes in the location of the multiple antennae used to create diverse signals. Some of the procedures commonly employed in cellular systems are examples of macrodiversity. The soft hand-off procedure instituted in IS 95–based systems is a form of macrodiversity (Pape 1995, Hum 2000). In this procedure, the

MU is contact with two base stations as the power and interference levels are being continuously monitored. It has also been shown that the macrodiversity concepts can be employed very easily in traditional TDMA- or FDMA-based systems. One typical geometry is shown in Figure C.3.1, where three base stations (at the corners of the hexagonal cell) are used as the diversity elements (Pape 1995, Haas 1998, Weis 1999, Hum 2000). It is possible to see that in the case of severe shadowing or lognormal fading, the presence of multiple base stations able to service the MU reduces the chances of outage. It is also possible to see that in a scenario where the presence of lognormal fading makes it impossible for the user to get enough signal strength even to get connected to the network, the presence of multiple base stations increases the likelihood of being connected.

The macrodiversity unit (MDU) receives the diverse signals and sends commands to the base stations regarding the decisions to be made. In the uplink (reverse channel), the three base stations collect the signals coming from the MU. The base station transmits these signals to the MDU, which combines them using some algorithm to create the best possible reconstruction of the data from the MU. These algorithms could involve selection combining, maximal ratio combining, or equal gain combining. In the downlink (forward channel), the MDU makes a decision to send information to the MU from a single BS based on the estimated best propagation conditions (form of selection diversity). The MDU may also choose to have all three base stations communicate simultaneously, or simulcast (Witt 1991). In this case, it may be possible to handle these multiple signals using equalizers or a RAKE receiver.

The base stations need not be set up specifically for the purpose of employing macrodiversity. The existing omnidirectional antenna or directional antenna systems could be utilized for macrodiversity. Two typical geometries using omnidirectional antennae are shown in Figure C.3.2. Using the omnidirectional

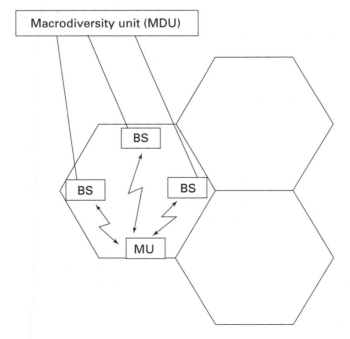

FIGURE C.3.1 Concept of macrodiversity.

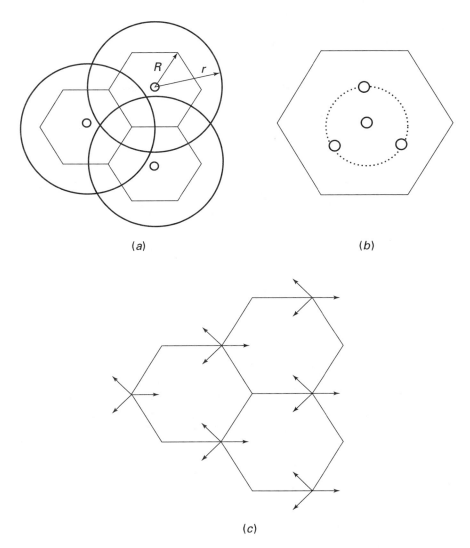

(a)

(b)

(c)

FIGURE C.3.2 (*a*) Overlapping coverage areas shown by circles. The base stations are at the center of the hexagons. (*b*) Four base stations. Three base stations are on a circle at a distance of *R* / *2* from the center base station. (*c*) Macrodiversity using sector antennae.

antennae, macrodiversity can be introduced by having overlapping coverage areas (Jake 1974, Klei 1996, Weis 1999) as shown in Figure C.3.2*a*. The overlap factor is given by r/R, which determines how many antennae can participate in the macrodiversity. The second arrangement uses installations of base stations (Jake 1974, Weis 1999) within the hexagonal cell as shown in Figure C.3.2*b*. Another arrangement, using existing 120° sector antennae (Weis 1999), is shown in Figure C.3.2*c*. This arrangement employs sectorized antennae at every other corner of the hexagonal cell and links them to the MDU.

We will now show the possible improvement in performance with the use of macrodiversity systems. Consider the case of a signal subjected to lognormal fading or shadowing, as discussed in Chapter 2. The probability density function, $f(P_{dB})$, for the received power P_{dB} in dBm, can be expressed as (eq. (2.66))

$$f(P_{dB}) = \frac{1}{\sqrt{2\pi\sigma_{dB}^2}} \exp\left[-\frac{(P_{dB}-P_{av})^2}{2\sigma_{dB}^2}\right], \tag{C.3.1}$$

where σ_{dB} is the standard deviation (dB) of lognormal fading and P_{dB} is the average received power (dBm). If the threshold power for acceptable performance is P_{th}, the outage probability P_{out} can be written as

$$P_{out} = \int_{-\infty}^{P_t} \frac{1}{\sqrt{2\pi\sigma_{dB}^2}} \exp\left[-\frac{(P_{dB}-P_{av})^2}{2\sigma_{dB}^2}\right] dp_{dB} = \frac{1}{2} + \frac{1}{2}\,\text{erf}\left[\frac{P_{th}-P_{av}}{\sqrt{2\sigma_{dB}^2}}\right]. \tag{C.3.2}$$

For the purposes of our calculations, let us assume that the average power remains constant, and we will therefore define the signal-to-noise ratio in lognormal fading, SNR_{lgf}, as

$$\text{SNR}_{lgf} = \frac{P_{av} - P_{th}}{\sigma_{dB}}. \tag{C.3.3}$$

The equation for the outage probability, eq. (C.3.2), can now be rewritten as

$$P_{out} = \frac{1}{2} + \frac{1}{2}\,\text{erf}\left(-\frac{\text{SNR}_{lgf}}{\sqrt{2}}\right). \tag{C.3.4}$$

We can vary the signal-to-noise ratio by varying the standard deviation of fading σ_{dB}, indicating high values of the signal-to-noise ratio for low values of σ_{dB} (nonmetropolitan areas) and low values of signal-to-noise ratio for high values of σ_{dB}

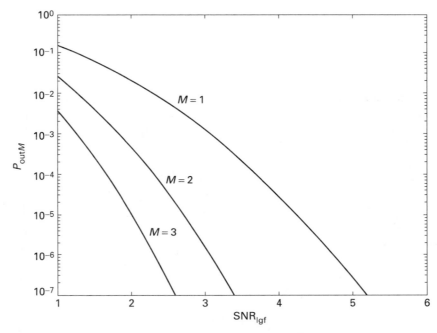

FIGURE C.3.3 Outage probability for different orders of diversity.

(metropolitan areas). We will now see how the performance of the wireless or mobile system is likely to improve with macrodiversity. Let us assume that we have M antennae in the MDU, each antenna independent of the others. We also assume that all of them receive the same average power. The standard deviation of fading σ_{dB} will not change, because we are looking at the same geographical region. The probability $P_{out\,M}$ that all the antennae participating in the MDU will be in outage at the same time is given by the product of the individual antenna outage probabilities. Because all the antennae are identical, the outage probability when M antennae participate in MDU is given by

$$P_{out\,M} = P_{out}P_{out}P_{out}\cdots(M \text{ terms}) = (P_{out})^M = \left[\frac{1}{2} + \frac{1}{2}\text{erf}\left(-\frac{\text{SNR}_{lgf}}{\sqrt{2}}\right)\right]^M. \qquad \text{(C.3.5)}$$

The outage probability is plotted in Figure C.3.3. We see that the performance improvement is dependent on the severity of fading (σ_{dB}) measured in terms of the lower values of the signal-to-noise ratio. The theoretical calculations plotted in Figure C.3.3 also show the improvement that may be obtained using macrodiversity. The actual improvement will depend on the type of modulation scheme employed, the type of receiver, and the correlation (if any) that exists between different diversity branches (Turk 1992, Wang 1999). The performance of cellular systems degraded by co-channel interference is also improved through macrodiversity techniques (Wang 1995, Li 1997).

Systems have also been proposed to implement both microdiversity and macrodiversity simultaneously. The schematic of such a setup is shown in Figure C.3.4. The microdiversity is undertaken at the base station with the multiple receiving antennae acting as the diversity receivers to mitigate Rayleigh fading. Macroscopic diversity is then performed by combining the outputs of these base stations to mitigate the effects of lognormal fading (Bern 1987, Turk 1991a).

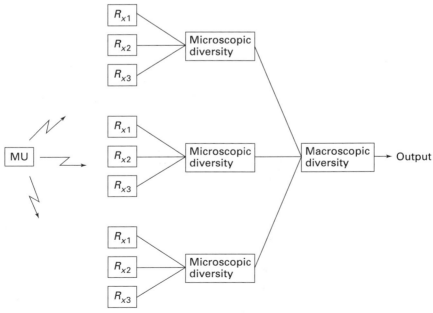

FIGURE C.3.4 Block diagram of the combined microdiversity and macrodiversity system. The R_{xi} are the receivers (antennae and processors) at each of the base stations.

ACRONYMS

A/D	Analog to digital
ACF	Autocorrelation function
ACI	Adjacent channel interference
ADC	American Digital Cellular
AM	Amplitude modulation
AMPS	Advanced Mobile Phone System
ARQ	Automatic repeat request
ASK	Amplitude shift keying
ATM	Asynchronous transfer mode
AWGN	Additive white Gaussian noise
BCH	Bose-Chaudhuri-Hoequenghem
BER	Bit error rate
BFSK	Binary frequency shift keying
B-ISDN	Broadband Integrated Services Digital Network
BPF	Bandpass filter
BPSK	Binary phase shift keying
BS	Base station
BSC	Base station controller
CC	Control channel
CCI	Co-channel interference
CDF	Cumulative distribution function
CDMA	Code division multiple access
CELP	Code excited linear predictor
CIN	Carrier-to-interference ratio
CNR	Carrier-to-noise ratio
CO	Central office
CPFSK	Continuous-phase frequency shift keying
CRC	Cyclic redundancy code
D/A	Digital to analog
dB	Decibel
dBm	Decibel relative to 1 mW
DCA	Dynamic channel allocation
DCS	Digital Communication System
DCS1800	Digital Communication System 1800
DECT	Digital European Cordless Telephone
DFE	Decision feedback equalization
DFM	Digital frequency modulation
DFT	Discrete Fourier transform
DM	Delta modulation
DPCM	Differential pulse code modulation
DPSK	Differential phase shift keying
DQPSK	Differential quadrature phase shift keying

DS	Direct sequence
DS/FHMA	Hybrid direct sequence/frequency-hopped multiple access
DSB	Double sideband
DSB-AM	Double-sideband amplitude modulation
DSB-SC	Double-sideband suppressed carrier
DSP	Digital signal processing
DS-SS	Direct-sequence spread spectrum
DUP	Duplexer
E/O	Electrooptic
EIRP	Effective isotropic radiated power
erf	Error function
erfc	Complementary error function
ETSI	European Telecommunications Standard Institute
FCC	Federal Communications Commission
FCC	Forward control channel
FDD	Frequency division duplex
FDMA	Frequency division multiple access
FEC	Forward error correction
FFH	Fast frequency hopping
FFSK	Fast frequency shift keying
FH	Frequency hopping
FH-SS	Frequency-hopping spread spectrum
FM	Frequency modulation
FOM	Fiberoptic mobile
FSK	Frequency shift keying
GFSK	Gaussian frequency shift keying
GMSK	Gaussian minimum shift keying
GOS	Grade of service
GSM	Global System for Mobile Communication
HPA	High-power amplifier
IDFT	Inverse discrete Fourier transform
IMT 2000	International Mobile Telecommunication 2000
IMTS	Improved Mobile Telephone Service
I-Q	Inphase and quadrature
IR	Infrared
IS 54	Interim standard 54
IS 95	Interim standard 95
IS 136	Interim standard 136
IM	Intermodulation
ISDN	Integrated Services Digital Network
ISI	Intersymbol interference
ISM	Industrial, scientific, and medical
JDC	Japanese Digital Cellular
LCR	Level crossing rate
LMS	Least mean square
LNA	Low-noise amplifier
LOS	Line of sight
LPC	Linear predictive coding
LPF	Low-pass filter

LSB	Lower sideband
LTE	Linear transversal equalizer
MAHO	Mobile-assisted hand-off
MC-CDMA	Multicarrier CDMA
MC-DS CDMA	Multicarrier direct-sequence CDMA
MCM	Multicarrier Modulation
MDU	Macrodiversity unit
MFSK	M-ary frequency shift keying
MLSD	Maximum-likelihood symbol detection
MLSE	Maximum-likelihood sequence estimation
MRC	Maximal ratio combining
MSC	Mobile switching center
MSK	Minimum shift keying
MT-CDMA	Multitone CDMA
MTSO	Mobile telephone switching office
MU	Mobile unit
MUX	Multiplexer
NAMPS	Narrowband Advanced Mobile Phone System
NBFM	Narrowband frequency modulation
N-LOS	Non-line of sight
NRZ	Non-return to zero
O/E	Optoelectronic
OFDM	Orthogonal frequency division multiplexing
OOK	On-off keying
OQPSK	Offset quadrature phase shift keying
P/S	Parallel to serial
PCS	Personal communication systems
PDA	Personal digital assistant
PDC	Pacific Digital Cellular
pdf	Probability density function
PLL	Phase-locked loop
PM	Phase modulation
PN	Pseudonoise
POTS	Plain old telephone system
PSD	Power spectral density
PSK	Phase shift keying
PSTN	Public switched telephone network
PU	Portable unit
QAM	Quadrature amplitude modulation
QPSK	Quadrature or quarternary phase shift keying
RC	Raised cosine
RF	Radio frequency
RS	Reed Solomon
RTT	Radio transmission technology
RZ	Return to zero
S/N	Signal-to-noise ratio
S/P	Serial to parallel
SAT	Supervisory audio tone
SELP	Stochastically excited linear predictor

SFH	Slow frequency hopping
SIR	Signal-to-interference ratio
SNR	Signal-to-noise ratio
SSB	Single sideband
ST	Supervisory tone
TDD	Time division duplex
TDMA	Time division multiple access
TIA	Telecommunications Industry Association
UMTS	Universal Mobile Telecommunication System
USB	Upper sideband
USDC	United States Digital Cellular
VAD	Voice activity detector
VLSI	Very large scale integration
VSB	Vestigial sideband
VSELP	Vector sum excited linear predictor
WBFM	Wideband frequency modulation
W-CDMA	Wideband CDMA

REFERENCES

[Abra 1995] Abrardo, A., Benelli, G., and Cau, G. R. "Multiple symbol differential detection of GMSK for mobile communications," *IEEE Trans. Veh. Technol.*, vol. 44, no. 3, pp. 379–389, Aug. 1995.

[Adac 1988a] Adachi, F., and Ohno, K. "Performance analysis of GMSK frequency detection with decision feedback equalization in digital land mobile radio," *IEE Proc.*, vol. 135, pt. F, no. 3, pp. 199–207, June 1988.

[Adac 1988b] Adachi, F., Feeney, M. T., and Parsons, J. D. "Level crossing rate and average fade duration for time diversity reception in Rayleigh fading conditions," *IEE Proc.*, vol. 135, pt. F, no. 6, pp. 501–506, Dec. 1988.

[Adac 1989] Adachi, F., and Parsons, J. D. "Error rate performance of digital FM mobile radio with post detection diversity," *IEEE Trans. Commun.*, vol. 37, no. 3, pp. 200–210, Mar. 1989.

[Adac 1991a] Adachi, F., and Ohno, K. "BER performance of QDPSK with post detection diversity reception in mobile radio channels," *IEEE Trans. Veh. Technol.*, vol. 40, no. 1, pp. 237–249, Feb. 1991.

[Adac 1991b] Adachi, F., and Ohno, K. "Postdetection MRC diversity for $\pi/4$-shift QDPSK mobile radio," *Electron. Lett.*, vol. 27, no. 18, pp. 1642–1643, Aug. 29, 1991.

[Adac 1993] Adachi, F. "Postdetection optimal diversity combiner for DPSK differential detection," *IEEE Trans. Veh. Technol.*, vol. 42, no. 3, pp. 326–337, Aug. 1993.

[Aghv 1993] Aghvami, A. H. "Digital modulation techniques for mobile and personal communication systems," *Electron. Commun. J.*, pp. 125–132, June 1993.

[Akai 1987] Akaiwa, Y., and Nagata, Y. "Highly efficient digital mobile communications with a linear modulation method," *IEEE J. Sel. Areas Commun.*, vol. SAC-5, no. 5, pp. 890–895, June 1987.

[Akai 1998] Akaiwa, Y. *Introduction to Digital Mobile Communication*. John Wiley & Sons, New York, 1998.

[Akki 1994] Akki, A. S. "Statistical properties of mobile to mobile land communication channels," *IEEE Trans. Veh. Technol.*, vol. 43, no. 4, pp. 826–831, Nov. 1994.

[Alhu 1985] Al-Hussaini, E. K., and Al-Bassiouni, A. M. "Performance of diversity systems for the detection of signals with Nakagami fading," *IEEE Trans. Commun.*, vol. COM-33, no. 12, pp. 1315–1319, Dec. 1985.

[Amor 1976] Amoroso, F. "Pulse and spectrum manipulation in the MSK format," *IEEE Trans. Commun.*, vol. 24, no. 3, pp. 381–384, Mar. 1976.

[Amor 1977] Amoroso, F., and Kivett, J. "Simplified MSK signaling technique," *IEEE Trans. Commun.*, vol. 25, no. 4, pp. 433–441, Apr. 1977.

[Amor 1980] Amoroso, F. "The bandwidth of digital data signals," *IEEE Commun. Mag.*, vol. 18, no. 11, pp. 13–24, Nov. 1980.

[Ande 1965] Anderson, R. R., and Salz, J. "Spectra of digital FM," *Bell Syst. Tech. J*, vol. 44, no. 6, pp. 1165–1189, July–Aug. 1965.

[Ande 1986] Anderson, J. B., Aulin, T., and Sundberg, C-E. *Digital Phase Modulation*. Plenum Press, New York, 1986.

[Ande 1999] Anderson, J. B. *Digital Transmission Engineering*. IEEE Press, New York, 1999.

[Anna 1998] Annamalai, A., Tellambura, C., and Bhargava, V. K. "A simple and accurate analysis of digital communication systems with diversity reception in different fading environments," *Proc. Ninth IEEE Intl. Symp. on Personal, Indoor, and Mobile Radio Communcations*, vol. 3, pp. 1055–1060, 1998.

[Anna 1999] Annamalai, A., and Tellambura, C. "Analysis of maximal ratio and equal gain diversity systems for *M*-ary QAM on generalized fading channels," *Proc. 1999 IEEE Intl. Conf. on Communications: ICC'99*, vol. 2, pp. 848–852, 1999.

[Ariy 1989] Ariyavisitakul, S., and Liu, T.-P. "Characterizing the effects of nonlinear amplifiers on linear modulation for digital portable radio communications," *Proc. IEEE GLOBECOM '89*, Nov. 27–30, 1989, Dallas, pp. 448–453.

[Arre 1973] Arredondo, G. A., Chriss, W. H., and Walker, E. H. "A multipath fading simulator for mobile radio," *IEEE Trans. Commun.*, vol. 21, no. 11, pp. 1325–1328, Nov. 1973.

[Atal 1967] Atal, B. S., and Schroeder, M. R. "Predictive coding of speech signals," *Proc. 1967 IEEE Conference on Communications and Signal Processing*, pp. 360–361, 1967.

[Atal 1970] Atal, B. S., and Schroeder, M. R. "Adaptive predictive coding for speech signals," *Bell Syst. Tech. J.*, vol. 49, pp. 1973–1986, 1970.

[Atan 2000] Atanassov, K., Narayanan, R. M., Stoiljkovic, V., and Kadambi, G. R. "Mobile station polarization diversity reception in handheld devices at 1.8 GHz," *Proc. 2000 IEEE APS Intl. Symposium*, vol. 1, pp. 290–293, 2000.

[Auli 1979] Aulin, T. "A modified model for the fading signal at the mobile radio channel," *IEEE Trans. Veh. Technol.*, vol. 28, no. 3, pp. 182–203, Aug. 1979.

[Auli 1982a] Aulin, T., and Sundberg, C.-E. "Calculating digital FM spectra by means of autocorrelation," *IEEE Trans. Commun.*, vol. COM-30, no. 5, pp. 1199–1208, May 1982.

[Auli 1982b] Aulin, T., and Sundberg, C.-E. "Minimum euclidian distance and power spectrum for a class of smoothed phase modulation codes with constant envelope," *IEEE Trans. Commun.*, vol. COM-30, no. 7, pp. 1721–1729, July 1982.

[Auli 1982c] Aulin, T., and Sundberg, C.-E. "Exact asymptotic behavior of digital FM spectra," *IEEE Trans. Commun.*, vol. COM–30, no. 11, pp. 2438–2449, Nov. 1982.

[Bake 1962] Baker, P. A. "Phase-modulation data sets for serial transmission at 2000, 2400 bits/second," *AIEE Trans. (Commun. Electron.)*, pp. 166–171, July 1962.

[Ball 1982] Ball, J. R. "A real-time fading simulator for mobile radio," *Radio Electron. Eng.*, vol. 52, no. 10, pp. 475–478, Oct. 1982.

[Beau 1991] Beaulieu, N. C., and Abu-Dayya, A. A. "Analysis of equal gain diversity on Nakagami fading channels," *IEEE Trans. Commun.*, vol. 39, no. 2, pp. 225–234, Feb. 1991.

[Beck 1962] Beckmann, P., "Statistical distribution of the amplitude and phase of a multiply scattered field," *J. Res. Natl. Bur. Stand.*, vol. 66D, no. 3, pp. 231–240, May–June 1962.

[Bell 1963a] Bello, P. A. "Characterization of randomly time variant linear channels," *IEEE Trans. Commun. Syst.*, vol. CS-11 (COM-11), pp. 360–393, Dec. 1963.

[Bell 1963b] Bello, P. A. and Nelin, B. D. "The effect of frequency selective fading on the binary error probabilities of incoherent and differentially coherent matched filter receivers," *IEEE Trans. Commun. Syst.*, vol. CS-11 (COM-11), pp. 170–186, June 1963.

[Bene 1999] Benedetto, S., and Biglieri, E. *Principles of Digital Transmission with Wireless Applications.* Kluwer, Norwell, MA, 1999.

[Berl 1987] Berlekamp, E. R., Peile, R. E., and Pope, S. P. "The application of error control to communications," *IEEE Commun. Mag.*, vol. 25, no. 4, pp. 44–57, Apr. 1987.

[Bern 1987] Bernhardt, R. C. "Macroscopic diversity in frequency reuse radio systems," *IEEE J. Sel. Areas Commun.*, vol. SAC-5, no. 5, pp. 862–870, June 1987.

[Berr 1977] Berry, L. A. "Spectrum metrics and spectrum efficiency: Proposed definitions," *IEEE Trans. Electromagn. Compat.*, vol. EMC-29, no. 3, pp. 254–264, Aug. 1967.

[Bigl 1998] Biglieri, E., Caire, G., Tarrico, G., and Ventura-Traveset, J. "Computing error probabilities over fading channels: A unified approach," *Eur. Trans. Telecommun.*, vol. 9, no. 1, pp. 15–25, Jan.–Feb. 1998.

[Bing 1990] Bingham, J. A. C. "Multicarrier modulation for data transmission: An idea whose time has come," *IEEE Commun. Mag.*, vol. 28, no. 5, pp. 5–14, May 1990.

[Biya 1995] Biyari, K. H., and Lindsey, W. C. "Error performance of DPSK mobile communication systems over non-Rayleigh fading channels," *IEEE Trans. Veh. Technol.*, vol. 44, no. 2, pp. 211–219, May 1995.

[Bodt 1982] Bodtmann, W. F., and Arnold, H. W. "Fade duration statistics of a Rayleigh distributed wave," *IEEE Trans. Commun.*, vol. COM-30, no. 3, pp. 549–553, Mar. 1982.

[Bour 1993] Bouras, D. P., Mathiopoulos, P. T., and Makrakis, D. "Optimal detection of coded differentially encoded QAM and PSK signals with diversity reception in correlated fast Rician fading channels," *IEEE Trans. Veh. Technol.*, vol. 42, no. 3, pp. 245–258, Aug. 1993.

[Brau 1991] Braun, W. R., and Dersch, U. "A physical mobile radio channel model," *IEEE Trans. Veh. Technol.*, vol. 40, no. 2, pp. 472–482, May 1991.

[Bren 1959] Brennan, D. G. "Linear diversity combining techniques," *Proc. IRE*, vol. 47, pp. 1075–1102, June 1959.

[Bull 1977] Bullington, K. "Radio propagation for vehicular communications," *IEEE Trans. Veh. Technol.*, vol. 26, no. 4, pp. 295–308, Nov. 1977.

[Bult 1983] Bultitude, R. J. C. "A study of coherence bandwidth measurements for frequency selective radio channels," *Proc. 33rd IEEE Vehicular Technology Conf.*, 25–27 May 1983, Toronto, Canada, pp. 269–278.

[Bult 1987] Bultitude, R. J. C. "Measurement, characterization and modeling of indoor 800/900 MHz radio channels for digital communications," *IEEE Commun. Mag.*, vol. 25, no. 6, pp. 4–12, June 1987.

[Bult 1989] Bultitude, R. J. C., Mahmoud, S. A., and Sullivan, W. A. "A comparison of indoor radio propagation characteristics at 910 MHz and 1.75 GHz," *IEEE Trans. Sel. Areas Commun.*, vol. 7, no. 1, pp. 20–30, Jan. 1989.

[Casa 1990] Casas, E., and Leung, C. "A simple digital fading simulator for mobile radio," *IEEE Trans. Veh. Technol.*, vol. 39, no. 3, pp. 205–212, Aug. 1990.

[Chan 1992] Chan, G. K. "Effects of sectorization on the spectrum efficiency of cellular radio systems," *IEEE Trans. Veh. Technol.*, vol. 41, no. 3, pp. 217–225, Aug. 1992.

[Chan 1994a] Chan, N. L. B. "Multipath propagation effects on a CDMA cellular system," *IEEE Trans. Veh. Technol.*, vol. 43, no. 4, pp. 848–855, Nov. 1994.

[Chan 1994b] Chang, C. S., and McLane, P. J. "Bit-Error-Probability for non-coherent orthogonal signals in fading with optimum combining for correlated branch diversity," *Proc. IEEE GLOBECOM '94*, pp. 1–7, 1994.

[Chen 1991] Chennakeshu, S., and Saulnier, G. J. "Differential detection of $\pi/4$-shifted DQPSK for digital cellular radio," *Proc. 41st IEEE Conf. on Vehicular Technology*, May 19–22, 1991, St. Louis, MO, pp. 186–191.

[Chen 1995] Chennakeshu, S., and Anderson, J. B. "Error rates for Rayleigh fading multichannel reception of MPSK signals," *IEEE Trans. Commun.*, vol. 43, no. 2/3/4, pp. 338–346, Feb./Mar./Apr. 1995.

[Chen 1999] Chen, K.-C., and Wu, S.-T. "A programmable architecture for OFDM-CDMA," *IEEE Commun. Mag.*, vol. 39, no. 11, pp. 76–82, Nov. 1999.

[Cheu 1998] Cheung, K., Sau, J. H. M., and Murch, R. D. "A new empirical model for indoor propagation prediction," *IEEE Trans. Veh. Technol.* vol. 47, no. 3, pp. 996–1001, Aug. 1998.

[Chou 1993] Chouly, A., Brajal, A., and Jourdan, S. "Orthogonal multicarrier techniques applied to direct sequence spread spectrum CDMA systems," *Proc. IEEE GLOBECOM '93*, Houston, Nov. 1993, pp. 1723–1733.

[Chu 1991] Chu, T. S., and Gans, M. J. "Fiber optic microcellular radio," *IEEE Trans. Veh. Technol.*, vol. 40, no. 3, pp. 599–606, Aug. 1991.

[Chun 1984] Chung, K. S. "Generalized tamed frequency modulation and its application for mobile radio communications," *IEEE J. Sel. Areas Commun.*, vol. SAC-2, no. 4, pp. 487–497, July 1984.

[Cimi 1985] Cimini, L. J. "Analysis and simulation of digital mobile channel using orthogonal frequency division multiplexing," *IEEE Trans. Commun.*, vol. 33, no. 7, pp. 665–675, July 1985.

[Clar 1968] Clarke, R. H. "A statistical description of mobile radio reception," *Bell Syst. Tech. J.*, vol. 47, pp. 957–1000, July–Aug. 1968.

[Clar 1985] Clark, A. P. "Digital modems for land mobile radio," *IEE Proc.*, vol. 132, pt. F., no. 5, pp. 348–362, Aug. 1985.

[Coop 1979] Cooper, G. R., Nettleton, R. W., and Grybos, D. P. "Cellular land mobile radio: Why spread spectrum?" *IEEE Commun. Mag.*, vol. 17, no. 3, pp. 17–23, Mar. 1979.

[Coop 1986] Cooper, G. R., and McGillem, C. D. *Modern Communications and Spread Spectrum*. McGraw-Hill, New York, 1986.

[Couc 1997] Couch, L. W., II. *Digital and Analog Communication Systems*. Prentice-Hall, Englewood Cliffs, NJ, 1997.

[Coul 1998] Coulson, A. J., Williamson, A. G., and Vaughan, R. G. "A statistical basis for lognormal shadowing effects in multipath fading channels," *IEEE Trans. Commun.*, vol. 46, no. 4, pp. 494–502, Apr. 1998.

[Cox 1972] Cox, D. C. "Time and frequency domain characterizations of multipath propagation at 910 MHz in a suburban mobile environment," *Radio Sci.*, vol. 7, no. 12, pp. 1069–1077, Dec. 1972.

[Cox 1975] Cox, D. C., and Leck, R. P. "Correlation bandwidth and delay spread multipath propagation statistics for 910-MHz urban mobile radio channels," *IEEE Trans. Commun.*, vol. 23, no. 11, pp. 1271–1280, Nov. 1975.

[Cox 1982] Cox, D. C. "Cochannel interference considerations in frequency reuse small coverage area radio systems," *IEEE Trans. Commun.*, vol. COM-30, no. 1, pp. 135–141, Jan. 1982.

[Cox 1995] Cox, D. C. "Wireless personal communications: What is it?" *IEEE Pers. Commun.*, pp. 20–35, Apr. 1995.

[Dava 1989] Davarian, F., and Sumida, J. T. "Multipurpose digital modulator," *IEEE Commun. Mag.*, vol. 27, no. 2, pp. 35–45, June 1989.

[Dave 1958] Davenport, W. B., Jr., and Root, W. L. *An Introduction to the Theory of Random Signals and Noise*. McGraw-Hill, New York, 1958.

[Davi 1971] Davis, B. R. "FM noise with fading channels and diversity," *IEEE Trans. Commun. Technol.*, vol. COM-19, no. 6, pp. 1189–1200, Dec. 1971.

[deJa 1978] de Jager, F., and Dekker, C. B. "Tamed frequency modulation, a novel method to achieve economy in digital transmission," *IEEE Trans. Commun*, vol. COM-26, no. 5, pp. 534–542, May 1978.

[Deli 1985] Delisle, G. Y., Lefevre, J., Lecours, M., and Chouinard, J. "Propagation loss prediction:

A comparative study with application to mobile radio channels," *IEEE Trans. Veh. Technol.*, vol. 34, no. 2, pp. 86–94, May 1985.

[Ders 1994] Dersch, U., and Zollinger, E. "Propagation mechanisms in microcell and indoor environments," *IEEE Trans. Veh. Technol.*, vol. 43, no. 4, pp. 1058–1066, Nov. 1994.

[Divs 1982] Divsalar, D., and Simon, M. "The power spectral density of digital modulators transmitted over nonlinear channels," *IEEE Trans. Commun.*, vol. COM-30, no. 1, pp. 142–151, Jan. 1982.

[Dixo 1994] Dixon, R. C. *Spread Spectrum Systems with Commercial Applications*. John Wiley & Sons, New York, 1994.

[Elno 1986a] Elnoubi, S. M. "Analysis of GMSK with discriminator detection in mobile radio channels," *IEEE Trans. Veh. Technol.*, vol. 35, no. 2, pp. 71–76, May 1986.

[Elno 1986b] Elnoubi, S. M. "Analysis of GMSK with differential detection in land mobile radio channels," *IEEE Trans. Veh. Technol.*, vol. 35, no. 4, pp. 162–167, Nov. 1986.

[El-Sa 1996] El-Saigh, A. I., and Macario, R. C. V. "Co- and adjacent channel interference performance of GSM in Rayleigh fading environment," *Proc. IEE Colloquium on Radio Communication at Microwave and Millimeter Wave Frequencies*, pp. 9/1–9/6, 1996.

[El-Ta 1998] El-Tanany, M., Wu, Y., and Hazy, L. "Impact of adjacent channel interference on the performance of OFDM systems over frequency selective channels," *IEEE VTC '98*, pp. 2241–2245, 1998.

[Eng 1996] Eng, T., Kong, N., and Milstein, L. B. "Comparison of diversity combining techniques for Rayleigh-fading channels," *IEEE Trans. Commun.*, vol. 44, no. 9, pp. 1117–1129, Sept. 1996.

[Evan 2000] Evans, M., Hastings, N., and Peacock, B. *Statistical Distributions*, 3rd ed. John Wiley & Sons, New York, 2000.

[Fech 1993] Fechtel, S. A. "A novel approach to modeling and efficient simulation of frequency selective fading radio channels," *IEEE J. Sel. Areas Commun.*, vol. SAC-11, pp. 422–431, Apr. 1993.

[Fehe 1995] Feher, K. *Wireless digital communications: Modulation and Spread Spectrum*

Applications. Prentice-Hall, Englewood Cliffs, NJ, 1995.

[Fleu 1996] Fleury, B. H., and Leuthold, P. E. "Radio wave propagation in mobile communications: An overview of European research," *IEEE Commun. Mag.*, vol. 34, no. 2, pp. 1–13, Feb. 1996.

[Forn 1973] Forney, D. G. "The Viterbi algorithm," *Proc. IEEE*, vol. 61, no. 3, pp. 268–278, Mar. 1973.

[Fren 1979] French, R. C. "The effect of fading and shadowing on channel reuse in mobile radio," *IEEE Trans. Veh. Technol.*, vol. 28, no. 3, pp. 171–180, Aug. 1979.

[Fuka 1996] Fukasawa, A., Sato, T., Tazikawa, Y., Kato, T., Kawabe, M., and Fisher, R. E. "Wideband CDMA system for personal radio communications," *IEEE Commun. Mag.*, pp. 116–123, Oct. 1996.

[Fung 1986] Fung, V., Rappaport, T. S., and Thoma, B. "Bit error simulation for $\pi/4$ DQPSK mobile radio communications using two-ray and measurement-based impulse response models," *IEEE J. Sel. Areas Commun.*, vol. 11, no. 3, Apr. 1993.

[Gagl 1988] Gagliardi, R. M. *Introduction to Communications Engineering*. John Wiley & Sons, New York, 1988.

[Gane 1989a] Ganesh, R., and Pahlavan, K. "On arrival of paths in fading multipath indoor radio channels," *Electron. Lett.*, vol. 25, no. 12, pp. 763–765, June 8, 1989.

[Gane 1989b] Ganesh, R., and Pahlavan, K. "On the modeling of fading multipath indoor radio channels," *Proc. IEEE GLOBECOM '89*, Nov. 27–30, 1989, Dallas, pp. 1346–1350.

[Gane 1991] Ganesh, R., and Pahlavan, K. "Statistical modeling and computer simulation of indoor radio channel," *IEE Proc.*, vol. 138, pt. I, no. 3, pp. 153–161, June 1991.

[Gans 1972] Gans, M. J. "A power spectral theory of propagation in the mobile radio environment," *IEEE Trans. Veh. Technol.*, vol. VT-21, no. 1, pp. 27–38, Feb. 1972.

[Garg 1996] Garg, V. K., and Wilkes, J. E. *Wireless and Personal Communications Systems*. Prentice-Hall, Englewood Cliffs, NJ, 1996.

[Garg 1997] Garg, V. K., Smolik, K., and Wilkes, J. E. *Applications of CDMA in Wireless/Personal Communications*. Prentice-Hall, Englewood Cliffs, NJ, 1997.

[Garg 1999] Garg, V. K., Halpern, S., and Smolik, K. F. "Third generation (3G) mobile communications systems," *Proc. 1999 IEEE Intl. Conf. on Personal Wireless Communications*, Jaipur, India, pp. 39–43, Feb. 1999.

[Garg 2000] Garg, V. K. *IS-95 CDMA and cdma2000 Cellular/PCS System Implementation*. Prentice-Hall, Englewood Cliffs, NJ, 2000.

[Gibs 1996] Gibson, J. D. (ed.). *The Mobile Communications Handbook*. IEEE Press/CRC Press, Boca Raton, FL, 1996.

[Gibs 1997] Gibson, J. D. (ed.). *The Telecommunications Handbook*. IEEE Press/CRC Press, Boca Raton, FL, 1997.

[Gilb 1965] Gilbert, E. N. "Energy reception for mobile radio," *Bell Syst. Tech. J.*, vol. 44, pp. 1779–1803, Oct. 1965.

[Gilb 1969] Gilbert, E. N. "Mobile radio diversity reception," *Bell Syst. Tech. J.*, vol. 48, pp. 2473–2492, Sept. 1969.

[Gilb 2000] Gilb, J. P. K. "Bluetooth radio architectures," *Proc. 2000 IEEE Radio Frequency and Integrated Circuits Symposium*, pp. 3–6, 2000.

[Gilh 1991] Gilhousen, K. S., Jacobs, I. M., Padovani, R., Viterbi, A., Weaver, L. A., Jr., and Wheatley, C. E., III. "On the capacity of a cellular CDMA system," *IEEE Trans. Veh. Technol.*, vol. 40, no. 2, pp. 303–312, May 1991.

[Glov 1998] Glover, I. A., and Grant, P. M. *Digital Communications*. Prentice-Hall, Englewood Cliffs, NJ, 1998.

[Gole 1994] Golestaneh, S., Hafez, H. M., and Mahmoud, S. A., "The effect of adjacent channel interference on the capacity of FDMA cellular systems," *IEEE Trans. Veh. Technol.*, vol. 43, no. 4, pp. 946–954, Nov. 1994.

[Good 1990] Goode, S. H., Kazecki, H. L., and Dennis, D. W. "A comparison of limiter-discriminator, delay and coherent detection for $\pi/4$ QPSK," *Proc. 40th IEEE Conf. on Vehicular Technology*, May 6–9, 1990, Orlando, FL, pp. 687–694.

[Good 1991] Goodman, D. J. "Second generation wireless information networks," *IEEE Trans. Veh. Technol.*, vol. 40, no. 2, pp. 366–374, May 1991.

[Good 1997] Goodman, D. J. *Wireless Personal Communications Systems*. Addison-Wesley, Reading, MA, 1997.

[Grad 1979] Gradshteyn, I. S., and Ryzhik, I. M. *Table of Integrals, Series, and Products*. Academic Press, Orlando, FL, 1979.

[Gron 1976] Gronemeyer, S. A., and McBride, A. L. "MSK and Offset QPSK modulation," *IEEE Trans. Commun.*, vol. COM-24, no. 8, pp. 809–820, Aug. 1976.

[Gunt 1996] Gunther, C. G. "Comment on estimate of channel capacity in Rayleigh fading environment," *IEEE Trans. Veh. Technol.*, vol. 45, no. 2, pp. 401–403, May 1996.

[Guo 1990] Guo, Y., and Feher, K. "Performance evaluation of differential π/4-QPSK systems in a Rayleigh fading/delay spread/CCI/AWGN environment," *Proc. 40th IEEE Vehicular Technology Conf.*, pp. 420–24.

[Gupt 1985] Gupta, S. C., Viswanathan, R., and Muammar, R. "Land mobile systems: A tutorial exposition," *IEEE Commun. Mag.*, vol. 23, no. 6, pp. 34–45, June 1985.

[Haar 2000] Haartsen, J. C., and Mattisson, S. "Bluetooth: A new low power radio interface providing short range connectivity," *Proc. IEEE*, vol. 88, no. 10, pp. 1651–1661, Oct. 2000.

[Haas 1998] Haas, Z., and Li, C.-P. " The multiply detected macrodiversity scheme for wireless systems," *IEEE Trans. Veh. Technol.*, vol. 47, no. 2, pp. 506–530, May 1998.

[Hahn 1994] Hahn, G. J., and Shapiro, S. S. *Statistical Models in Engineering*. John Wiley & Sons, New York, 1994.

[Hamm 1998] Hammuda, H. *Cellular Mobile Radio Systems*. John Wiley & Sons, Chichester, UK, 1998.

[Hans 1977] Hansen, F., and Meno, F. I. "Mobile fading: Rayleigh and lognormal superimposed," *IEEE Trans. Veh. Technol.*, vol. 26, no. 4, pp. 332–335, Nov. 1977.

[Hanz 1994] Hanzo, L., and Steele, R. "The Pan European mobile radio system, Part 1 and Part 2," *Eur. Trans. Telecommun.*, vol. 5, no. 2, pp. 245–276, Mar.–Apr. 1994.

[Har 1999] Har, D., Xia, H. H., and Bertoni, H. "Path–loss prediction model for microcells," *IEEE Trans. Veh. Technol.*, vol. 48, no. 5, pp. 1453–1461, Sept. 1999.

[Hara 1993] Hara, S., Okada, M., and Morinaga, N. "Multicarrier modulation technique for wireless local area networks," *Proc. Fourth European Conference on Radio Relay Systems*, pp. 33–38, Oct. 1993.

[Hara 1997] Hara, S., and Prasad, R. "Overview of multicarrier CDMA," *IEEE Commun. Mag.*, pp. 126–133, Dec. 1997.

[Harl 1989] Harley, P. "Short distance attenuation measurements at 900 MHz and 1.8 GHz using low antenna heights for microcells," *IEEE J. Sel. Areas Commun.*, vol. 7, no. 1, pp. 5–11, Jan. 1989.

[Hash 1979] Hashemi, H. "Simulation of urban radio propagation channel," *IEEE Trans. Veh. Technol.*, vol. 28, no. 3, pp. 213–225, Aug. 1979.

[Hash 1993] Hashemi, H. "The indoor radio propagation channel," *Proc. IEEE*, vol. 81, no. 7, pp. 943–968, July 1993.

[Hata 1980] Hata, M. "Empirical formulae for propagation loss in land mobile radio services," *IEEE Trans. Veh. Technol.*, vol. 29, no. 3, pp. 317–325, Aug. 1980.

[Hata 1985] Hata, M., Kinoshita, K., and Hirade, K. "Radio link design of cellular land mobile communication systems," *IEEE Trans. Veh. Technol.*, vol. 31, no. 1, pp. 25–31, Feb. 1985.

[Hatf 1977] Hatfield, D. N. "Measures of spectral efficiency in land mobile radio," *IEEE Trans. Electromagn. Compat.*, vol. EMC-19, no. 3, pp. 266–268, Aug. 1977.

[Haug 1994] Haug, T. "Overview of GSM: Philosophy and results," *Int. J. Wireless Inf. Networks*, vol. 1, no. 1, pp. 7–16, 1994.

[Hayk 2001] Haykin, S. *Communication Systems*, 4th ed. John Wiley & Sons, New York, 2001.

[Hira 1979a] Hirade, K., and Murota, K. "A study of modulation for digital mobile telephony," *Proc. 29th IEEE Conf. on Vehicular Technology*, Mar. 27–30, 1979, Arlington Heights, IL, pp. 13–19.

[Hira 1979b] Hirade, K., Ishizuka, M., Adachi, F., and Ohtani, K. "Error rate performance of digital FM with differential detection in land mobile radio channels," *IEEE Trans. Veh. Technol.*, vol. 28, no. 3, pp. 204–212, Aug. 1979.

[Hiro 1981] Hirosaki, B. "Orthogonally multiplexed QAM system using the discrete Fourier transform," *IEEE Trans. Commun.*, vol. COM-29, no. 7, pp. 982–989, July 1981.

[Hiro 1984] Hirono, M., Miki, T., and Murota, K. "Multilevel decision method for bandlimited digital FM with limiter-discriminator detection," *IEEE J. Sel. Areas Commun.*, vol. SAC-2, no. 4, pp. 498–506, July 1984.

[Ho 1999] Ho, C.-L., and Su, W.-P. "Performance calculation of cellular radio systems over fading channel," *IEE Proc. Commun.*, vol. 146, no. 3, pp. 201–207, June 1999.

[Hodg 1990a] Hodges, M. R. L. "The GSM radio interface," *Br. Telecommun. Res. Lab. J.*, vol. 8, no. 1, pp. 31–43, Jan. 1990.

[Hodg 1990b] Hodges, M. R. L., and Jensen, S. A. "Laboratory testing of digital cellular systems," *Br. Telecommun. Res. Lab. J.*, vol. 8, no. 1, pp. 57–66, Jan. 1990.

[Hoff 1960] Hoffman, W. C. (ed.). *Statistical Methods on Radio Wave Propagation.* Pergamon Press, New York, 1960 (pp. 3–36, "The M-distribution: A general formula of intensity distribution in rapid fading" by Nakagami, M.).

[Hong 1986] Hong, D., and Rappaport, S. S. "Traffic model and performance analysis for cellular mobile radio telephone systems with prioritized and non-prioritized handoff procedures," *IEEE Trans. Veh. Technol.*, vol. 35, no. 3, pp. 77–92, Aug. 1986.

[Hori 1990] Horikoshi, J., and Shimura, S. "Multipath distortion of differentially encoded GMSK with 2-bit differential detection in bandlimited frequency selective mobile radio channel," *IEEE Trans. Veh. Technol.*, vol. 39, no. 4, pp. 308–314, Nov. 1990.

[Huan 1992] Huang, C.-C., and Khayata, R. "Delay spreads and channel dynamics measurements at ISM bands," *1992 Intl. Commun. Conf.*, Chicago, 1992.

[Hugh 1985] Hughes, C. J., and Appleby, M. S. "Definition of a cellular mobile radio system," *IEE Proc.*, vol. 132, pt. F, no. 5, pp. 416–424, Aug. 1985.

[Hum 2000] Hum, V. S., Host-Madsen, A., and Okoniewski, M. "Improving reverse link capacity of CDMA systems using macrodiversity," *Proc. 2000 Antennas and Propagation Soc. Symp.*, pp. 936–939, 2000.

[IEEE 1988] IEEE Vehicular Technology Society Committee on Radio Propagation. "Coverage prediction for mobile radio systems operating in the 800/900 MHz frequency range," *IEEE Trans. Veh. Technol.*, vol. 37, no. 1, pp. 3–44, Feb. 1988.

[Ikeg 1980] Ikegami, F., and Yoshida, S. "Analysis of multipath propagation structure in urban mobile radio environments," *IEEE Trans. Antennas Propag.*, vol. AP–28, no. 4, pp. 531–537, July 1980.

[Ikeg 1991] Ikegami, F., Tekeuchi, T., and Yoshida, S. "Theoretical prediction of mean field strength for urban mobile radio," *IEEE Trans. Antennas Propag.*, vol. 39, no. 3, pp. 299–302, Mar. 1991.

[Jake 1971] Jakes, W. C. "A comparison of specific space diversity techniques for the reduction of the fast fading in UHF mobile radio systems," *IEEE Trans. Veh. Technol.*, vol. 20, no. 4, pp. 81–92, Nov. 1971.

[Jake 1974] Jakes, W. C. (ed.). *Microwave Mobile Communications.* IEEE Press, Los Alamitos, CA, 1974.

[Jenk 1972] Jenks, F. G., Morgan, P. D., and Warren, C. S. "Use of four level phase modulation for digital mobile radio," *IEEE Trans. Electromagn. Compat.*, vol. EMC-14, no. 4, pp. 113–128, Nov. 1972.

[Jung 1993] Jung, P., Baier, P. W., and Steil, A. "Advantages of CDMA and spread spectrum techniques over FDMA and TDMA in cellular mobile radio applications," *IEEE Trans. Veh. Technol.*, vol. 42, no. 3, pp. 357–364, Aug. 1993.

[Kaas 1998] Kaasila, V. P., and Mämmelä, A. "Bit error probability for an adaptive diversity receiver in a Rayleigh-fading channel," *IEEE Trans. Commun.*, vol. 46, no. 9, pp. 1106–1108, Sept. 1998.

[Kaji 1996] Kajiya, S., Tsukamoto, K., and Komaki, S. "Proposal of fiber-optic radio highway networks using CDMA method," *IEICE Trans. Electron.*, vol. E79–C, no. 1, pp. 111–117, Jan. 1996.

[Kcha 1993] Kchao, C., and Stüber, G. L. "Analysis of a direct-sequence spread-spectrum cellular radio system," *IEEE Trans. Commun.*, vol. 41, no. 10, pp. 1507–1516, Oct. 1993.

[Kenn 1969] Kennedy, R. S. *Fading Dispersive Communication Channels.* John Wiley & Sons, New York, 1969.

[Kerr 1993] Kerr, R. "CDMA digital cellular," *Appl. Microwave Wireless,* pp. 30–41, Fall 1993.

[Kim 1993] Kim, K. I. "CDMA cellular engineering issues," *IEEE Trans. Veh. Technol.,* vol. 42, no. 3, pp. 345–350, Aug. 1993.

[Klei 1996] Klein, A., Steiner, B., and Steil, A. "Known and novel diversity approaches as a powerful means to enhance the performance of cellular mobile radio systems," *IEEE J. Sel. Areas Commun.,* vol. 14, no. 9, pp. 1784–1795, Dec. 1996.

[Knis 1998a] Knisely, D. N., Kumar, S., Laha, S., and Nanda, S. "Evolution of wireless data services: IS-95 to cdma2000," *IEEE Commun. Mag.,* pp. 140–149, Oct. 1998.

[Knis 1998b] Knisely, D. N., Li, Q., and Ramesh, N. S. "cdma 2000: A third generation radio transmission technology," *Bell Labs Tech. J.,* pp. 63–78, July/Sept. 1998.

[Kohn 1995] Kohno, R., Meidan, R., and Milstein, L. B. "Spread spectrum access methods for wireless communications," *IEEE Commun. Mag.,* vol. 35, no. 1, pp. 58–67, Jan. 1995.

[Korn 1991] Korn, I. "Error floors in the satellite and land mobile channels," *IEEE Trans. Commun.,* vol. 39, no. 6, pp. 833–837, June 1991.

[Kosh 1997] Koshy, B. J., and Shankar, P. M. "Efficient modeling and evaluation of fiber-fed microcellular networks in a land mobile channel using a GMSK modem scheme," *IEEE J. Sel. Areas Commun.,* vol. 15, no. 4, pp. 694–706, May 1997.

[Kosh 1999] Koshy, B. J., and Shankar, P. M. "Spread-spectrum techniques for fiber-fed microcellular networks," *IEEE Trans. Veh. Technol.,* vol. 48, no. 3, pp. 847–857, May 1999.

[Kuch 1999] Kuchar, A., Taferner, M., Tangemann, M., and Hoek, C. "Field trial with GSM/DCS 1800 smart antenna base station," *Proc. 50th IEEE Vehicular Technology Conf., VTC '99,* pp. 42-46, 1999.

[Le Fl 1989] Le Floch, B., Halbert-Lassalle, R., and Castelain, D. "Digital sound broadcasting to mobile receivers," *IEEE Trans. Consum. Electron.,* vol. 35, pp. 493–503, Aug. 1989.

[Lee 1967] Lee, W. C. Y. "Statistical analysis of the level crossings and duration of fades of the signal from an energy density mobile radio antenna," *Bell Syst. Tech J.,* vol. 47, pp. 417–448, Feb. 1967.

[Lee 1971a] Lee, W. C. Y., "Antenna spacing requirements for a mobile radio base–station diversity," *Bell Syst. Tech J.,* vol. 50, no. 6, pp. 1859–1877, July–Aug. 1971.

[Lee 1971b] Lee, W. C. Y. "A study of the antenna array configuration of an *M*-branch diversity combining mobile radio receiver," *IEEE Trans. Veh. Technol.,* vol. 20, no. 4, pp. 93–104, Nov. 1971.

[Lee 1980] Lee, W. C. Y. "Studies on base station antenna height effects on mobile radio," *IEEE Trans. Veh. Technol.,* vol. 29, pp. 252–260, May 1980.

[Lee 1985] Lee, W. C. Y. "Estimate of local average power of a mobile radio signal," *IEEE Trans. Veh. Technol.,* vol. 34, no. 1, pp. 22–27, Feb. 1985.

[Lee 1986] Lee, W. C. Y. "Elements of cellular mobile radio systems," *IEEE Trans. Veh. Technol.,* vol. 35, no. 2, pp. 48–56, May 1986.

[Lee 1990] Lee, W. C.Y. "Estimate of channel capacity in Rayleigh fading environment," *IEEE Trans. Veh. Technol.,* vol. 39, no. 3, pp. 187–189, Aug. 1990.

[Lee 1991a] Lee, W. C. Y. "Overview of cellular CDMA," *IEEE Trans. Veh. Technol.,* vol. 40, no. 2, pp. 291–302, May 1991.

[Lee 1991b] Lee, W. C. Y. "Smaller cells for greater performance," *IEEE Commun. Mag.,* vol. 29, pp. 19–23, Nov. 1991.

[Lee 1993] Lee, W. C. Y. *Mobile Communications Design Fundamentals.* John Wiley & Sons, New York, 1993.

[Lee 1997] Lee, W. C. Y. *Mobile Communications Engineering.* McGraw-Hill, New York, 1997.

[Lehn 1987] Lehnert, J. S., and Pursley, M. B. "Multipath diversity reception of spread-spectrum multiple access communications," *IEEE Trans. Commun.,* vol. COM-35, no. 11, pp. 1189–1198, Nov. 1987.

[Li 1997] Li, C.–P., and Haas, Z. J. "Macrodiversity techniques for improvement in BER in wireless systems," *Electron. Lett.,* vol. 33, no. 7, pp. 556–557, Mar. 27, 1997.

[Liu 1991a] Liu, C., and Feher, K. "Bit error rate performance of $\pi/4$-DQPSK in a frequency selective fast Rayleigh fading channel,"

IEEE Trans. Veh. Technol., vol. 40, no. 3, pp. 1–8, Aug. 1991.

[Liu 1991b] Liu, C., and Feher, K. "π/4-QPSK modems for satellite sound/data broadcast systems," *IEEE Trans. Broadcast.,* vol. 37, no. 1, pp. 1–8, Mar. 1991.

[Lots 1992] Lotse, F., Berg, J.-E., and Bownds, R. "Indoor propagation measurements at 900 MHz," *Proc. 42nd IEEE Vehicular Technology Conf.,* pp. 629–632, 1992.

[MacD 1979] MacDonald, V. H. "The cellular concept," *Bell Syst. Tech. J.,* vol. 58, no. 1, pp. 15–41, Jan. 1979.

[MacW 1976] MacWilliams, F. J., and Sloane, N. J. "Pseudo-random sequences and arrays," *Proc. IEEE,* vol. 64, no. 12, pp. 1715–1729, Dec. 1976.

[Maki 1967] Makino, H., and Morita, K. "Design of space diversity receiving and transmitting systems for line-of-sight microwave links," *IEEE Trans. Commun.,* vol. 15, no. 4, pp. 603–613.

[Malm 1997] Malm, P., and Maseng, T. "Adjacent channel separation in mobile cellular systems," *Proc. 47th IEEE Vehicular Technology Conf.,* pp. 642–646, 1997.

[Mase 1985a] Maseng, T. "Digitally phase modulated (DPM) signals," *IEEE Trans. Commun.,* vol. 33, no. 9, pp. 911–918, Sept. 1985.

[Mase 1985b] Maseng, T. "The power spectrum of digital FM as produced by digital circuits," *Signal Process.,* vol. 9, pp. 253–261, 1985.

[Mehr 1999] Mehrnia, A., and Hashemi, H. "Mobile satellite propagation channel: Part 1—A comparative evaluation of current models," *Proc. 50th IEEE Vehicular Technology Conf.,* vol. 5, pp. 2775–2779, 1999.

[Mela 1986] Melancon, P., and Le Bel, J. "A characterization of the frequency selective fading of the mobile radio channel," *IEEE Trans. Veh. Technol.,* vol. 35, no. 4, pp. 153–161, Nov. 1986.

[Meli 1993] Melin, L., Rönnlund, M., and Angbratt, R. "Radio wave propagation—a comparison between 900 and 1800 MHz," *Proc. IEEE VTC '93,* pp. 250–252, 1993.

[Miya 1997] Miya, K., Watanabe, M., Hayashi, M., Kitade, T., Kato, O., and Homma, K. "CDMA/TDD cellular systems for the 3rd generation mobile communication," *Proc.*

47th IEEE Vehicular Technology Conf., vol. 2, pp. 820–824, 1997.

[Molk 1991] Molkdar, D. "Review on radio propagation into and within buildings," *IEE Proc.,* vol. 138, pt. H, no. 1, pp. 61–73, Feb. 1991.

[Mons 1980] Monsen, P. "Fading channel communications," *IEEE Commun. Mag.,* vol. 18, no. 1, pp. 16–25, Jan. 1980.

[Moul 1995] Mouly, M., and Pautet, M.-B. "Current evolution of the GSM systems," *IEEE Pers. Commun.,* vol. 2, no. 5, pp. 9–19, Oct. 1995.

[Muro 1981] Murota, K., and Hirade, K. "GMSK modulation for digital mobile radio telephony," *IEEE Trans. Commun.,* vol. COM–29, no. 7, pp. 1044–1050, July 1981.

[Muro 1985] Murota, K. "Spectrum efficiency of GMSK land mobile radio," *IEEE Trans. Veh. Technol.,* vol. 34, no. 2, pp. 69–75, May 1985.

[Naga 1987] Nagata, Y., and Akaiwa, Y. "Analysis of spectrum efficiency in single cell trunked and cellular mobile radio," *IEEE Trans. Veh. Technol.,* vol. VT-35, no. 3, pp. 100–113, Aug. 1987.

[Naka 1990] Nakajima, N., Kuramoto, M., Kinoshita, K., and Utano, T. "A system design for TDMA mobile radios," *Proc. 40th IEEE Vehicular Technology Conf.,* May 6–9, 1990, Orlando, FL, pp. 295–298,

[Ng 1993] Ng, C. S., Tjhung, T. T., Adachi, F., and Lye, K. M. "On the error rates of differentially detected narrowband π/4-DQPSK in Rayleigh fading and Gaussian noise," *IEEE Trans. Veh. Technol.,* vol. 42, no. 3, pp. 259–265, Aug. 1993.

[Ng 1994] Ng, C. S., Tjhung, T. T., and Adachi, F. "Effects of non-matched receiver filters on π/4-DQPSK bit error rate in Rayleigh fading," *IEICE Trans. Commun.,* vol. E77-B, no. 6, June 1994.

[Nogu 1986] Noguchi, T., Daido, Y., and Nossek, J. A. "Modulation techniques for microwave digital radio," *IEEE Commun. Mag.,* vol. 24, no. 10, pp. 21–30, Oct. 1986.

[Oett 1979] Oetting, J. D. "A comparison of modulation techniques for digital radio," *IEEE Trans. Commun.,* vol. COM-27, no. 12, pp. 1752–1762, Dec. 1979.

[Oett 1983] Oetting, J. D. "Cellular mobile radio: An emerging technology," *IEEE Commun. Mag.,* vol. 21, pp. 10–15, Nov. 1983.

[Ohno 1990] Ohno, K., and Adachi, F. " Post detection diversity reception of QDPSK signals under frequency selective Rayleigh fading," *Proc. 40th IEEE Vehicular Technology Conf.,* 1990.

[Okum 1968] Okumura, Y., Ohmuri, E., Kawano, T., and Fukuda, K. "Field strength and its variability in VHF and UHF land mobile radio service," *Rev. Elect. Commun. Lab.,* vol. 16, pp. 825–873, Sept.–Oct. 1968.

[Ono 1991] Ono, S., Kondoh, N., and Shimazaki, Y. "Digital cellular system with linear modulation*," Proc. 39th IEEE Vehicular Technology Conf.,* May 1–3, 1991, San Francisco, pp. 44–49.

[Padg 1995] Padgett, J. E., Gunther, C. G., and Hattori, T. "Overview of wireless personal communications," *IEEE Commun. Mag.,* vol. 33, no. 1, pp. 28–41, Jan. 1995.

[Pado 1994] Padovani, R. "Reverse link performance of IS 95 based cellular systems," *IEEE Pers. Commun.,* vol. 1, no. 3, pp. 28–34, 1994.

[Pahl 1995] Pahlavan, K., and Levesque, A. H. *Wireless Information Systems.* John Wiley & Sons, New York, 1995.

[Pale 1991] Palestini, V. "Evaluation of overall outage probability in cellular systems," *Proc. 39th IEEE Vehicular Technology Conf.,* May 1–3, 1991, San Francisco, pp. 625–630.

[Pani 1996] Panicker, J., and Kumar, S. "Effect of system imperfections on BER performance of a CDMA receiver with multipath diversity combining," *IEEE Trans. Veh. Technol.,* vol. 45, no. 4, pp. 622–630, Nov. 1996.

[Pape 1995] Papen, W. "Improved soft handoff and macrodiversity for mobile radio," *Proc. ICC '95,* pp. 1828–1833, 1995.

[Papo 1991] Papoulis, A. *Probability, Random Variables, and Stochastic Processes,* 3rd ed. McGraw-Hill, New York, 1991.

[Park 1978] Park, J. H. "On binary PSK detection," *IEEE Trans. Commun.,* vol. COM-26, no. 4, pp. 484–486, Apr. 1978.

[Pars 1983] Parsons, J. D., and Ibrahim, M. F. "Signal strength prediction in built areas," *IEE Proc.,* vol. 130, pt. F, no. 5, pp. 385–391, Aug. 1983.

[Pars 1992] Parsons, D. *The Mobile Radio Propagation Channel.* John Wiley & Sons, New York, 1992.

[Pasu 1979] Pasupathy, S. "Minimum shift keying: A spectrally efficient modulation," *IEEE Commun. Mag.,* vol. 17, no. 7, pp. 14–22, July 1979.

[Peeb 1987] Peebles, P. Z. *Digital Communication Systems.* McGraw-Hill, New York, 1987.

[Peri 1998] Perini, P. L., and Holloway, C. L. "Angle and space diversity comparisons in different mobile environments," *IEEE Trans. Antennas Propag.,* vol. 46, no. 6, pp. 764–775, June 1998.

[Peri 1999] Perini, P. L. "Angle and space diversity comparisons in different mobile radio environments," *1999 IEEE Aerospace Conference Proc.,* vol. 3, pp. 211–218, 1999.

[Pete 1992] Petersen, B. R., and Falconer, D. D. "Suppression of adjacent-channel interference in digital radio by equalization," *Proc. ICC '92,* pp. 657–661, 1992.

[Pick 1982] Pickholtz, R. L., Schilling, D. L., and Milstein, L. B. "Theory of spread spectrum communications: A tutorial," *IEEE Trans. Commun.,* vol. 30, no. 5, pp. 855–884, May 1982.

[Pick 1991] Pickholtz, R. L., Milstein, L. B., and Schilling, D. L. "Spread spectrum for mobile communications," *IEEE Trans. Veh. Technol.,* vol. 40, no. 2, pp. 313–322, May 1991.

[Pier 1960] Pierce, J. N., and Stein, S. "Multiple diversity with nonindependent fading," *Proc. IRE,* vol. 48, no. 1, pp. 89–104, Jan. 1960.

[Pras 1999] Prasad, N. R. "GSM evolution towards third generation UMTS/IMT 2000," *Proc. 1999 IEEE Intl. Conf. on Personal Wireless Communications,* Jaipur, India, Feb. 1999, pp. 50–54.

[Pric 1958] Price, R., and Green, P. E. "A communication technique for a multipath channel," *Proc. IRE,* vol. 46, no. 3, pp. 555–570, 1958.

[Proa 1994] Proakis, J. G., and Salehi, M. *Communication Systems Engineering.* Prentice-Hall, Englewood Cliffs, NJ, 1994.

[Proa 1995] Proakis, J. G. *Digital Communications.* McGraw-Hill, New York, 1995.

[Purs 1977] Pursley, M. B. "Performance evaluation for phase-coded spread-spectrum multiple-access communication—Part I: System analysis," *IEEE Trans. Commun.,* vol. COM-25, no. 8, pp. 795–799, Aug. 1977.

[Rahm 1998] Rahman, T. A., Burok, H., and Geok, T. K. "The cellular phone industry in Malaysia: Toward IMT-2000," *IEEE Commun. Mag.,* pp. 154–156, Sept. 1998.

[Rait 1991] Raith, K., and Uddenfeldt, J. "Capacity of a digital cellular system," *IEEE Trans. Veh. Technol.,* vol. 40, no. 2, pp. 323–332, May 1991.

[Rao 1999] Rao, Y. S., and Kripalani, A. "cdma2000 mobile radio access for IMT," *Proc. 1999 IEEE Intl. Conf. on Personal Wireless Communications,* Jaipur, India, Feb. 1999, pp. 6–15.

[Rapp 1995] Rappaport, T. S. (ed.) *Cellular Radio & Personal Communications,* vol. 1. IEEE Press, Los Alamitos, CA, 1995.

[Rapp 1996a] Rappaport, T. S. (ed.). *Cellular Radio & Personal Communications*, vol. 2. IEEE Press, Los Alamitos, CA, 1996.

[Rapp 1996b] Rappaport, T. S. *Wireless Communications: Principles & Practice.* Prentice-Hall/ IEEE Press, Englewood Cliffs, NJ, 1996.

[Razb 1972] Razbel, M., and Pasupathy, S. "Spectral shaping in MSK type signals," *IEEE Trans. Commun.,* vol. 26, no. 1, pp. 189–195, Jan. 1972.

[Ricc 1997] Ricci, F. J. *Personal Communications Systems and Applications.* Prentice-Hall, Englewood Cliffs, NJ, 1997.

[Robe 1994] Roberts, J. A., and Bargallo, J. M. "DPSK performance for indoor wireless Rician fading channels," *IEEE Trans. Commun.,* vol. 42, no. 2/3/4, pp. 592–596, Feb./ Mar./Apr. 1994.

[Rowe 1975] Rowe, H. E., and Prabhu, V. K. "Power spectrum of a digital frequency modulation signal," *Bell Syst. Tech. J.,* vol. 54, no. 4, pp. 1095–1125, July–Aug. 1975.

[Rumm 1986] Rummler, W. D., Coutts, R. P., and Liniger, M. "Multipath fading channel models for microwave radio," *IEEE Commun. Mag.,* vol. 24, no. 11, pp. 30–41, Nov. 1986.

[Salt 1998] Saltzberg, B. R. "Comparison of single carrier and multitone digital modulation for ADSL applications," *IEEE Commun. Mag.,* vol. 38, no. 11, pp. 114–121, Nov. 1998.

[Samp 1993] Sampei, S., Leung, P., and Feher, K. "High capacity cell configuration strategy in ACI and CCI conditions," *Proc. IEEE VTC '93,* pp. 185–188, 1993.

[Samp 1997] Sampei, S. *Applications of Digital Wireless Technologies to Global Wireless Communications.* Prentice-Hall, Englewood Cliffs, NJ, 1997.

[Samu 1998] Samukic, A. "UMTS universal mobile telecommunications system: Development of standards for the third generation," *IEEE Trans. Veh. Technol.,* vol. 47, no. 7, pp. 1099–1104, Nov. 1998.

[Sasa 1998] Sasaki, A., and Inada, S. "The current situation of IMT-2000 standardization activities in Japan," *IEEE Commun. Mag.,* pp. 145–153, Sept. 1998.

[Saun 1999] Saunders, S. R. *Antennas and Propagation for Wireless Communication Systems.* John Wiley & Sons, New York, 1999.

[Schi 1991] Schilling, D. L., Milstein, L. B., Pickholtz, R. L., Bruno, F., Kanterakis, E., Kullback, M., Erceg, V., Biederman, W., Fishman, D., and Salerno, D. "Broadband CDMA for personal communications systems," *IEEE Commun. Mag.,* pp. 86–93, Nov. 1991.

[Schi 1994] Schilling, D. L. "Wireless communications going into the 21st century," *IEEE Trans. Veh. Technol.,* vol. 43, no. 3, pp. 645–652, Aug. 1994.

[Schn 2000] Schneiderman, R. "Bluetooth's slow dawn," *IEEE Spectrum,* pp. 61–66, Nov. 2000.

[Scho 1977] Scholtz, R. A. "The spread spectrum concept," *IEEE Trans. Commun.,* vol. 25, pp. 748–755, Aug. 1977.

[Scho 1982] Scholtz, R. A. "The origins of spread spectrum," *IEEE Trans. Commun.,* vol. 30, pp. 822–854, May 1982.

[Schr 1966] Schroeder, M. R. "Vocoders: Analysis and synthesis of speech," *Proc. IEEE,* vol. 54, pp. 720–734, 1966.

[Schr 1985] Schroeder, M. R. "Linear predictive coding of speech: Review and current directions," *IEEE Commun. Mag.,* vol. 23, no. 8, pp. 54–61, Aug. 1985.

[Schw 1996] Schwartz, M., Bennett, W. R., and Stein, S. *Communication Systems and Techniques.* IEEE Press, Los Alamitos, CA, 1996.

[Shaf 1998] Shafi, M., Sasaki, A., and Jeong, D.-G. "IMT-2000 developments in the Asia Pacific region," *IEEE Commun. Mag.,* p. 144, Sept. 1998.

[Shan 1949] Shannon, C. E. "A mathematical theory of communication," *Bell Syst. Tech. J.,* vol. 27, no. 3, pp. 379–423, July 1948.

[Shan 1979] Shanmugam, K. S. *Digital and Analog Communication Systems.* John Wiley & Sons, New York, 1979.

[Shap 1994] Shapira, J. "Microcell engineering in CDMA cellular networks," *IEEE Trans. Veh. Technol.,* vol. 43, no. 4, pp. 817–825, Nov. 1994.

[Shib 1993] Shibutani, M., Kanai, T., Domom, W., Emura, K., and Namiki, J. "Optical fiber feeder for microcellular mobile communication systems (H–015)," *IEEE J. Sel. Areas Commun.,* vol. 11, no. 7, pp. 1118–1126, Sept. 1993.

[Shum 1998] Shumin, C. "Current development of IMT-2000," *IEEE Commun. Mag.,* pp. 157–159, Sept. 1998.

[Sidd 1963] Siddiqui, M. M., and Weiss, G. H. "Families of distributions for hourly median power and instantaneous power of received radio signals," *J. Nat. Bur. Stand. (D–Radio Propag.),* vol. 67-D, no. 6, pp. 753–762, Nov./Dec. 1963.

[Simo 1976] Simon, M. K. "A generalization of minimum shift keying (MSK)–type signaling based upon input data symbol pulse shaping," *IEEE Trans. Commun.* vol. COM-24, no. 8, pp. 845–856, Aug. 1976.

[Simo 1978] Simon, M. K. "Comments on binary PSK detection," *IEEE Trans. Commun.,* vol. COM-26, no. 10, pp. 1477–1478, Oct. 1978.

[Simo 1984] Simon, M. K., and Wang, C. C. "Differential detection of Gaussian MSK in a mobile radio environment," *IEEE Trans. Veh. Technol.,* vol. 33, no. 4, pp. 307–320, Nov. 1984.

[Simo 1992] Simon, M. K., and Divsalar, D. "On the implementation and performance of single and double differential detection schemes, "*IEEE Trans. Commun.,* vol. 40, pp. 278–291, Feb. 1992.

[Skla 1988] Sklar, B. *Digital Communications: Fundamentals and Applications.* Prentice–Hall, Englewood Cliffs, NJ, 1988.

[Skla 1993] Sklar, B. "Defining, designing, and evaluating digital communication systems," *IEEE Commun. Mag.,* vol. 31, no. 11, pp. 92–101, Nov. 1993.

[Slim 1998] Slimane, S. B. "OFDM schemes with non-overlapping time waveforms," *Proc.*

48th IEEE Vehicular Technology Conf., 1998, pp. 2067–2071.

[Smit 1994] Smith, W. S., and Wittke, P. H. "Differential detection of GMSK in Rician fading," *IEEE Trans. Commun.,* vol. 42, no. 2/3/4, Feb./Mar./Apr. 1994.

[Span 1994] Spanis, A. S. "Speech coding: A tutorial review," *Proc. IEEE,* vol. 82, no. 10, pp. 1541–1582, 1994.

[Stee 1989] Steele, R. "The cellular environment of lightweight handheld portables," *IEEE Commun. Mag.,* vol. 27, no. 6, pp. 20–29, June 1989.

[Stee 1992] Steele, R., and Nofal, M. "Teletraffic performance of microcellular personal communication networks," *IEE Proc.,* vol. 139, pt. I, no. 4, pp. 448–461, Aug. 1992.

[Stee 1994] Steele, R. "The evolution of personal communications," *IEEE Pers. Commun.,* vol. 1, no. 2, pp. 6–11, 1994.

[Stee 1999] Steele, R., and Hanzo, L. (eds.). *Mobile Radio Communications,* 2nd ed. IEEE Press, Los Alamitos, CA, 1999.

[Stei 1964] Stein, S. "Unified analysis of certain coherent and non-coherent binary communications systems," *IEEE Trans. Inf. Theor.,* vol. IT-10, no. 1, pp. 43–51, Jan. 1964.

[Stei 1987] Stein, S. "Fading channel issues in systems engineering," *IEEE J. Sel. Areas Commun.,* vol. SAC-5, no. 2, pp. 68–89, Feb. 1987.

[Stub 1996] Stuber, G. L. *Principles of Mobile Communication.* Kluwer Academic Press, Norwell, MA, 1996.

[Sull 1972] Sullivan, W. A. "High capacity microwave system for digital data transmission," *IEEE Trans. Commun.,* vol. 20, no. 6, pp. 466–470, June 1972.

[Sund 1986] Sundberg, C.-E. "Continuous phase modulation," *IEEE Commun. Mag.,* vol. 24, no. 4, pp. 25–38, Apr. 1986.

[Suzu 1977] Suzuki, H. "A statistical model for urban radio propagation," *IEEE Trans. Commun.,* vol. COM-25, no. 7, pp. 673–680, July 1977.

[Tant 1998] Tantaratana, S., and Ahmed, K. M. *Wireless Applications of Spread Spectrum Systems.* IEEE Press, Los Alamitos, CA, 1998.

[Tara 1988] Tarallo, J. A., and Zysman, G. I. "Modulation techniques for digital cellular systems," *Proc. 38th IEEE Vehicular*

Technology Conf., June 15–17, 1988, Philadelphia, PA, pp. 245–249.

[Taub 1986] Taub, H., and Schilling, D. L. *Principles of Communication Systems,* 2nd ed. McGraw-Hill, New York, 1986.

[Tayl 1998] Taylor, D. P., Vitetta, G. M., Hart, B. D., and Mammela, A. "Wireless channel equalization," *Eur. Trans. Telecommun.,* vol. 9, no. 2, pp. 117–143, Mar.–Apr. 1998.

[Toma 1998] Tomasi, W. *Advanced Electronic Communications Systems*, 4th ed. Prentice-Hall, Englewood Cliffs, NJ, 1998.

[Turi 1972] Turin, G. L., Clapp, F. D., Johnston, T. L., Fine, S. B., and Lavry, D. "A statistical model of urban multipath propagation," *IEEE Trans. Veh. Technol.,* vol. VT-21, no. 1, pp. 1–9, Feb. 1972.

[Turk 1990] Turkmani, A. M. D., and de Toledo, A. F. "Time diversity for digital mobile radio," *Proc. 40th IEEE Vehicular Technology Conf.,* pp. 576–581, 1990.

[Turk 1991a] Turkmani, A. M. D. "Performance evaluation of a composite microscopic plus macroscopic diversity system," *IEE Proc.,* vol. 138, pt. I, no. 1, pp. 15–20, Feb. 1991.

[Turk 1991b] Turkmani, A. M. D., and de Toledo, A. F. "Radio transmission at 1800 MHz into, and within, multistory buildings," *IEE Proc.,* vol. 138, pt. I, no. 6, pp. 577–584, Dec. 1991.

[Turk 1992] Turkmani, A. M. D. "Probability of error for *M*-branch macroscopic selection diversity," *IEE Proc.,* vol. 139, pt. I, no. 1, pp. 71–78, Feb. 1992.

[Vars 1993] Varshney, P., Salt, J. E., and Kumar, S. "BER analysis of GMSK with differential detection in a land mobile channel," *IEEE Trans. Veh. Technol.,* vol. 42, no. 4, pp. 683–688, Nov. 1993.

[Vaug 1988] Vaughn, R. G. "On optimum combining at the mobile," *IEEE Trans. Veh. Technol.,* vol. 37, no. 4, pp. 181–188, Nov. 1988.

[Vaug 1990] Vaughn, R. G. "Polarization diversity in mobile communications," *IEEE Trans. Veh. Technol.,* vol. 39, no. 3, pp. 177–186, Aug. 1990.

[Vite 1971] Viterbi, A. J. "Convolutional codes and their performance in communication systems," *IEEE Trans. Commun. Technol.,* vol. COM-19, no. 5, pp. 751–772, Oct. 1971.

[Vite 1979] Viterbi, A. J. "Spread spectrum communications: Myths and realities," *IEEE Commun. Mag.,* vol. 17, pp. 11–18, May 1979.

[Vite 1985] Viterbi, A. J. "When not to spread spectrum: A sequel," *IEEE Commun. Mag.,* vol. 23, no. 1, pp. 12–17, Apr. 1985.

[Walf 1988] Walfisch, J., and Bertoni, H. L. "A theoretical model of UHF propagation in urban environments," *IEEE Trans. Antennas Propag.,* vol. 36, no. 12, pp. 1788–1796, Dec. 1988.

[Walk 1966] Walker, W. F. "Baseband multipath fading simulator," *Radio Sci.,* vol. 1, no. 7, pp. 763–767, July 1966.

[Walk 1983] Walker, E. H. "Penetration of radio signals into buildings in the cellular radio environment," *Bell Syst. Tech. J.,* vol. 62, no. 9, pp. 2719–2734, Nov. 1983.

[Wang 1993] Wang, J., and Yongaçoglu, A. "Direct sequence CDMA employing combined modulation schemes," *Proc. IEEE GLOBECOM '93,* Houston, pp. 1729–1733, Nov. 1993.

[Wang 1995] Wang, L.-C., and Lea, C.-T. "Macrodiversity cochannel interference analysis," *Electron. Lett.,* vol. 37, no. 8, pp. 614–616, Apr. 13, 1995.

[Wang, 1999] Wang, L.-C., Stüber, G. L., and Lea, C-T. "Effects of fading and branch correlation on a local mean based macrodiversity cellular system," *IEEE Trans. Veh. Technol.,* vol. 48, no. 2, pp. 429–436, Mar. 1999.

[Way 1993] Way, W. I. "Optical fiber–based microcellular systems: An overview," *IEICE Trans. Electron.,* vol. E76-B, no. 9, pp. 1091–1102, Feb. 1993.

[Weav 1962] Weaver, C. S. "A comparison of several types of modulation," *IRE Trans. Commun. Syst.,* vol. CS-10, pp. 96–101, Mar. 1962.

[Wee 1998] Wee, K. J., and Shin, Y. S. "Current IMT-2000 R&D status and views in Korea," *IEEE Commun. Mag.,* pp. 160–164, Sept. 1998.

[Weis 1999] Weiss, U. "Designing macroscopic diversity cellular systems," Proc. 49th *IEEE Vehicular Technology Conf.,* pp. 2054–2058, 1999.

[Wese 1998] Wesel, E. K. *Wireless Multimedia Communications: Networking Video, Voice, and Data.* Addison-Wesley, Reading, MA, 1998.

[Whip 1994] Whipple, D. P. "The CDMA standard," *Appl. Microwave Wireless,* pp. 24–39, 1994.

[Wils 1996] Wilson, S. G. *Digital Modulation and Coding.* Prentice-Hall, Englewood Cliffs, NJ, 1996.

[Wint 1984] Winters, J. H. "Optimum combining in digital mobile radio with co-channel interference," *IEEE Trans. Veh. Technol.,* vol. VT-33, no. 3, pp. 144–155, Aug. 1984.

[Witt 1991] Wittneben, A. "Basestation modulation diversity for digital simulcast," *Proc. IEEE Vehicular Technology Conf.,* pp. 848–853, 1991.

[Wong 1997] Wong, D., and Lim, T. J. "Soft hand-offs in CDMA mobile systems," *IEEE Pers. Commun.,* pp. 6–17, vol. 4, no. 6, Dec. 1997.

[Wu 1995] Wu, G., Jalali, A., and Mermelstein, P. "On channel model parameters for microcellular CDMA systems," *IEEE Trans. Veh. Technol.,* vol. 44, no. 3, pp. 706–711, Aug. 1995.

[Yama 1989] Yamao, Y., Saito, S., Suzuki, H., and Nojima, T. "Performance of $\pi/4$ QPSK transmission for mobile radio applications," *Proc. IEEE GLOBECOM '89,* Nov. 27–30, 1989, Dallas, pp. 443–447.

[Yang 1971] Yang, J., and Feher, K. "An improved $\pi/4$-QPSK with non-redundant error correction for satellite mobile communication systems," *IEEE Trans. Broadcast.,* vol. 37, no. 1, pp. 9–16, Mar. 1971.

[Yeun 1996] Yeung, K. L., and Nanda, S. "Channel management in microcell/macrocell cellular systems," *IEEE Trans. Veh. Technol.,* vol. 45, no. 4, pp. 601–612, Nov. 1996.

[Youn 1979] Young, W. R. "Advanced mobile phone service: Introduction, background, and objectives," *Bell Syst. Technol. J.,* vol. 58, no. 1, pp. 1–14, Jan. 1979.

[Yue 1983] Yue, O. C. "Spread spectrum mobile radio, 1977–1982," *IEEE Trans. Veh. Technol.,* vol. 32, no. 1, pp. 98–105, Feb. 1983.

[Zeng 1999] Zeng, M., Annamalai, A., and Bhargava, V. K. "Recent advances in cellular wireless communications," *IEEE Commun. Mag.,* vol. 37, no. 9, pp. 128–138, Sept. 1999.

[Ziem 1995] Ziemer, R. E., and Tranter, W. H. *Principles of Communications: Systems, Modulation and Noise.* John Wiley & Sons, New York, 1995.

[Zurb 2000] Zurbes, S. "Considerations on link and system throughput of Bluetooth network," *Proc. PIMRC 2000,* pp. 1315–1319, 2000.

INDEX